Johann Ludwig Casper

A Handbook of the Practice of Forensic Medicine, Based upon

Personal Experience

Vol. 4

Johann Ludwig Casper

A Handbook of the Practice of Forensic Medicine, Based upon Personal Experience
Vol. 4

ISBN/EAN: 9783337778071

Printed in Europe, USA, Canada, Australia, Japan

Cover: Foto ©berggeist007 / pixelio.de

More available books at **www.hansebooks.com**

A

HANDBOOK

OF THE PRACTICE OF

FORENSIC MEDICINE,

BASED UPON PERSONAL EXPERIENCE.

BY

JOHANN LUDWIG CASPER, M.D.,

LATE PROFESSOR OF FORENSIC MEDICINE IN THE UNIVERSITY OF BERLIN,
FORENSIC PHYSICIAN TO THE COURTS OF JUSTICIARY OF BERLIN AND MEMBER OF THE ROYAL CENTRAL
MEDICAL BOARD OF PRUSSIA.

VOL. IV.

BIOLOGICAL DIVISION.

TRANSLATED FROM THE THIRD EDITION OF THE ORIGINAL BY
GEORGE WILLIAM BALFOUR, M.D., St. Andrews,
FELLOW OF THE ROYAL COLLEGE OF PHYSICIANS, EDINBURGH.

THE NEW SYDENHAM SOCIETY,
LONDON.

MDCCCLXV.

BIOGRAPHICAL NOTICE

BY THE TRANSLATOR.

JOHANN LUDWIG CASPER was the son of Joachim Casper, a merchant in Berlin, and was born upon the 11th of March, 1796. He was educated in the Cöllnische Gymnasium, at Berlin, and in it he also learned the art of an apothecary; in the spring of 1816, he commenced to study the science of medicine in the University of Berlin; in 1817, he continued these studies at Göttingen; and, in 1818, he went to Halle, where he graduated in 1819. In 1820 he travelled through France and Britain, and subsequently settled in Berlin, as a practising physician, when, having passed through the preliminary position of private teacher, he was appointed, in 1824, Extraordinary Professor of Medicine, Medical Councillor, and Member of the Medical College for the province of Brandenburg. In 1834 he was appointed Privy Superior Medical Councillor, Member of the Superior Medico-scientific College, and of the Royal Scientific Commission for the Administration of Ecclesiastical, Educational and Medical Affairs. In 1839, he was appointed Professor of Forensic Medicine in the University at Berlin, and, in 1841, he was also appointed Forensic Physician to the Courts of Justiciary at Berlin, and Director of the School of Practical Medicine. On the 23rd of February, 1864, while yet in the full vigour both of body and mind, he suddenly ceased to be mortal.

Professor Casper's literary labours commenced at a somewhat early age, and were continued uninterruptedly to the close of his life. The year before he graduated, he published a small pamphlet, entitled, " Modest Doubts in regard to the New Clairvoyante at Carlsruhe ; with a Few Thoughts upon Animal Magnetism in General,"* and this,

* Bescheidene Zweifel gegen die neue Hellseherinn in Carlsruhe ; mit einigen Gedanken über den thierisichen Magnetismus überhaupt. Leipzig, 1818.

his earliest work, seems to have attracted some attention in his native country, as it was reviewed in several of the German periodicals; beyond this I have no information regarding it. Casper's inaugural dissertation, "De Phlegmatia Alba Dolente," Halle, 1819, was also well received both at home and abroad, and an interesting and careful review of it will be found in the *Edinburgh Medical and Surgical Journal* Vol. XXXII., 1829, p. 422-31, in which it is characterized as giving a distinct account and judicious enumeration of the existing opinions as to this disease, as well as a minute and detailed account of three cases, with the *sectio cadaveris* of one of them. The view adopted by Casper of this disease was, that it was an inflammation, mainly affecting the absorbents, but implicating, to some extent, the cellular tissue also—a view of the disease which is probably pretty generally maintained even at the present day. The writer in the *Conversations Lexicon* states that, at the commencement of his medical career, Professor Casper published anonymously several literary works, as well as many non-professional articles in various journals, which exhibited such a talent for humorous and comic description as would certainly have gained for him a wide and extended reputation, had he not preferred leaving off everything which could possibly interfere with his endeavours to improve and advance medical science. In 1822, Casper published an extremely interesting account of the observations made by him during his tour through France and England,* including an admirable critique of Broussais and his doctrine, a sketch of the various physicians, hospitals, lunatic asylums, and modes of treatment of the sick and the insane, both in Paris and London, along with an account of all that is interesting in the sanitary regulations, or remarkable in the forms of quackery in these two cities. The whole work forms a complete *repertorium* of the state of medicine and medical hygiene, for the time being, in these two countries, and is, besides, doubly interesting as exhibiting the source of much of the information which we find cropping out in his subsequent works. It also shows the careful manner in which Professor Casper educated himself for the discharge of his professional duties, the diligence with which he made use of every available opportunity of acquiring information, and the spirit which he brought to the study, and subsequently carried out in the practice of his profession, and to which much of his fame as a medical author is unquestionably due. In his

* Characteristik der Französichen Medizin, &c. Leipzig, 1822.

preface he says, " Whoever seeks throughout this book for what is new, and only what is new, will find himself occasionally at fault. For I hold that in an empirical science, such as medicine is, it is of far less importance to discover and put in circulation new modes of cure, than continually to confirm by fresh evidence, the efficacy of the old, tried, and proved methods. I have frequently seen that important medical dogma, ' *non nocére*,' sacrificed for the sake of novelty, and I have omitted detailing many new things when I was not convinced that they were not only true but really practical."*

With these views Professor Casper commenced his professional career, and every work which has fallen from his pen has proved how steadily, throughout his long life, he acted up to them, and the more learning and knowledge of the subject we bring to the investigation, the more evidently do we see how carefully he has avoided mere novelty, how diligently and laboriously he has accumulated fresh evidence in favour of what was tried and found good in the olden views, and how sedulously he has set aside whatever was injurious or unpractical, whether it were new or old. I cannot omit mentioning that, when speaking in this work of William Lawrence, Casper takes the opportunity of paying a high compliment to the cordiality and true-heartedness of the English generally, in spite of their apparently cold and inaccessible exterior.†

In 1824, Casper published a pamphlet, showing the influence of vaccination upon the general mortality of Berlin.‡ He showed that formerly one out of twelve died from small-pox in Berlin, then only one out of one hundred and sixteen; that of one hundred children born fifty-one died formerly in childhood, then only forty-three, in spite of the greater frequency of ordinary children's complaints; that formerly one death occurred out of every twenty-eight inhabitants, and then only one out of every thirty-four. In 1825, the first volume of his contributions to medical statistics§ was published, and for this important work the King of Prussia presented Professor Casper with a valuable diamond ring. This work contains articles upon suicide and its increase in our day, upon the poor and sick of Paris, and upon the mortality of children in Berlin. The reviewer in

* *Op. cit.* Preface, p. xxi.   † *Op. cit.* p. 477.
‡ De vi atque efficacitate insitionis variolæ vaccinæ in mortalitatem civium Berolinensium hucusque demonstrata. Berlin, 1824.
§ Beiträge zur Medicinischen Statistik und Staatsarzneikunde. Berlin 1825.

the *Edinburgh Medical and Surgical Journal* (July 1826) characterises this as "one of the most interesting contributions to medical statistics that has appeared during the present century; there is not much novelty in it, but it establishes upon satisfactory evidence many points which have been merely matters of doubtful opinion, and it settles some important questions regarding which there has been a keen controversy among physicians and political economists;" and he adds, in conclusion, "In bidding adieu to Dr. Casper, we may venture to predict, that if we possessed upon the other topics of medical police a few such monographs as his, it would not long remain so neglected as it now is; but that, as it is one of the most elevated, so it would likewise soon become one of the most fascinating and popular of all the departments of medicine." The second volume of these Statistics* was published in 1835, and is chiefly remarkable for a paper on the probable duration of life in man, the result of ten years' laborious research. In 1846 appeared his "Memorabilia,"† forming a third volume of Medical Statistics, and containing, among other important papers, most interesting communications on the influence of the weather upon the health and mortality of mankind, upon the geography of crime, on the influence of the period of the day upon the birth and death of man, and an essay upon Pyromania. What may be termed the physiological, in opposition to the medico-legal portion of this work, was subjected to an able critical analysis by a writer in the *British and Foreign Medical Review* (Vol. XXIV., 1847), in the course of which he states :—" This is a remarkable work from the pen of a man who, by his extensive and elaborate contributions to medical literature, has long since deservedly acquired a European reputation. We know no one so well fitted as Professor Casper to undertake a work of this kind, involving, as it does, immense labour, and requiring a mind well disciplined by the experience of years devoted to the study of medical philosophy in its higher and more recondite ramifications; and although some may be disposed to differ from the author in his conclusions, and to dispute the practical utility of his researches, there is no one, we think, who will refuse to admit that he deals with his subject fairly, candidly, and with an earnest desire

* Beiträge zur Medicinischen Statistik und Staatsarzneikunde. Berlin, 1835.

† Denkwürdigkeiten zur Medicinischen Statistik und Staatsarzneikunde. Berlin, 1846.

to improve medical science and practice." The writer concludes by stating—" it would be difficult to point to any researches more accurate, more extensive, or more elaborate than those of the learned author of these essays; and there is no doubt, if ever it be permitted to the medical philosopher to determine the exact influence of the weather on the salubrity of places, and on the healthy condition of the human species, it will be by pursuing the course so well marked out by Casper."

In 1851 Casper published the first volume of his Medico-legal Dissections; * to this a second was shortly afterwards added, and the two were subsequently developed into the Thanatological portion of his Handbook of Forensic Medicine, which was first published under that title in October, 1856, and was reprinted, unaltered, in December, 1857. To it was added a Biological division in September, 1858. Another edition of the whole work was issued in April, 1860, in which it was thus for the first time presented to the public in a complete condition. Such was the estimation in which this most valuable work was held that, long before the year was out, a fourth edition was called for. It was also translated into French, Italian, and Dutch, and when Professor Casper became aware that the New Sydenham Society intended to translate it into English, he most kindly offered, *unasked*, to take the very great trouble of revising each sheet of the translation before it went to press, so as to bring it abreast of the very latest German edition. He thus ensured its accuracy, whatever may be the defects of its literary execution. As the translator of this important contribution to medical scientific literature, I owe Professor Casper a deep debt of gratitude for much pleasing literary intercourse, and it has been a labour of love with me to endeavour to exhibit to my fellow members of the New Sydenham Society how continuously, energetically, and successfully, Professor Casper has laboured for the advance of medical science.

In 1863 Casper published, under the title of Clinical Novelties in Forensic Medicine,† a volume in which are discussed many questions which either had been omitted from the Handbook entirely, or which had not been treated in it with that fulness of detail which their importance demanded, and these are illustrated by three hundred and thirty-nine new cases. This volume, therefore, may be regarded as, properly speaking, a supplement to the Handbook, which is constantly

* Gerichtliche Leichenöffnungen. Berlin, 1851.
† Klinische Novellen zur gerichtlichen Medicin. Berlin, 1863.

referred to throughout it. Two of the most important questions discussed in this work are entirely new, namely, first, a comparison of
Pæderastia with Rape, that is, a treatise on the question whether it
can be enforced by one man on another, against the will of the latter;
three cases are related bearing upon this, the most important being
one of imputed crime. Casper decided that, under such circumstances, the act was impossible. The second important novelty is
the question of priority in the causation of death, a question of the
utmost consequence in diagnosing which of two wounds or injuries has been the actual occasion of death, or whether death has
arisen from bodily disease or from injury, evidence of both being
found at the autopsy. It is obvious that this question may be of
great importance in relation to which of two men is to be considered
the murderer, both having wounded the deceased in a scuffle, but
with different weapons. Although this question is not formally raised
in the Handbook, yet a considerable number of cases actually involving it are to be found carefully reasoned in § 41 of Vol. I. of this
translation. Amongst these novelties, the reader will also find *in
extenso* the anonymous communication referred to at page 330 of
Vol. III. of this translation, which Tardieu declared to be either a
myth or a mystification, and which is published as in itself affording
unequivocal proofs of its authenticity.

But these works, large and important as they are, and seemingly
enough to occupy in their production the life-time of one man, by no
means represent the whole of Professor Casper's literary labours.
From 1823 he was co-editor of Rust's *Magazin für die gesammte
Heilkunde*, and 1829 he became sole editor of it. This magazine
was merged, in 1833, in Casper's well-known *Wochenschrift*,* which
was characterized by the late Sir John Forbes, in the *British and
Foreign Medical Review*, as by far the best and most valuable of all
the German medical journals. Casper also edited a quarterly
journal of forensic medicine † from 1852, and in these journals he
published, during the course of years, an enormous number of most
interesting and valuable contributions on medicine, surgery, medical
hygiène, and medical jurisprudence, very many of which were subsequently published separately, and had a large independent circulation.
During the prevalence of cholera in 1831, Casper was made Director

* Wochenschrift für die gesammte Heilkunde. Hirschwald, Berlin.
† Vierteljahrschrift für gerichtliche und offentliche Medicin. Hirschwald,
Berlin.

of the Cholera Hospital, and published numerous admirable observations on this disease in the *Berliner Cholera-Zeitung*, which he edited. At that time, steam apparatus, warm baths, and violent counter irritants were largely employed in the treatment of cholera. Professor Casper, on the other hand, was extremely successful in the employment of the external application of cold, and he published his results as a pamphlet.* All Professor Casper's works are distinguished by a clearness and vivacity of description, and a writer in the *Conversations Lexicon*, a work which must be acknowledged to be a competent authority, characterizes his style as distinguished by a correct and fluent manipulation of his mother tongue; while, as already stated, there is not one of his many works which does not seem based upon an earnest desire for the good of his fellowmen and the improvement of his adopted science, or which is not thoroughly permeated by evidence of a most earnest and pains-taking search after truth. And yet, Professor Casper's life, in spite of his voluminous authorship, was by no means solely devoted to literary labours. For nearly forty years he lectured on forensic medicine to large classes, with whom he also, each session, performed a series of experimental investigations in regard to the various injuries and accidents which may befall the human frame. During all that time he was also largely engaged in practical medico-legal investigations; and, on the completion of his thousandth autopsy, he was publicly *fêted* in Berlin, not long before his death. He was also largely employed to give evidence as a medical expert both in his own courts and in those of other countries, and from his official position as a member of the highest and last professional court, every medico-legal case in the kingdom, in which the opinion of the medical college of the province was disputed, came under his cognizance for review and revisal. Besides all this, Professor Casper enjoyed, to a large extent, the confidence of the public in his more private capacity, and had a very extensive practice as a family and consulting physician. Professor Casper was not only the most distinguished man of the day in his own department, but he was also well acquainted with science and literature generally. He understood and spoke French, Italian and English, and his judgment in social and political matters was as correct, and as much sought after, as in medical. Eminently sincere, affable, and social himself, he was one of the most esteemed

* Die Behandlung der Asiatischen Cholera durch Anwendung der Kälte. Berlin, 1832.

men in Berlin, and his drawing-room, to the last day of his life, was the resort of the most distinguished members of society in that city, having among its frequenters all the most famous men of the day, whether in science, diplomacy, or statecraft.   During the course of his long life, Professor Casper amassed a very handsome fortune, and, with that earnest desire to do good and be useful, which was so eminently characteristic of his life, he provided in his last will that, after the decease of his widow and daughters, the University of Berlin should inherit his fortune, to be employed in the relief of such students, and widows and unmarried daughters of professors, as should require pecuniary assistance.

18, LYNEDOCH PLACE, EDINBURGH,
January, 1865.

CONTENTS.

SPECIAL DIVISION.—PART FOURTH.

PART FIFTH.

PART SIXTH.

CHAPTER FIRST.

CHAPTER SECOND.

FIRST SECTION.

SECOND SECTION.

ERRATA.

VOL. I.—Page 79, line 18 from top, *instead of* "ilea," *read* "ilia."
,, 79, line 21 from top, *instead of* "ilea," *read* "ilia."
,, 85, line 13 from top, *instead of* "exhibiting in it all that the external senses have taken cognizance of," *read* "causing to be shown to him everything that can be taken cognizance of by the external senses."
,, 110, bottom line, *instead of* "ileum," *read* "ilium."

VOL. II.—Page 12, line 8 from top, *instead of* "sulphuret of gold," *read* "golden sulphuret of antimony."
,, 90, line 3 from top, *instead of* "41° F," *read* "43°.25 F."
,, 136, line 14 from bottom, *instead of* "67°.75 F," *read* "70°.25 F."
,, 151, line 18 from top, *instead of* "24°.5," *read* "27°.5."
,, 202, line 17 from bottom, *instead of* "STANGULATION," *read* "STRANGULATION."
,, 207, line 14 from top, *instead of* "20°.75 F," *read* "20°.25 F."
,, 276, line 5 from top, *instead of* "30°.25 to 19° F," *read* "— 1°.75 to — 13 F."
,, 281, line 14 from bottom, *instead of* "16° F," *read* "16° R."
,, 281, line 15 from bottom, *instead of* ".11°.75," *read* "12°.25."
,, 283, line 21 from bottom, *instead of* "*Elaychlorür?*" *read* "Dutch liquid ($C^4 H^4 Cl^2$)."

VOL. III.—Page 64, top line, *instead of* "INFILTRATION," *read* "INFLATION."
,, 122, line 17 from bottom, *instead of* "50° F," *read* "14° F."
,, 369, line 12 from bottom, *instead of* "three hundred," *read* "thirty."

CASPER

ON

FORENSIC MEDICINE.

BIOLOGICAL DIVISION.

SPECIAL DIVISION.—PART FOURTH.

DISPUTED RESULTS OF INJURIES AND ILL-TREATMENT UNATTENDED BY FATAL CONSEQUENCES.*

STATUTORY REGULATIONS.

GENERAL COMMON LAW, TIT. VI., PART I., § 115. *If the injury inflicted is such as completely to prevent the person injured from discharging the functions of his office or employment as previously, then the party who inflicted it is liable in damages to the amount of the profits of which the person injured is thereby deprived.*

§ 119. *Whenever the person injured, in spite of his injury is able to obtain a livelihood by means of his bodily or mental exertion, this must be deducted from the damages recovered (under § 115-117).*

§ 120. *If the person injured is only temporarily rendered unable to pursue his employment by the infliction of the injury, he can only sue for the reimbursement of his loss.*

§ 121. *The amount of indemnification thus claimed must, however, in accordance with the principles laid down in § 115, be determined in relation to the length of time during which the injury received prevented the person injured from pursuing his employment.*

§ 122. *When the person injured has not been rendered wholly in-*

* For the consideration of *fatal injuries*, and the medico-legal definition of an injury, as well as for the regulation of the Penal Code concerning them, *vide* Vol. I., p. 236, &c.

capable of discharging the functions of his office or employment, but their discharge has been rendered more difficult and expensive, the Judge must assess the damages in accordance with the principles laid down, and with due regard to the injurious influence which the injury received is likely to have upon the circumstances of the person injured.

§ 123. *When an unmarried female has been so disfigured by a bodily injury, as to render it more difficult for her to get married, she may demand a dowry from the party who inflicted it.*

§ 128. *Moreover, if the worldly prosperity of any one is hindered by any deformity occasioned intentionally, or through gross carelessness, he is entitled to reasonable (&c.) indemnification.*

PENAL CODE, § 188. *When slight bodily injuries or ill-treatment are immediately repeated, the Judge shall, &c.* (has reference to the punishment to be awarded).

§ 189. *When the private complaint regarding the infliction of slight bodily injuries or ill-treatment is withdrawn, &c.* (refers to the amount of punishment).

§ 192 a. *If intentional ill-treatment or bodily injury has produced important injury to the health or limbs of the person injured, or has resulted in a long-continued incapacity for work, the punishment is to be imprisonment for not less than six months.*

§ 193. *If the person intentionally injured or ill-treated has been mutilated, deprived of his speech, sight, hearing, or procreative power, or has had any mental affection produced, the punishment is to be penal servitude for not more than fifteen years.*

§ 195. *When a man is killed in a brawl, or in an assault perpetrated by several persons, or when a severe (§ 193) or important (§ 192 a) bodily injury or ill-treatment has been thus inflicted on him, every one who has taken part, &c.* (refers to the amount of punishment).

§ 233. *Robbery is punished with penal servitude for life:—1, &c.; 2, when the robbery has been attended with the torturing of any person, or their mutilation or deprivation of speech, sight, hearing, or the power of procreation, or the induction of any mental affection by ill-treatment or bodily injury; 3, when the robbery has been attended by the death of any person from ill-treatment or bodily injury.*

§ 43. GENERAL.

There is no question in which, as experience has taught me, there · is so much uncertainty as to the limits of competence (both judicial and medical), as in that regarding the results of injuries during life; in none are there such remarkable variations in opinion in regard to these limits among individual judges, public prosecutors, and courts of justice. Whilst in one place, the forensic physician is *asked* by the court, as is constantly and without exception the case in the forensic practice in Berlin, whether for instance, N., in consequence of the injury received, has suffered an "important injury to his health or limbs;" and whether the injury is to be regarded as important in the sense of § 192 *a* of the Penal Code, or as severe in the sense of § 193 ? A proof that our judges recognize the competence of the medical definition of the ideas contained in the portions of the Penal Code referred to ; in another place, the forensic physician will be informed by the court, that his duty is only to describe the disturbance to the health arising from the injury inflicted, and to leave to the Judge, to whom the interpretation of the statutes belongs, to declare how far he understands these results to be comprised under the "important injuries" of the one or the other paragraph, &c. Whilst our superior Scientific Medical Board, in consequence of a requisition from a Prussian court, has found occasion to give a definition of the expression "capacity for work" contained in the Penal Code, the competence of medical experts to give this explanation being thus judicially acknowledged, the superior courts of justice have on their part, given an independent and perfectly different interpretation of the same expression (§ 50) ; while the same national Medical Board, upon the requisition of another indigenous court, has taken occasion in another *super-arbitrium* to explain the expression "mutilation" in the Penal Code, medical competence in this matter being thus again recognized, I have subsequently had occasion to hear a very high judicial authority state, that this interpretation belongs only to the Judge. I myself have constantly experienced in the course of my official connexion with the courts here (Berlin), the judicial members of which are constantly changing, as also with jury courts outside of Berlin, that there is in this respect not the slightest uniformity of opinion among public prosecutors, Judges, or advocates. And this is the reason why I have, after much consideration, treated this question as I have in the following pages, as I have thought that by so doing I could best be of use to medico-legal practice.

The Prussian statutes, as the regulations quoted above show, distinguish, 1. from the civil law point of view, and with respect to actions for damages, injuries—*a*, which produce a complete incapacity for procuring a livelihood by work or service; *b*, which produce a temporary incapacity for procuring a livelihood; *c*, which render the person injured partially incapable of procuring his livelihood; *d*, which render it more difficult to procure a livelihood; and *e*, which mutilate the person injured. 2. From a penal point of view the Prussian statutes distinguish; *a*, severe, *b*, important, and *c*, slight injuries; and in regard to this mode of classification all the German Statute Books, the *Code Pénal* (Art. 309, &c.), and all analogous Italian Statute Books, only take cognizance of the actual results to the person injured, and not, as formerly, of the mere possibility of the occurrence of these results, the mere danger to be apprehended from the injury. Therefore, in the present state of the matter, every forensic physician has a much more certain basis for his opinion, and a firmer position than formerly, when he was obliged to tread on the slippery ground of prognosis, and to state whether a given injury *might* have this or the other result, as defined by the statute? When we remember the very uncertain prognosis of cranial injuries for instance, we see at once that the present regulations of the Penal Code, however numerous the difficulties may be which they often in individual cases oppose to the medico-legal opinion, have yet brought about a most delightful change in regard to it. The medical man is no longer so placed, as to be obliged, contrary to his own internal convictions, to over-rate very highly the danger of an injury, because the Statute Book, with its bare possibilities, left him no other choice. In April, 1847, there was a great riot in the streets of Berlin, during which many of the police, both rank and file, were injured by stones thrown at them. Among those examined by us there were twelve who had received blows from stones upon their heads, faces, or eyes. In every one, more or less, insensibility had been produced; a few had to keep their room for one or two days, but all were, however, completely restored in a few days. Nevertheless, no one can deny the blows from stones, such as there described, "*might*" have produced "important injuries to the health or limbs;" and in accordance with the then state of the law, we were obliged to declare all these persons, who had been actually only slightly, to have been severely injured. A hundred times have I been in similar cases obliged to turn and wind in order to bring a formally correct opinion into unison with the precepts of experience, and of a sound understanding. As the Prussian Statute Book has classified injuries into severe, important, and slight, so the other extra Prussian Statute Books have also adopted a substantially similar division. In the first place, however, I may point out that the different countries have adopted in making their classification various criteria, which the Prussian Penal Code does *not* recognize. Among these are :—

" Permanent enfeeblement" (not merely the "deprivation ") " of the speech, hearing, and mind."—In the Austrian Penal Code,

" permanent ill-health"—*ibidem ;*

" incurable disease "—*ibidem ;*

"permanent " (not merely temporary) " unfitness for the duties of his calling."—In Austria, Bavaria, Saxony, Hanover, Oldenburg,

" incurable deprivation of the use of a limb."—In Bavaria,

" permanent " (not merely "important ") "injury to the health." —In Saxony,

"considerable," or also "remarkable " mutilations or disfigurements, and such as are "of slight degree," are included in most of the German Statute Books, as in those of Hanover, Saxony, Wirtemberg, Hesse, Baden, Weimar, Anhalt, &c. : whilst the Prussian Statute Book does not include two different classes of mutilations, and says nothing at all of disfigurements (§ 44).

Finally, the Hessian Statute Book mentions " the deprivation of any sense," thus including both smell and taste, none of these senses being even named by any one of the other Statute Books. I will by and by recur to other differences, meanwhile I may only remark, that not one of the more recent Statute Books contains such concise and simple regulations in regard to this matter as the Prussian, which in this respect is only surpassed by the *Code Pénal.* This simplicity may be preferable for penal purposes to those more complex regulations which enter more into the minutiæ of the results of injuries; of this the physician can form no opinion; but his relation to these questions is made decidedly more easy, when a large selection of definitions is placed at his disposal. In this case he will be more rarely placed in such a position as to be compelled to make such forced assumptions, as, under the Prussian law, he is frequently obliged to do (*vide* the illustrative cases following under § 52).

The conditions which, in the earlier penal laws were connected with the determination of the fatal character of injuries, the so-called degrees of lethality, have, as is well known (*vid.* Vol. I., *loc. cit.*) been fortunately deleted from all the penal codes of our day. It is singular, however, that the new Prussian, as well as almost all the other Statute Books, also makes no mention of these conditions in regard to injuries which have not proved fatal, that is, that it lays down no regulations in regard to the idiosyncracy of the person injured, the nature or degree of surgical assistance afforded him, or the necessary or unnecessary nature of the consequences which have resulted, &c., and as it were *implicite* excludes the idea of the possibility of the occurrence of any other result, inasmuch as it keeps steadily in view only the actual results which have occurred. Only the penal code of Hanover, so far as I know, expressly states (art. 241) that (in determining the amount of punishment) no regard shall be paid, " whether the consequences (which have occurred) have been prevented in other cases by proper assistance; whether this might probably have been the case had proper assistance been timely afforded in the case in question; whether these consequences have been the direct result of the injury, or of intermediate causes, which the injury has set in action; and whether these have only arisen from the constitutional peculiarity of the person injured." Almost precisely the same words which the Prussian Penal Code has employed (in § 185) in relation to the decision respecting injuries which have proved *fatal*. Nevertheless, however, all these conditions occur in the *living subject*, and come in question, too, *in judicial cases*, in spite of the position taken up by the Statute Book. The person injured is *now*, that is, at the time when the opinion of the forensic physician is demanded respecting him, indubitably " mutilated," for he wants an arm, and the case is unquestionably one of " severe bodily injury" according to § 193; but the injury has not caused this " mutilation," but the amputating knife of the surgeon. Has this amputation been necessary? Was it impossible to avert this " mutilation" from the person injured? Or the ill-treatment of a woman in the third or fourth month of pregnancy, blows with a stick upon the back and belly have produced an abortion with all its injurious results; such, for instance, as a long period of " incapacity for work," and the injury must be declared to be an important one in the sense of § 192 *a*. The "constitutional peculiarity," however, of the person injured, inasmuch as she happened to be pregnant at the time, has certainly aided in producing these results ! Cases similar to these are of frequent occurrence, and are constantly coming before us officially. The advocate for the defence in such cases, in spite of the silence of the penal code respecting such intermediate causes, is always in the habit of bringing forward queries in regard to them. The position of the medical jurist in such a case is quite plain. We declare in all such cases that in our opinion the *majus* in § 185 of the Penal Code includes the *minus*, that is, that since the medical jurist, according to the paragraph quoted, is no longer entitled, even in cases of injury which have proved fatal, to take these intermediate causes into consideration in order to show the injury to be a " fatal" one, he is still less justified in considering them in order to point out that the injury has been " severe" or " important." Next we unravel all the peculiarities of the individual case, and show why, for instance, the " mutilating" surgical operation has been from internal causes a necessary result of the original injury or the reverse, &c. and then we leave the interpretation of the case and its decision to the court.

§ 44. SEVERE BODILY INJURY.—1. MUTILATION.

There are several conditions laid down in the Statute Books, particularly in § 193 of the Prussian one already quoted, as constituting a " severe" injury ; these are as follow :—

1. *Mutilation.*—The etymology of the (German) word from the vernacular for 'trunk' or ' stump' shows that it signifies the loss of any part of the body. The Prussian Statute Book does not point out any particular part, while other German Statute Books specially mention the arm, the hand, or the foot. Nevertheless it is evident that the Prussian Statute Book does not intend the loss of any and every part of the body—a hair, for instance—to be included under the head of mutilation ; since the connexion of the definition " mutilation" with the most grievous results that can befal a person who has been injured, and yet escaped with life, shows that in relation to mutilation, the loss itself is not so much regarded as the consequences of that loss to the person injured. And here we must in the first place consider, whether these results can be averted by the medical art ? If this be the case under ordinary circumstances, then the results of such an injury cannot be regarded as a " mutilation ;"

for instance, the mere alteration in the position of internal organs,
and in particular, and in regard to the possible occurrence of such
accidents from ill treatment or bodily injury, *ruptures* or *prolapsus
uteri* cannot be regarded as mutilations, as they have occasionally
been, and this all the more that in such cases there is actually no
loss of any part. It is another question, however, whether the re-
placement of the part lost, by an artificial substitute, is to be regarded
as a kind of cure, and whether, therefore, the loss of a hand or of a
row of teeth ought any longer to be regarded as a " mutilation,"
since the person injured is no longer mutilated when he wears an
artificial hand or set of teeth ? This question must be answered
in the negative, quite irrespective of the fact that the greater number
of mankind are not in a position to procure any such artificial substitute, for the dead substitute can never replace the living member,
can never bring about an actual and complete *restitutio in integrum*,
and that alone is truly a cure. In general also, two other important
results of ill-treatment or injury—*deformity* and *crippling*—cannot
be cured. I have already stated that (whilst the Prussian Civil
Code makes mention of " deformity," and the damages to be assigned
for it) the Prussian Penal Code recognizes neither of these definitions,
whilst most of the other German Penal Codes mention deformities,
and even several kinds of them, and a few of these Penal Codes mention crippling, without exactly saying the word. Nevertheless, it
would be an error to conclude that our Statute Book intended to include deformities and disfigurements under the head of " mutilation,"
it would be an error, because deformities can be just as properly included among the " important injuries to the health or limbs" of
§ 192 a, which most indubitably they are. And the same is true in
regard to crippling. A man crippled by blows has indubitably suffered an " important injury to his health and limbs," and is also
usually for ever, or for " a long period rendered incapable of work."
Therefore, such cases combine both the conditions mentioned in
§ 192 a, and there has been an " important" injury, but no mutilation, for by the latter definition the actual loss of a part of the body
is always intended. But now we come to the point, that the
actual loss of a part of the body, is *in itself* not sufficient
to bring the injury within this category. And the reason
for this is that the Penal Code threatens mutilation with so
severe a punishment, because it places it in the same category
with those results of injuries which involve the most serious con-

sequences which a man can suffer from an injury and yet survive, and which set him more or less aside from all human society, namely, " the deprivation of speech, of sight, of hearing, of the power of procreation, or the production of any affection of the mind." We must, consequently look for some criterion which shall be common to all those injuries to health which are included in § 193, for each of them, as well as a mutilation, constitutes a " severe" injury. And such a common criterion we find in the important, severe, or incurable disturbance of some bodily or mental function, such a functional disturbance being openly expressed in the other conditions mentioned in the remaining paragraphs, as " deprivation of speech," &c. Since, therefore, the legislator has placed "mutilation" in the same category, it must be defined in a practical medico-legal sense, as *the loss by violence of a portion of the body, whereby an important, incurable disturbance of some function is occasioned*. We cannot avoid seeing that this definition involves many apparent anomalies and inconsequences. The loss of *one tooth* will be, according to it, no mutilation, while the loss of a whole row of teeth must, on the other hand, be certainly acknowledged to be a mutilation* (*vide* also case CXVII.). When we consider, however, I repeat, the position in which the word " mutilation" is placed in the Statute Book, and the nature of

* The Superior Court of Justice of Prussia has expressed a similar opinion, in a decision recently (24th June, 1858) delivered; yet it would appear from it as if the Court could only recognize the loss of all the teeth to be a mutilation. In the case referred to, two incisors were broken off close to the gums, and another tooth loosened. The Judge, on appeal, declared this to be a mutilation, since it was a lasting injury of a not unimportant part of the body; and in common language, to mutilate is to cause the loss of any limb, or portion of the human body belonging to it in its normal condition, and in any measure contributing to its wellbeing and the performance of its functions. In the annulling decision of the Superior Tribunal, the occurrence of a mutilation was denied, inasmuch as it cannot be assumed that the legislator has included under this term the loss of only two teeth, solely because such a loss inflicts a lasting " injury" on an important part of the body, and the teeth are "an external portion of the body that in some measure " contribute to its wellbeing and the performance of its functions; for this reason, that such an injury *bears not the slightest relation to the other severe injuries* included in § 193 along with mutilation; and such important differences exist between the loss of an essential part of the body (the teeth) and the "injury" of the same, that the latter can neither in common language, nor according to statute, be so expanded, without at once bringing into view these differences in connexion with the intention of the legislator. (Archiv. für Preuss. Strafr. 1859, vii. 1, s. 104.)

the matter, as in this case, for instance, the function of the teeth in relation to speech and digestion, the apparent anomaly disappears. A more important objection would be found in the cutting off an entire ear or the point of the nose. Both, in common language, are mutilations; according to the definition given here, only the loss of an entire ear is a mutilation, for the external ear is, as is well known, an important aid to hearing, which is materially injured by the loss of an entire ear. The loss of the cartilaginous nose, on the other hand, would be (a horrible *disfigurement* indeed, but) according to our definition, no mutilation, since the sense of smelling would not be injured by it. When we consider, moreover, that the legislator, for evident and certainly very proper reasons, has not mentioned the sense of smelling at all in § 193, the apparent anomaly, even in this instance, is found to disappear.

I have unfolded these views in an explanatory judicial superarbitrium, which was required from the Superior Scientific Medical Board, and they have received the sanction of that Board.*

§ 45. CONTINUATION.—2. DEPRIVATION OF SPEECH.

2. *Deprivation of Speech* also, according to the Prussian and other German Penal Codes, involves the idea of a "severe" injury. Austria, including Wirtemberg and others, also mention "a permanent debility of speech." Both of these horrible results of ill-treatment or injury can, from the physiology of the matter, be but of very rare occurrence. Nevertheless, among the very numerous cases of injury which have come before me, I have seen one instance of this nature (*vide* Case CXVIII.). A "permanent debility of speech," with us is included among the "important injuries to health or limbs," of section 192 *a*, and can consequently be only reckoned as an "important" injury by the medical jurist, a debility of this character may certainly persist after injuries to the head, accompanied by compression of the brain, just as it is constantly observed as a residual paralysis after spontaneous cerebral affections. In most cases there will be no difficulty in determining

---

* *Vide* the Superarbitrium of the Royal Scientific Medical Commission, in Casper's *Vierteljahrschft.* xi. 2, 1857, s. 193; and also *ibidem*, xv. 1, 1859, s. 162, where there is another opinion of our Superior Board of Justice than that given above, and in which also there is the most perfect agreement with the interpretation of the word "mutilation" given above.

its existence, and whenever there is any doubt, from suspected simulation, I urgently advise, as in every other case of disputed results
of ill-treatment and injuries, that the medical jurist should first of
all, by requesting an inspection of the documentary evidence,—
which, in Berlin, is generally sent to us in the first place—make
himself accurately acquainted with the manner in which the injury
was inflicted, the implement employed, &c., &c., since the statements of the injured person who is to be examined are never to be
trusted.   Thus instructed as to what has actually happened, he must,
in the next place, inquire whether it is physiologically possible that a
debility of speech should be produced by violence of this character:
whereby he must consider, not merely the material influences brought
to bear, but also the excessive terror, and other emotional excitement;
and then the medical jurist must proceed carefully to investigate the
result.   What, however, is a "deprivation" of speech?   This
question was forced upon my attention, in connexion with the judicial
case which came before me, as one perfectly novel and not easily
answered.   Of course, in the first place, it could not mean the inability to speak loud, and so as to be heard; for if so, the words,
voice, and speech would have been confounded, and we must term a
man, who had lost his voice from hoarseness, one deprived of speech!
But even those wholly without voice can communicate their thoughts
to their fellow-men by means of articulate sounds which can be
understood, and only when this power ceases does "deprivation of
speech" commence.   Even a deaf mute gives a vent to tones, and even
sounds, yet he is unquestionably "deprived of speech;" because his
sounds are inarticulate, only understood at the most by those
acquainted with himself and his gestures, but conveying no meaning
whatever to most men.   The case I have to relate proves that such
a deprivation of speech may be produced by violence.   And this
case has also been instructive in another point of view, namely, as to
prognosis.   As in this case, so in any similar one, we can never
declare with any certainty that the unfortunate results of the injury,
so evident at the time of the examination, would be subsequently
cured.   Yet in the case referred to a complete cure ensued, and
that in a comparatively short space of time.   Is, or was a man so
subsequently cured, "deprived of speech?"   This question must
be answered by the Judge.   The medical jurist will, after a careful
and repeated examination of the case continued for weeks, describe
the facts as he now finds them ; he will, if there be any hope, declare

the possibility of a cure, and propose a fresh examination at a future period, nevertheless declaring the person injured to be at present "deprived of speech," and must then leave the case for the decision of the Judge. In all the "deprivations" of the other paragraphs of the Penal Code, of the sight—of the hearing, and of the procreative power—this consideration ceases to have any influence, because any improvement or cure of the unfortunate subject of any actual "deprivation" of such a nature is no longer in the power of possibility. The experienced medical jurist will also beware of a simulated deprivation of speech, though, indeed, such a deceit will not readily occur to any injured person, and would be extremely difficult properly to carry out for any length of time; and this very difficulty will render the simulation easy to be detected by the medical man, provided he possesses some adroitness and practice.

§ 46. Continuation—3. Deprivation of Sight or Hearing.

It is precisely in regard to morbid affections of the organs of the senses that the medical jurist frequently feels so strongly, and is embarrassed by the fact of the Prussian Penal Code containing, as I have already pointed out, so few intermediate degrees between the most trifling and the most severe consequences, the result of violence unattended by a fatal issue. It frequently occurs, as may be readily imagined, that a man receives a cranial injury, be it from a blow of any kind, a fall, or a cut, &c., and as its result retains some defect in one or other of his senses, such, for instance, as singing or deafness in one or both ears, feeble or indistinct vision of both eyes, &c., although he, nevertheless, still hears or sees, so that it is impossible to declare that he is "deprived" of either sight or hearing. Other German Penal Codes give the medical jurist in this respect much freer scope. How many of these defects of the senses can be suitably comprised under the heads of "permanent ill-health," "incurable disease" (Austria), "protracted disease" (Bavaria, Baden), "permanent injury" (Saxony, Brunswick, Weimar, &c.), finally, even in certain circumstances, "disfigurement" (Austria, Saxony, Würtemberg, Brunswick, Hanover, Hesse, Baden, &c.) : some even of the Statute-Books go yet farther, and explicitly mention what I now refer to : thus, that of Austria mentions at least a "difficulty of hearing," if not of seeing also ; and that of Baden, which expressly mentions a *limitation* in the use of a limb, *or of the organ of any of the senses.* Consequently, in these cases, which are of such frequent occurrence, the Prussian medical jurist has no other choice but, when he is required so to do, to place the individual case under §§ 192 *a* or 193; to describe the existing defective condition of the sight or hearing, after he has satisfied himself that this is not merely simulated (§ 53, &c.) ; and further, to declare that it constitutes " an important injury to the health " (§ 192 *a*) ; or, his best plan in such a case, when not fettered by the mode of putting the judicial queries, is to describe correctly the actual state of matters, and leave it to the public prosecutor or the Judge to decide under which section of the Penal Code it ought to be included.

In other cases, six of which (Cases CXXI. to CXXIII., and CXXVI. to CXLVIII.) I shall relate presently, a consideration of a different kind arises. A man, for instance, is unquestionably " deprived " of sight or hearing, ay, completely deprived, but—only of *one* eye or ear. In regard to this also, other Statute-Books are more comprehensive than the Prussian ; Austria lays down regulations in regard to the " loss of one eye ;" Wurtemberg, in regard to the loss of sight, " or of one eye ;" Baden, in regard to the deprivation of a sense, or " of one of the organs of the senses." With us it is always questionable whether the loss of only one eye, &c., is to be regarded as a " deprivation of sight ? " For forensic medicine it is perfectly superfluous to enter upon etymological considerations regarding the signification of the word " deprivation," as I myself have done before the true state of the matter was made clear to me in the actual transactions of a court of justice; it is also superfluous to point out that there are possessions, such as health, goods, &c., of which we can be partially deprived, and others whose nature excludes the idea of partial deprivation—such as life and honour. For, whenever the medical jurist enters upon any such considerations, and endeavours to assert his own opinion in regard to them—as that a man with one eye knocked out, is not therefore to be regarded as " deprived of sight," or the reverse, he will be placed in direct opposition, in one case to the public prosecutor, and in another to the advocate for the defence. It is specially requisite, therefore, for the physician, in such cases, to hold fast to the old rule, and to leave the legal experts to fight out the dispute as to the interpretation of the statutes, and to content himself with describing the actual facts of the case—such as, that the injured person has been deprived of the sight of one eye, or the hearing of one ear, and how this has happened; or that one of the organs of the senses is weakened or temporarily diseased and its function disturbed, though it may still be restored, &c., and to consider that having done this he has fulfilled his duty.*

§ 47. CONTINUATION.—4. DEPRIVATION OF THE POWER OF PROCREATION.

All the German penal codes have included the loss of the procreative power from violence or injury, in the category of " severe" bodily injury. I have already (Vol. III., p. 238, § 1) described the difficulty of ascertaining the actual facts in cases of this character. Accusations of this nature are, however, of very rare occurrence, because the instinct of the public not incorrectly supposes, that the procreative power can only be injured by great violence locally applied; because, also, the genital organs are more rarely injured in brawls, sudden attacks, &c., than other parts of the body; and because, finally, other infirmities are more easily made plausible, by lies and simulation, than this, &c. Of all the causes of incapacity for procreation already (*loc. cit.*) described, only the following can specially be considered as likely to be produced by injuries or violence :—1. *In the male*, such injuries of the penis as may by their results—of which amputation may possibly be one—render that organ apparently unfit for the discharge of its function. A drunken crew, from pure cannibal wantonness, drew forward the penis of one of their comrades who lay senseless among them, laid it in a small frying-pan taken out of a set of child's playthings, and held lighted shavings beneath it to roast it!! A serious illness, ending in amputation of the organ, was the result of the infamous deed. I need not repeat, that where there are slight accidental injuries of the penis, the greatest caution must be exercised; just as in every case where the procreative capacity is disputed because of the condition of this organ. (*Vide* Vol. III., pp. 238 and 239, §§ 1 and 2.) Injuries of the testicles are also of rare occurrence in practice. When

* The High Court of Justiciary in Prussia has recently recognized (in a decision given March 3, 1859), " that the loss of the sight of one eye is not to be regarded as a deprivation of the sight in the sense of § 193 of the Penal Code, since it is unmistakable that the expression there employed, ' deprived of sight,' can only be understood to mean the total, and not merely the partial, loss of the power of seeing."—*Vide* Archiv. für Preuss. Strafrecht vii. 3, s. 390.

the most careful examination does not reveal the slightest material alteration in a testicle said to be, for instance, crushed by the tread of a foot, we must in the first place obtain permission, by giving our reasons, to defer giving an opinion for some time, so that any inflammation, effusion of blood, or the like, which may possibly be present, may have time to be removed; and we must also, and always remember the possibility of its being a mere case of simulation. The case is different when the injury has resulted in the castration of one testicle,* where its extirpation has been rendered necessary subsequent to, and in consequence of, the injury received. This is a case precisely similar to the loss of sight or hearing in one eye or one ear. The physician cannot speak physiologically of the "deprivation of the procreative power" where only one testicle is lost (Vol. III., p. 255, § 7); he must therefore certify, in his opinion, that this loss has actually occurred, and leave it to the Judge to refer it to its proper statutory category; in such a case he will probably prefer the definition "mutilation."—In one case an inguinal hernia, alleged to have been produced by violence, has been asserted to me as the cause of subsequent incapacity for procreation. It cannot be doubted that hernia may possibly render a man impotent (Vol. III., p. 245, § 4). But this only in the case of very large, old, scrotal herniæ which are irreducible, or cannot be retained, and which completely enclose the organ. Moreover experience, in opposition to popular belief and the opinion of many medical men, teaches us that ruptures are very rarely produced by injuries, ill treatment, blows, treads, falls caused by violence, &c., least of all ruptures of the kind described, which, when observed, may be always with certainty regarded as of many years' duration. Accordingly, any case such as that referred to is easily decided. 2. *In the female,* a local injury may possibly, by setting up inflammation, or by the surgical assistance rendered necessary, occasion such an adhesion of the walls of the vagina as to render the act of procreation physically impossible, and the woman would consequently be "deprived of the power of procreation." The fact itself would be easily ascertained; but in respect to its results, we must take into consideration all that has been already said as to the physiological operation of atresia of the vagina (*Vide* Vol. III., p. 243, § 3).—A much more frequent

* The *Code Pénal* expressly mentions "the crime of castration," and threatens it with penal servitude for life, and with capital punishment if death should occur within forty days.

cause of inability to conceive, namely, the ante- and retro-flexion of the uterus (Vol. III., p. 259, § 8) may be produced by violence acting on the lower part of the abdomen. But the curability of this affection excludes the idea of a "deprivation" of the power of procreation. Finally, at the risk of appearing superfluous, I will, because of my own forensic experience, mention what every one knows, that an abortion (which has itself been produced by violence) is no cause of future incapacity to conceive; wherefore, in the (three) cases which came before me of ill treatment, which must be acknowledged, from the circumstances, to have been in each case the cause of an abortion, it was not recognized to be "severe," that is, such as would occasion a "deprivation of the power of procreation."

§ 48. CONTINUATION.—5. THE PRODUCTION OF ANY MENTAL AFFECTION.

The German Penal Codes not only reckon the mutilation of the bodies of men as a "severe injury," but also, and very properly, that of their minds; for the latter even more than the former deprives him of communication with his fellow-men, and excludes him more or less from their society. And in regard to this also the Prussian Penal Code (and along with it only that of Brunswick) is distinguished, by a summary conciseness, from all the others, which may indeed be of advantage to the Judge, in giving him personally more freedom of action, with which we have no business, but which is, on the other hand, productive of more restraint for the medical jurist consulted. Whilst Austria, Wirtemberg, Hanover, Saxony, Hesse, and Baden, in the case of a mental affection produced by violence, distinguish in regard to the punishment applicable, whether there exists any probability of recovery or not, the Prussian Statute does not recognise this difference, but only asks the physician if any "mental affection" does exist? Whilst Austria, Wirtemberg, Hesse, and Baden speak of a "destruction of the mind," and Bavaria goes yet further, and especially mentions by name "furious madness, insanity, idiocy, or other similar mental diseases," Prussia only mentions the perfectly general term, "mental disease." What, however, is mental disease? and in asking this question, I mean, in reference to the statutes in regard to injuries—what is the boundary betwixt mental health and mental "disease?" This question is here also, not merely in relation to a state of disputed accountability, of great practical importance. For example, it frequently happens, after injuries or violence of any kind which has affected the head, that those injured assert, on examination, " My head feels always so stupid;" or, " I am so weak in the head, that I often do not know what I am doing;" or, "My thoughts are so confused that I act very stupidly," &c.; and these statements are also confirmed and found to be true. Is such a condition that of one " mentally diseased?" The mind is certainly not " destroyed;" and there is just as little question of its being a state of "furious madness, insanity, idiocy, or other similar mental disease." But such a condition cannot even be termed a " mental disease," for a psychical disease only exists where there are delusions. Again, therefore, there is nothing for the medical jurist but to determine the state of the case at the time of his examination, and to give a reasoned account of it in his opinion, after which it remains with the Judge to determine whether he will regard any such intermediate mental state only as an " important injury to the health " (§ 192 a, Penal Code), and thus include the injury that has caused it under the head of "important;" or whether he, on his own responsibility, nevertheless will regard it as a " mental disease " (§ 193, Penal Code), and the causal injury, consequently, as "severe." On the other hand, there will be no hesitation in deciding in regard to those, fortunately very rare, cases which exhibit the characteristics of perfectly indubitable specific morbid mental affections. But in these cases we meet with the second difficulty, already hinted at, one moreover, which also equally affects all those intermediate mental affections just referred to; I mean those frequent, very frequent, cases in which, in consequence of local violence, we, after the lapse of weeks or months, are able to discover the affection of the mind in the person injured, but in which we must confess that there does exist that " probability," or "reasonable expectation" of recovery mentioned in other German Statute Books, though unnoticed in the Prussian one. In such cases, a " mental disease" does indeed exist now, but will not probably in six months, or in one or two years. It is evident that the modern course of administration of justice, which will no longer permit a case to be hung up for a whole year, prevents the employment, in such cases, of an expedient to which we are compelled, both by conscience and experience, to resort in many other cases, namely, to reserve our opinion at the time, and to move for permission to make a fresh examination, after the lapse of a sufficient length of time. There is nothing, however, not even in the concise construction of the paragraphs of our Penal Code, to prevent the medical jurist from reasoning out, and laying before the Judge his opinion as to the curability of the present anormal mental condition, stating the possibility or probability, or very great probability of the subsequent restoration of the patient, and then leaving it to him once more to interpret and apply the statute properly.

§ 49. IMPORTANT BODILY INJURY.—1. IMPORTANT INJURY TO THE
HEALTH OR LIMBS.

All other Penal Codes assume, along with that of Prussia, the existence of an intermediate condition, intermediate between the more severe and the slighter injuries; and in the Prussian Statutes injuries belonging to this category are termed "important" (Penal Code, § 295, Vol. IV. p. 2). But in their regulations, which daily affect the medical jurist and his opinions, they deviate materially from one another, and particularly from the most recent construction of the Prussian Statutes. In the first place I may remark, that the Penal Codes of Austria, Bavaria, Brunswick, Wirtemberg, Baden, Hesse, Hanover, and the Saxon Dukedoms, speak of a "disease" as the result of injury, while phrases which express limitations or modifications of this idea, such as "disturbance of the health" or "chronic illness" are found in the Austrian code, and "injury to the health" in the Bavarian Penal Code. These expressions give a firmer hold to the physician; while every forensic practitioner must have felt with us the embarrassment frequently provided for him by the use of a word of so many meanings as "disease." The Prussian Penal Code also formerly spoke of "disease" as the result of injury, while now, since its recent reconstruction on the 14th of April, 1856, it no longer recognizes a disease as such a result. For who is "diseased" and who is healthy? There is scarcely such a thing anywhere to be found as absolute health, but only relative, and the physician can consequently not say that the individual (who has been injured) is absolutely healthy, but he can only ascertain and determine whether his state of comparative health be similar to that which he enjoyed previous to the receipt of the injury? And even with this necessary limitation, experience has taught me that cases do occur which are very difficult to decide medico-legally, in which a man was already diseased previous to the receipt of the injury, was indubitably diseased and also continued to suffer from the same illness at the time of the examination, such as chest complaint, chronic inflammation of the eyes, paralysis, &c., consequently is not healthy (is diseased), suffers from a disease which, however unquestionably, so to speak, has nothing to do with the matter, since it has nothing in common with the injury inflicted, and in these cases it is often extremely difficult to determine whether the (old) disease has not perhaps been made worse by the injury; that is, in other words, whether there is not now present as the result of the injury a *quasi* new disease, a greater amount of disease than formerly? But the word "disease" in the sense in which it is employed by the Penal Code requires yet another limitation. It is impossible to suppose that the legislator intended to avenge with such severe punishment—and no physician can or will understand the following to be a "disease" the result of an injury—when, for instance, the person injured at the time of the examination, for example, after the receipt of blows with a stick, still exhibits a yellowish-green mark or a little pain on pressure on the part struck, but is otherwise in perfectly good general health. Such cases as that in manifold variety are of frequent occurrence. In such a case there may indeed be an "injury to the health," but there is certainly no disease. No physician, and also no Judge, will call such a man diseased, although the results of the injury unquestionably still persist. Therefore, whenever and wherever the Penal Code speaks of "disease," a forensic definition of the idea thus expressed must be sought for and completely abstracted from that usual in medicine. And in this sense *disease* must be termed *such a disturbance of the health as gives rise to a general affection of the body, such as fever, violent pains affecting the whole body, a condition of general debility, &c., or, even when this is not the case, by which any function of the body is materially disturbed*, as, for example, the movement of any of the limbs or of the whole body, digestion, respiration, &c. This *medicoforensic* definition of the word "disease" also agrees with the general and popular idea, which I regard as by no means quite unimportant. A man suffering from fever, or inflammatory disease, or violent and constant pain, who has, in short, any general affection of the body, or a man who cannot move a limb, or who can eat and digest nothing, or who cannot make his water, or who suffers from difficulty of breathing, &c., in short, in whom any bodily function is materially disturbed, such a man will be termed "diseased," and rightly so, both by the medical and non-medical public; but not so a man who goes about his business perfectly healthy, but has a blood-

shot eye or blue welts upon his back.* In order to meet the difficulty just referred to, by far the greater number of Penal Codes have
set limits to the duration of the disease, and have proportioned the
punishment according as the " disease" occasioned by the injury has
been of longer or shorter duration. This also was done in § 193 of
the Prussian Penal Code previous to its reconstruction, and when the
" disease" resulting was " of more than twenty days' duration," the
injury was reckoned a "severe" one. The Penal Code of Austria
speaks of a duration of the disease of twenty or thirty days; Bavaria,
of it lasting one month or longer; and Brunswick, of a disease
lasting longer than three months; Baden, of one lasting longer than
two months, and Hanover and Hesse speak only indefinitely of " a
disease lasting for a long time." The German Penal Codes, therefore, vary in their Regulations from twenty up to more than ninety
days! Here there is evidently, from the very nature of the matter,
an absence of any guiding principle. Therefore, and because thereby
many practical disadvantages, as I myself have felt often enough, could
not be avoided, because the mere chance of better or worse nursing,
proper or improper treatment of the sick person, &c. might readily
cause a similar disease to cease in one case a day sooner, and in
another not for one or two days longer, but to exceed the statutory
limits of duration of the disease by even one day involved the most
important difference in the penal treatment of the matter, and therefore the Prussian Legislature was certainly justified, when, fortified
by an experience collected during five years, it omitted the definition
originally taken from the *Code Pénal* (art. 309) as to the " duration
of the disease of more than twenty days " from this statute, as recon-
structed in 1856, and completely set aside both the idea and the
word " disease." Besides—and this is another great difficulty which
was formerly much felt by us in medico-legal matters, and must still
be so in every other part of Germany and in France—medical men must
in these countries always object to their Judicial Boards, that health
and disease are not so distinctly defined in nature as in the Statute
Books, that in most cases these are separated by a stage of *convales-
cence*, which, intermediate betwixt both, can neither be properly
termed disease nor perfect health. The person injured, therefore,
may be, after twenty, thirty, or ninety days no longer " diseased,"

---

* *Vide* the superarbitrium of the Scientific Commission for Medical
Affairs, in Casper's Vierteljarschrift, III. s. 185, &c.

but not yet healthy, only convalescent. Further, every limitation of the duration of a disease implies regard to its curability or incurability. This possible curability is, moreover, explicitly expressed in the Penal Codes of Austria, Wirtemberg, Saxony, Brunswick, Hesse, Baden, and the Saxon Dukedoms, but not in the Prussian Penal Code. For this has, after completely setting aside all idea of "disease," pursued quite a different method, inasmuch as it reckons that to be an "important injury," when, in the first place, the person injured has thereby received an "*important injury to his health or limbs.*" Saxony and Hanover alone have similar definitions, inasmuch as the former speaks of "injurious consequences to the health" and of a "permanent injury," and the latter of a "trifling hurt." I do not need to examine the reasons of the Legislators, but I must point out what I have learned from much experience, namely, that a wide field is thus opened up for the most arbitrary individual estimation. What is an *important* injury to the health? remembering that in this instance those injuries which are most severe and consequential—as mutilation, loss of sight, &c.— are already excluded, since injuries of this character are included in § 193 (Severe Injuries). What one physician calls in any given case "an important injury," another will, with just as good reason, decline to consider important. An eyetooth was knocked out during a brawl; I found after several months a slight stiffness of the left ring finger, the result of a bite, and the curability of this was at the time just as uncertain as that of a small fistula on the right arm, which was the result of a punctured wound. I found that a buzzing sound in the head remained (unless it was simulated) subsequent to a blow from a stone. Similar cases are of extremely frequent occurrence, and the illustrative cases following will supply a number of other similar examples. Were these "important injuries to the health?" The answer may be either affirmative or negative from a medical point of view. In this matter also, therefore, it is proper for the medical jurist to take up the position which has been so often already recommended to him, that is, to describe the facts as he has found them on examination, and to leave it to the Judge to decide whether the injury found and described is to be considered "important" or not. If, however, the physician should be expressly asked, as according to my experience indeed almost always happens, to state his opinion as to the importance of the injury, then he must estimate the damage to the best of his conviction, and with due regard to all the other

circumstances of the individual case, and—content himself, should
the Judge (public prosecutor, or advocate for the defence) have a
different opinion, as is unquestionably their right, and his view is
not the one adopted.

§ 50. CONTINUATION :—2. LONG-CONTINUED INABILITY FOR WORK.

An injury is also, however, to be reckoned important according to
§ 192 a, of our Penal Code, when it has produced " a long-continued
inability for work," a state of matters very often combined with an
" important injury to the health or limbs," since the latter condition
is very likely to produce the former.   In the original construction of
the Penal Code, it contained (following the *Code Pénal*) a limitation
of the time, as in regard to " disease," to " an inability for work of
more than twenty days' duration," precisely as is still to be found in
the Penal Codes of Austria, Bavaria, Brunswick, and Baden, which
respectively define the duration of this condition as of from twenty
to thirty days, or one or several months, of more or less than three
months, and of more than two months, whilst the Wirtemberg Code
lays down an indirect limitation of time, when it defines the inability
for labour, as " temporary or permanent."   Only the Hanoverian
and Hessian Penal Codes use the same expression as the Prussian
one, and speak of a longer or shorter inability for work ; evidently
another extremely arbitrary definition on the part of the legislator,
and conveying a perfectly relative idea.   But in relation to the
medical jurist, this definition has this advantage over that conveyed
in an "important injury to the health," that there is nothing scientific in it.   The physician is thus, either in regard to this or other
matters, never in the right when he declares that, for scientific
reasons, he is justified in assuming that the person examined must,
from his bodily or mental condition, have been "unable to work"
for eight, fourteen, or thirty days, &c., and that he leaves it to others
to determine whether that is a long or a short time, as that is not
the duty of an expert.   If, nevertheless, he should be asked his
opinion, which, I repeat, is usually the case, since the opinion of the
forensic physician is only taken quite generally in relation to the
applicability of the statutory paragraphs, and he is usually asked
whether the injury has been a "severe," or only an "important"
one, &c., in the sense of the statutes ?   He may then declare his
own individual opinion, without putting any particular value upon

it. For my part, in these cases, which have been of such frequent occurrence, I have, in order to have some practical rule for guidance, been in the habit of defining a period of more than three weeks as " a long-continued inability to work," and in this I think I have the support both of the former statutory regulation and of the popular opinion; I always, however, make the reservation already referred to.

It takes far more consideration, however, to define what is " capable for work," than merely to estimate the period of time. . For there is, as I have already pointed out in regard to health, no absolute capability for labour, but only a relative one. Different ages, sexes, and positions have different standards of capacity for work; consequently, in any question as to the capacity for work of any injured person, only that relative capacity can be meant, which was possessed previous to the receipt of the injury, in relation to the subject as well as the object. The learned man, who, from an injury to his head, has become weak in the mind and half idiotic, and must give up his literary or teaching labour, the violin-player, who has lost one finger of his left hand, and can no longer play the instrument which fed him, the professional flute-player, whom the same fate has overtaken from an injury to his tongue, have all become unable for work; and to assume the contrary, because these three men can still plait straw or strip quills, would be to give the legislator credit for an absurdity. Just as little can it be assumed that the definition, " unable for work," refers only to a mere livelihood, the bodily necessaries only, for in that case it must be assumed that the legislator must have intended that children, for instance, and persons of independent means, &c., should never suffer an important injury in the sense of the Penal Code. " Capacity for work" is rather to be defined as, " the capacity for exerting the ordinary bodily or mental power in its ordinary measure." In this sense, even a child can become incapable of performing its " labours," to go to school, &c., and thus prepare itself for a future position, and the man of property also, even though his ordinary " labours " consist in looking after his means, in taking daily walks for the preservation of his health, in the usual mental employments, &c. Wherever the relative capacity, thus defined, is cancelled by an injury, and for " a long time" remains less than it was, there an " important " injury, in the sense of § 192 a, of the Penal Code, must be assumed to have existed. This opinion of mine is also acknowledged to be that of the highest scientific Medical Board, in the superarbitrium already quoted (Vol. IV., p. 20, *note*); it is not so recognised, however, by the highest Prussian Court of Justice. It agrees with the *travail personnel* of the French Penal Code and the expression "professional labour," employed by all the German Penal Codes, except those of Prussia and Brunswick. Nevertheless, the Royal Obertribunal has seen it right to state (Decisions, vol. xxviii.), "he is not to be regarded as incapable of labour, who is not able for his usual amount, but can still work a good deal; neither is he, who, though he is unfit for his professional avocations, is yet able to perform other ordinary bodily labour. Incapacity for labour must not be understood as meaning every diminution of the power to labour, and not the inability to discharge professional duty, but only the incapacity to perform ordinary bodily labour not requiring special skill." It is neither requisite, nor do I possess sufficient knowledge of the matter to enable me to examine and criticise the expressed opinion of so august a judicial Board. In justification, however, of an opinion, which is not only mine, but also that of the highest Medical Board in the kingdom, I must be permitted to point out the extraordinary hardship involved in such an interpretation- in regard to those, by no means selected instances, already given, to point out that it stands opposed to that of all the other Statutes which relate to "professional avocation;" that, further, it stands opposed to the Prussian Civil Statutes themselves, which everywhere (*vide* Vol. IV., p. 1) refer distinctly and expressly to the *travail personnel* (office or trade); and this is of great practical importance, because very frequently the criminal action on account of long-continued "inability to work," is followed by a civil action for damages. According to the interpretation given by our Obertribunal, a little child cannot receive an "important" injury, in the sense of the Penal Code, since, even previous to the receipt of the injury, it was "incapable of performing ordinary bodily labour (that did not even require special skill);" the injury has not, therefore, deprived it of anything of which it was previously possessed!

But even in a practical, medico-legal point of view such a definition of "incapacity for labour," is an extremely restricting one. The physician can, indeed, in any given case, make himself acquainted with the several muscles or parts of the body which are chiefly employed, or what injuries to health may arise from any trade or craft in question, and employ the knowledge thus obtained in forming his

opinion; but he could not be expected, for it lies beyond the domain
of his science, to make himself acquainted with the hundred different
" ordinary bodily labours not requiring special skill," to which a
learned man, an artist, an official, a tradesman, a coachman, or an
agriculturist, &c., may turn their hand, when rendered unfit for the
performance of their " professional avocation " by the effects of an
injury.   I make this statement also on my own experience, for I
have often enough had to deal with this difficulty in civil actions for
damages.   Another defect in our Penal Code is still more frequently
felt in practice.   It says nothing at all, namely, in regard to the
amount of capacity for labour.   And yet it daily occurs that a man
(formerly twenty days, now a "long time") after the receipt of an
injury can do indeed something by dint of great exertion or with
frequent interruption, but cannot do a full day's work.   I have,
therefore, considered it right, in the definition of my idea of capacity
for labour (Vol. IV., p. 23), to have regard to the "usual amount."
The explanatory interpretation by the Obertribunal of the paragraph
of the Penal Code referred to, takes a different view in regard to this
also, inasmuch as it expressly excludes the mere " diminution of the
power to labour;" thus, for example, holding as capable of labour
the teacher, who by great exertion can teach one hour, instead of
seven or eight, as formerly; or the labourer who, with difficulty, can
execute a fourth part of his former day's work.   This also is opposed
to the views expressed in several of the other Penal Codes (Bavaria,
Wirtemberg, Saxony, Hanover, and the Saxon Dukedoms), which
speak of a *complete* (consequently also of an incomplete) unfitness for
the discharge of professional duty, and thus their medical jurists have
a basis according to which to decide each individual case, which we
in Prussia have not; our position in regard to the question of the
disputed capacity of a man for labour is in general as follows:
whenever a definite question is put by the Judge, this must be, as
in every case, as far as possible answered literally, in accordance with
the actual state of matters found, or reasons must be given why this
cannot be done.   When this is not done, but the Judge, as in by far
the larger proportion of cases, presupposing (and rightly so) the
forensic physician to be acquainted with the statutes applicable, merely
asks, for instance, whether the injury is to be regarded as a severe one
according to § 193, or as an important one, according to § 192 a?
Then the physician may, in a course of suitable reasoning, unfold his
opinion in regard to the "long-continued incapacity for labour,"—

and the expressed opinion of the highest Medical Board has certainly just as little statutory influence on the formation of this opinion as that of the highest judicial court—and leave it to the Judge to accept this opinion or not, as he pleases, the material and essential part of the medical report, the actual facts found upon examination, remaining, nevertheless, unchallenged; and this is here, as always, that which the Judge, in a court of examination, specially requires to be determined by the medical jurist. I refer for this matter, as well as for all this chapter, to the illustrative cases following.

§ 51. SLIGHT BODILY INJURY.

Besides severe and important injuries, the Penal Code, in the paragraph already quoted (Vol. IV., p. 2), also makes mention of " slight bodily injury." Of course all the other Statute Books contain similar regulations. The Austrian one speaks of injuries " in themselves slight;" that of Saxony, of injuries " without danger or injurious consequences;" that of Wirtemberg, of "injuries of less severity ; " that of Hanover, of injuries attended by " the danger of only a slight permanent blemish, or of a permanent disfigurement of trifling importance," or by the results of "a disease curable in a short time, or a short time's incapacity for professional duty; " the Hessian Penal Code speaks of "a trifling bodily injury attended by a short illness or unfitness for work ; " and so does that of Baden, which moreover, enumerates as diagnostic signs of the condition meant, " the slightly remarkable deformity or the mere limitation in the use of a limb, or of one of the organs of sense," &c. The Prussian differs from all these German Penal Codes in not vouchsafing one single word to say what it wishes to be understood by the expression, a " trifling " injury, in this also following the example of the *Code Pénal*, the regulations of which are equally negative in construction. But this negative definition is perfectly sufficient for the practical purposes of forensic medicine. Because to the medical jurist *every bodily injury* is, by logical necessity, a *" slight" one, which has not produced any of the consequences detailed in* §§ 193 *or* 192 *a*, and which consequently cannot be termed either a severe or unimportant one. As always in determining the alleged results of violence or injury, so I must repeat (Vol. IV., p. 11), that the medical jurist must specially exert the most scrupulous caution in investigating all disturbances of the health which have arisen subsequent to injuries which were apparently only trifling and unimportant; and to this also the personal experience of every one will very soon urge him. Mutilations, deprivation of sight, &c. &c., cannot be easily simulated with any certainty of result, and this, therefore, is but seldom attempted, even when there is an abundance of wrath against the perpetrator of the injury, or of desire to extort money from him, &c. Nothing is, however, easier or more likely to occur, even to the malevolent possessed of but little cunning, than to simulate subjective disturbances and ailments of every kind, pains, sleeplessness, debility, giddiness and the like, or to exaggerate extremely any such affections which may actually exist, in order to deceive the medical man, and to get him to give an opinion favourable to the prosecutor. The more, therefore, at the first glance, on examining the condition of the body, only a "trifling" injury seems to exist, so much the more must I, taught by my experience, say to him, " *Cave!*"

§ 52. Illustrative Cases.*

A.   Severe Bodily Injuries.

I. MUTILATION.

Case CXV.—Bite of the Finger.—Amputation.

On the evening of the 9th of August the journeyman locksmith, P., bit, in a brawl, the little finger of the left hand of the turner, P. On the 14th, that is after five days, the surgeon H. saw the wounded person, and found on the finger a "gangrenous inflammation," as evinced by "bluish-black colouration," separation of the cuticle, and redness and swelling of the surrounding parts. The appropriate remedies did not produce the desired result. The gan-

* Many of the cases here given occurred before the introduction of the present Regulations of the Penal Code on the 14th of April, 1856; that is to say, either while the old Penal Code (Common Law, Tit. 20, Part II.) was still in force, or under the *régime* of the Penal Code of 1851, previous to its most recent reform. This occasions and explains the various modes of deciding the cases, according to the form of the judicial queries. When we were asked as to a "severe" bodily injury previous to 1856, it meant something quite different from what it now does, as I have already pointed out. In self-justification, and for the proper understanding of the cases, those which have occurred most recently (since the 14th of April, 1856) are marked with an *.

grenous soft parts did indeed become partially separated, but the middle joint of the finger was opened into after destruction of the capsular ligament; and on the 31st of August, twenty-two days after the bite, the finger had to be taken off at this joint. The surgeon had not been able to see the marks of the teeth upon the gangrenous finger, yet he inferred "pretty certainly" the truth of the statement of the person injured, "since the capsule of the joint was injured." The accused had in the same brawl bit the fingers of *two* other men, and was, from the evidence of eye-witnesses, "perfectly mad" from drink and excitement. "At the present time (October), the amputation wound is perfectly healed, but the injured person is *mutilated*, inasmuch as he is deprived of the little finger of the left hand, and the use of the left hand is thereby much injured. It is a positive truth, based upon medical experience, that the most dangerous wounds may result from the bite of a man excited with wrath. And it must also, therefore, be confessed, that a bite, followed by inflammation, and gangrene, with all its results up to the final necessity of amputating the limb, as in the case before us, might have occurred, and the entire history of the case, as described by the person injured and by the surgeon, presents nothing improbable." Accordingly, and since a "mutilation" did exist, we answered the question, is this a "severe" injury? affirmatively.

CASE CXVI.—A TOOTH-STUMP KNOCKED OUT.—IS THIS A
MUTILATION?

By no means an easy case! On the 7th of August, ten weeks before my examination, the accused struck the woman examined with his fist upon the face, upon the head, and on the breast. Besides this, according to her statement to me, he also gave her a blow with the fist on the left side, about the region of the spleen, upon which she slid down a few steps. Dr. B. had examined this woman, aged thirty-five, shortly after the violence had been inflicted, and found that her under lip was swollen, as well as the chin covered with coagulated blood. A piece of the cuticle of the upper lip, the size of a fourpenny-piece, was abraded, and there was some fluid blood visible at the left corner of the mouth. The gum was bloody; and blood trickled from the cavity from which the tooth had been forced. On the right cheek there were two ecchymoses, arising from violence applied with dirty fingers. The injured woman complained of pain,

particularly in the region of the left ovary, of headache and giddiness. ´ Her head was hot, her pulse quickened, beating ninety-two
times in the minute; great exhaustion and tendency to weep were
also evinced. The physician declared the injuries to be of considerable severity, because the woman N. had for years suffered
from hysterical convulsions, as well as from a tumour of the cervix
uteri and left ovary, and had been all that time under medical treatment. The injured woman had to keep her bed for twenty-three
days, and on the 3rd of September she was, according to the statement of Dr. B., still so weak that she could not leave her room;
"at present no traces of the unimportant external injuries just
described are any longer visible. But close to the old gaps of two
of the upper incisors there is a small depression, still reddened, in
the gum in the situation of the left eye-tooth, and the woman declares, that this has arisen from the stump of that tooth having been
knocked out by a blow with the fist. The appearances certainly do
not contradict this statement : nevertheless, the mere loss of the *last
remnant of a tooth* cannot be regarded as a "mutilation" (§ 193,
Penal Code), since the woman is not thereby either disfigured, or
deprived of the function of any organ, for the stump of a tooth cannot be regarded as such. Further, I have also found that there is a
prolapsus vaginæ. But the woman, N., does not deny that she
suffered from this previously, and moreover, this could not be
regarded as having been caused by the comparatively trifling violence
inflicted. The same may also be said in regard to an inguinal
hernia, which is to be felt in the left groin. Consequently we have
only further to consider the general disease under which the woman,
N., is said to have laboured as the result of the violence inflicted.
It is confessed, that she was previously an ailing woman. She
asserts, nevertheless, which is also medically very probable, that in
spite of her chronic ailments, she was still able to look after her
household affairs. To suppose that an ailing person cannot become
more ailing, that is, that a new disease cannot be superadded to the
old—cannot suffer from a severe injury, when the new disease lasts
longer than twenty days, is to impute an absurdity to the legislator.
In all these cases we have only to consider the relative and individual
health of the person injured, as it existed *previous* to the receipt of
the injury, and to compare it with his state of health subsequent to
the injury. As now the evidence of her medical attendant shows,
that the woman, N., was more ailing subsequent to the receipt of

the injury than she was previously, since she suffered in particular
from such great bodily debility that for more than twenty days she was
unable to leave either her room or her bed, or to discharge her household duties as formerly, so we must assume, that she has been for more
than twenty days after the infliction of the violence " ill and also
unfit for work," and, therefore—presupposing always that Dr. B.
has assured himself that his patient was not merely simulating—
we must answer the question put before us by saying, that the
injury in question has been a severe one* in the sense of § 193 of the
Penal Code.

Case CXVII.—Amputation of a Breast.—Is this a Mutilation?

This was a rare example of cases such as those referred to in
§ 44 (Vol. IV.). According to the definition of the word " mutilation" there given, the apparent anomaly must be held to be correct,
the loss of one tooth is no mutilation, but the loss of a whole row
of teeth is a mutilation; thus also the injury of the female breast
which may occasion its partial or complete extirpation, must be
termed in one case a mutilation, and in another not so, but only an
"important" bodily injury, according as the *function* of the breast may,
or may not be interfered with. Of this the present case is a proof.
The woman, B., already some fifty years of age, had been struck on
the left breast a year and a-half ago by a piece of metal, alleged to
be heavy, which was thrown at her. Pain instantly commenced and
continued. After from six to eight weeks she observed a hardness
of the breast, for which she used domestic remedies, till the increase
of the swelling and the violence of the pain forced her to consult
Dr. X., who removed a tumour the size of a goose-egg three months
before our examination. We only found the red line of the incision,
and the breast almost quite disappeared. There was no appearance of any glandular swelling, either in the neighbourhood or in the
axilla, &c., and her general health was perfect. At the preliminary
examination, we decided that it was "not improbable" that the
affection of the breast had been the result of the injury, since the
present appearances did not justify us in assuming the existence of
any carcinomatous diathesis, and the mode of development of the
tumour supported our idea. But we could not declare that this

* Now only an "important" one, according to § 192 *a*.

woman had met with a mutilation, because her age and constitution proved that she had passed the period of childbearing, consequently, she could no longer be a nurse. In this case, therefore, this injury was only an "important" one. In a woman still capable of bearing children a similar injury, attended by similar results, must have been declared to be a "severe" one. At the trial for the first time (1) the *corpus delicti*, a light piece of tin was produced, and we had to declare that it was still less likely than we had formerly supposed, that this body had produced the injury.

II. DEPRIVATION OF SPEECH.

CASE CXVIII.—TEMPORARY LOSS OF SPEECH SUBSEQUENT TO THE INFLICTION OF VIOLENCE.

Loss of speech is the rarest of all the ills that result from violence. The following case is therefore all the more remarkable (even in a pathological point of view!). The accused was a teacher, who had been previously insane, and was well known for the violence of his temper. On the 17th of July he had struck Eliza, aged twelve, with his fist on the breast and back, and had also "seized her by the throat." The child was brought home (on foot), and there livid patches were seen upon the neck and breast, and it was immediately discovered that the child had lost her speech. We examined her fourteen days after the accident. Not a trace of any violence could be any longer discovered, and the child was perfectly healthy. On the other hand, her speech, which was said to have been formerly normal, was most remarkably altered. For the child's voice consisted of inarticulate sounds, perfectly unintelligible, and when asked the simplest question, it was only with difficulty that what she wished to say could be puzzled out. But just as the inarticulate, howling, sudden bursts of sound from a deaf mute cannot be called "speech," so neither could the child's present mode of speaking be so termed, and I had made up my mind to declare that a "deprivation of speech" had in this case been the result of the injury, whether this had been caused by psychical impression or actual compression of the nerves affected. Partly, however, from the unheard-of rarity of the case, and partly to do away with any possibility of simulation, I still withheld my report, and continued to observe the child.   Fifteen days later, however, the child's condition was perfectly unchanged; she could not name one single consonant, and nothing but vowel sounds were heard.   The fact of there being a "deprivation of speech" was specially evident when the child was made to read aloud, for then nothing was heard but a perfectly unintelligible blatter of vowels.   For this reason, therefore, according to the statutes, the injury must be regarded as a "severe" one.   In the expectation, however, that this affection would probably disappear in time, I applied for and obtained permission to postpone my report. Four weeks subsequently I saw the child again, and found her perfectly restored and speaking with fluency and distinctness.   According to the statement of her family, this improvement had taken place gradually, and it was only eight days since she had been first observed to speak in her former normal manner.   The injury could not now be declared to be a "severe" one.

CASE CXIX.—NON-DEVÉLOPMENT OF THE SPEECH AS THE RESULT
OF GENERAL ILL-TREATMENT.

A boy was said to have been crippled by ill-treatment, and in particular to have had the development of his speech thereby prevented. We found this child, aged five, very backward both in bodily and mental development, small for his age, extremely thin, and the bones both of his chest and pelvis very much behind in their development. His appearance was extremely characteristic—half-animal, half-idiot— and his whole physiognomy had a most remarkable similarity with that of an ape.   And his whole mental condition was of a correspondingly low type; in particular, the child could certainly not yet speak, but only mumbled inarticulate sounds.   "Accordingly," we stated, in our opinion, "it is at once evident to every man of experience, that these mental and bodily anomalies bespeak some congenital defect, some fault in the primitive formation, some imperfect development, and not anything which has been acquired or depends upon external laws; consequently, any ill-treatment to which the child might have been subjected could not have produced such results; moreover, no trace of any ill-treatment could be observed on any part of the boy's body."

3. DEPRIVATION OF SIGHT.

CASE CXX.—LOSS OF BOTH EYES FROM QUICKLIME.

A most horrible case, and one to which the statutory definition, "Deprivation of sight," is most undeniably applicable! On the evening of the 17th of July, as the boy Hugo B., aged fifteen, was standing some little way off and watching the workmen, M. and K., slaking lime in the court, either intentionally or by accident—both were asserted, but which it mattered not as far as our opinion was concerned—he had some of the half-slaked lime thrown upon him, which burned him. Dr. R. examined him the same evening, and found a most intense inflammation of the conjunctiva of the cornea, and of the eyelids. On the 19th of the same month Dr. S. found the closed eyelids of both eyes swollen, inflamed, and in part suppurating. On separating the lids the conjunctiva was found to be swollen in both eyes, and the cornea so obscured that even then the entire loss of sight seemed probable. The boy himself has declared, that immediately after the accident he felt violent pain in his eyes, and could not open them. "The curative means hitherto employed have been of no avail, and the boy B. has been dismissed as incurable from the clinique of Dr. v. G., after fourteen days' treatment. The condition of his eyes, as observed by me, is as follows : the eyelids of the left eye are puffy, swollen, and reddened, those of the right one are much less so. On both eyes, however, the lids are so closely adherent, not only to each other but also to the whole eyeball, that they cannot be opened in the very slightest, and of course, it is therefore impossible that vision can exist. This adhesion must be the result of a foregoing very intense inflammation, such as was observed immediately after the receipt of the injury. At present the condition of either eyeball can be no longer ascertained. The extreme intensity of the inflammation is explained by reflecting that a hot, corrosive fluid was squirted, in considerable quantity, into the eyes." Of course, I had to declare that I quite agreed with Dr. v. G. as to the incurability of the mischief, and that there could not be a doubt that the case was one of "deprivation of sight" ("severe" bodily injury).

*CASE CXXI.—LOSS OF ONE EYE FROM SULPHURIC ACID.

A worthy companion to the foregoing, and a case of the vilest wickedness.

A young person had been scorned by her former sweetheart, H., and she resolved to be revenged on him. She waylaid him, and flung undiluted sulphuric acid over his whole face! The result was ineffaceable corrosion of his whole countenance. When requested to examine H., in the Hospital, "in regard to §§ 192 a, and 193, of the Penal Code," I could not at the first visit observe anything, because I was unwilling to remove the carefully applied dressings, and besides, I must at any rate wait for and see what the result of the injury would be. On a subsequent examination, when the swelling of the eyelids had somewhat lessened, and the state of the eyes could be investigated, I found that the vision of the right eye was not lost, but that the left eye had completely suppurated, and was consequently lost for ever. Three months subsequently, at the time of the jury trial, his condition was as follows,—His whole countenance was disfigured with cicatrices. From deficiency of skin, and from the cicatrices the lips could not be freely moved in every direction. On the right (the good) eye there was an ectropium of both eyelids, consequently this eye could not be shut. The left eye was completely closed by adhesion of the lids, and through them the globe could be distinctly felt as a mutilated stump. I declared and explained this state of matters to the jury, and—with the assent of the Judge, the public prosecutor, and the advocate for the defence—I left it to the court to decide whether there was in this case a "deprivation of sight," or "mutilation," or "severe injury," according to § 193, or whether there was only "important injury to the health, or long-continued inability to work"—only an important injury, according to § 192 a. The judicial decision of so remarkable a medico-legal case will be learned with some interest. The sentence was ten years' penal servitude, the case being regarded as one of severe injury intentionally inflicted, and which had resulted in "deprivation of sight of the left eye, and mutilation of the right one (?), which could not be shut, and its integrity was thus endangered."

*CASE CXXII.—LOSS OF AN EYE FROM A STAB OR BLOW.

A young watchmaker, in a perfectly trifling struggle with an older associate, was struck in the right eye by the hand of the latter. Whether, as the injured party alleged, there was a pointed instrument in the clenched fist, or whether the thumb alone projected from between the closed fingers and had struck the eye, as the accused asserted, stating that "this was his usual blow when any one insulted him" (!), remained unascertained, and was of no consequence, since the results of the injury were certain. According to the perfectly credible diagnosis in the certificate from the Eye-clinique, this consisted in a "separation of the retina with irrecoverable loss of sight." To this a complete cataract was subsequently conjoined, which at the time of my examination completely prevented any inspection of the interior of the eye. I explained the facts, and stated that there was "deprivation of the sight of one eye." In this case, however, there was no judicial interpretation, as the jury returned a verdict of "Not Guilty."

*CASE CXXIII.—LOSS OF AN EYE BY A BLOW.

The following truly lamentable case was very similar to the foregoing; in it a *deaf mute* child lost an eye! The boy, aged eight, received a blow on the left eye, from a willow wand thrown by another boy. In spite of the most careful attention in an hospital, the inflammation advanced to ulceration, perforation of the cornea took place, and the eye was lost. My opinion was given in precisely the same words as in the foregoing case, leaving it to the court to decide whether the deprivation of the sight of one eye should be included under the head of "severe" injury, or not. In this instance the court only declared the injury to be "important."

*CASE CXXIV.—ALLEGED LOSS OF AN EYE SUBSEQUENT TO BLOWS IN THE FACE.

W. displayed the commencement of a cataract in his left eye. It is perfectly credible, we stated, that, as he asserts he can only imperfectly see with this eye, as this affection is readily explicable by the advanced age of seventy years attained by the person examined, while the assumption that it has arisen from blows received in the

D 2

face five months ago is not supported by medical experience, the latter idea is all the less probable, as the surgeon L., who examined him four days after the infliction of the violence, did not observe any inflammation or anything else anormal about this eye. Moreover, W. also complains of weakness in the left arm, which he likewise attributes to the blows received. Objectively, there is nothing anormal to be discerned on this limb; this statement, therefore, is purely subjective. The surgeon also, at the time referred to, observed ecchymoses on the left breast, but nothing of the kind upon the left arm. But even if the arm had been struck by blows with the hand, such an "injury" is not of such a character as that any long-continued inability to work would be in general supposed to result from it, and when, moreover, the person examined asserts that even yet, after the lapse of more than five months, he cannot make use of his arm, the truth of this statement is liable to so much doubt that all that we can agree to is, that advanced age has rendered it less fit for performing laborious handiwork. Accordingly, I declared, in answer to the question put to me, that the injury could neither be regarded as " severe " nor " important."

CASE CXXV.—DREADED LOSS OF ANY EYE BY A BLOW FROM A WHIP.

This was another case which we had to report upon with express reference to § 193 of the (then) Penal Code. A boy, aged seven, had received a blow from a whip in the left eye. We found an active inflammation of the conjunctiva and an ulcer the size of a millet seed three lines from the edge of the cornea. Accordingly, we declared that " deprivation of the sight," even of one eye, was not to be feared under any proper treatment. But as it can be foreseen that the disease will not be cured even after the lapse of six days ($14 + 6 = 20$), and the injured boy cannot, consequently, follow his usual employment, go to school, &c.; therefore, since in this case we have to deal with "disease and inability to work for more than twenty days," the injury must be declared to be "severe," according to § 193. (Now-a-days it would only be termed " important.")

4. Deprivation of Hearing.

Case CXXVI.—Can One Ear be deprived of Hearing by the slap of a hand upon it ?

Whilst the earlier Penal Code (Gen. Com. Law, Tit. 20) continued to be in force, we had to deliver a superarbitrium in the form of an answer to the following question placed before us by the public prosecutor : " Is rupture of the drum of the ear, by means of a blow with the open hand, when given with special violence, a rare occurrence, and is not rather, in opposition to the opinion of Dr. K., such a result very likely to occur when such a blow is given directly upon the surface of the ear, and is not the accused, even if he cannot be supposed to be guilty of intention to injure, at least liable to be dealt with for punishable carelessness, in the sense of the General Common Law?" This case affords another proof of how often the interpretation of the statutes concerned is required from the medical jurist. We answered : " According to the documentary evidence, on the 30th of November, S. struck the journeyman musical instrument maker C., while working in the workshop, such a blow upon the left cheek with his open hand that it swelled up and blood ran out of the left ear, while C. could not work any more on account of violent pain in his head. Dr. K., who visited medically the injured person on the same day, declared in his certificate of the same date (and on account of which the injured party maintained that a lifelong injury had been done him, *since he would be for ever prevented from tuning musical instruments*), ' That no alteration of the ear was visible externally, but that the probe penetrated one line deeper into the left than into the right ear, and that air driven, while holding the nose, through the Eustachian tube escaped through the left ear, from which it could be indubitably concluded that there was an opening in the tympanum.' At the oral trial the journeyman joiner R., an eye-witness, confirmed the case so far as that he stated that he had not indeed seen, but had heard a blow, and that he had seen his left cheek not swollen ' but slightly reddened,' and the ear not bleeding. At the oral trial Dr. K. thus delivered his opinion, ' That complete deafness of the injured ear would not occur, but a certain amount of dulness of hearing would. It is, indeed, possible that the tympanum may be.

ruptured by a blow with the hand on the ear from the compression of the air thus induced; but such a result is one of the rarest occurrences, and one which cannot be foreseen. He did not observe any swelling of the cheek or traces of blood, and he leaves it undecided whether the aperture found in the tympanum is actually the result of the blow received or of some other and earlier cause.' The undersigned can only express his agreement with this opinion. Ruptures of the tympanum by external violence, not merely apertures caused by diseases of the internal ear, are phenomena of the rarest occurrence, and have only been observed when violent concussion or compression of the air in the auditory canal has been produced by explosions of powder, spent balls, and the like ; independent of openings or lacerations in the membrane caused by pointed weapons which, as produced by actual contact, have no reference to the present case. That, however, a rupture, so rare of itself, should be produced by so slight a concussion and compression of the air as that caused by a slap on the cheek, even though this should have struck ever so directly upon the surface of the ear, must be certainly a most rare occurrence, as daily experience teaches, when we consider the extremely frequent application of violence of this character compared with uncommon rarity of ruptures of the tympanum on the whole."

" With this statement I think I have also answered the second of the queries put before me, Whether the accused, though not guilty of intentional injury, is yet liable to be dealt with for punishable carelessness in the Common Law sense of the expression? Of all the various paragraphs referred to (Tit. 20, §§ 511, 691, 780, 28), § 28 seems to me to be specially referred to, according to which he, who by proper mindfulness and consideration might have foreseen the illegal results of his actions, has been guilty of punishable carelessness. With all mindfulness and consideration, however, in my opinion, only such results of illegal actions can be foreseen as are always, or for the most part, or as the Common Law says in another part, 'according to the usual course of things,' observed to occur, but not those which, in rare instances, occur by accident, as it were, as exceptions to the general rule. It has been, however, already stated, that a rupture of the tympanum can only be the result of a slap on the ear and cheek in the rarest cases ; but I cannot omit to add that I fully share the doubts expressed by Dr. K., as to whether the rupture of the tympanum in C. did actually result from the slap on the ear received. Accordingly, I have to state that my opinion in answer to the question put to me is as follows : That rupture of the tympanum by a blow with the open hand, even if given with great violence, must be one of the rarest accidents; that, also, such a result is not very likely to happen, even when the blow is given right on the ear, and no punishable carelessness, in the sense in which that expression is used in the General Common Law can therefore be attributed to the accused."

*CASE CXXVII.—HAS DEAFNESS OF ONE EAR BEEN PRODUCED
BY TWO SLAPS ON IT?

The young man, K., declared that on the 7th of March two slaps had been given him on his ear, after which he had become " quite deaf" of the left ear, and then a blow on the breast had " felled him senseless to the earth," when the blood ran out of his mouth. According to a medical certificate, he had on the day following the infliction of this violence laboured under the " evident phenomena of congestion of the brain," but without fever, and he complained of pain in the head, ringing in the ears, deafness of the left ear, and giddiness. On the second day he complained of cutting pains in the chest. The suspicion that a rib might be fractured was not confirmed. After one week's continuous medical attendance " he was again fit to attend to his business." On or in the ear the physician, even with the aid of an ear-speculum, could discover nothing anormal. At the examination made by me, six weeks subsequent to the violence, I found K. perfectly healthy. I also could find nothing anormal on or in the left ear in particular; as the examination proved, the tympanum was uninjured, and though K. still complained of difficulty of hearing with this ear, yet the tests to which he was subjected proved either that he was only simulating or that he was intentionally exaggerating a perfectly trifling dullness. Accordingly, I could not assume that K. had been " deprived of his hearing" (§ 193 Penal Code), nor that he had suffered an " important injury" (§ 192 a) to his health and limbs from the violence to which he had been subjected, nor even, finally, that he had been thereby rendered for " a long time " unfit to follow his usual employment (ibid. Penal Code), since he had been able to return to his work after only one week's absence.

*Case CXXVIII.—Alleged Deprivation of Hearing by a
          Blow with the Fist.

The following case was precisely similar. The bookbinder G.
declared that on the 15th of June he had received a blow with the
fist upon the left ear, and that he was thus stunned for an instant.
The practising physician, Dr. E., found next day, " on the external
auditory canal an extensive recent ecchymosis, with almost complete
deafness." Fourteen days afterwards the same physician found an
aperture in the tympanum three-quarters of a line in diameter, and
he further certified that the injured person could only indistinctly
hear the ticking of a watch, even when it was held close to his ear.
—" It is not, and it cannot be determined," I stated, " whether the
small opening in the tympanum did not already exist in the person
injured previous to the infliction of the violence ; this question is,
however, of comparatively no consequence, only the other questions
put before me are of importance : Has G. actually suffered from the
violence inflicted on him any ' important injury to his health or
limbs,' or has he been rendered ' for a long time unable to work ?'
(Penal Code, § 192 a) or, has he been indeed thereby ' deprived of
his hearing ? ' (§ 193). I must answer both of these questions
negatively. In the first place, as to the inability to work, the party
himself does not deny that he worked as usual the day after the
receipt of the injury. The small opening in the tympanum cannot
be regarded as an ' important injury,' since of itself it in general
injures the hearing but little or not at all. Now, G. has not only
asserted to me that he is quite deaf of the left ear, but also that the
hearing of his right ear is also much weakened, and he declared that
he could not hear the ticking of my watch. However, I do not
hesitate to declare that it is all a piece of simulation, which the party
does not even understand how to carry out properly. I intentionally
spoke with him on indifferent subjects, in his shop and workroom,
amid the noise of the streets and the court, quite low and standing
at a distance from him of from two to two and a half feet, and sometimes even with my head turned aside, yet he understood me perfectly well, and answered all my questions without exception quickly
and correctly, so as it would have been quite impossible for a man
deprived of his hearing to do. After this I must declare that the
violence inflicted on G. has neither produced an " important "
(§ 192 a) nor a " severe " bodily injury (§ 193).

5. Deprivation of the Power of Procreation.

* Case CXXIX.—Strangulation of the Penis.

The possibility of future impotence came in question in the following rare case. Human hair was tied round the penis, close behind the glans, of a boy, aged two years, out of wrath against his parents!! Three weeks subsequently I found the boy perfectly well in health, and with his penis quite normal, as was to be expected, seeing that the constriction of the member with the hair ligature only lasted for two hours, as Surgeon W. had immediately removed it, after swelling and severe pain had set in. Of course these results were at once allayed, and injurious consequences to the health neither actually did set in, nor was the future power of procreation at all threatened. The question however, " has there been an injury inflicted in the sense of § 193 of the Penal Code, and in how far might this treatment have proved injurious to the health of the child?—we, after negativing all idea of its being a " severe injury," had to answer as follows— That, if the strangulating ligature had remained on longer it might have produced an important and more or less persistent swelling of the prepuce, inflammation of the urethra, &c.; this " treatment, therefore, might possibly have proved dangerous to the health of the child."

Case CXXX.—Impotence alleged to have been caused by a Kick.

The labourer B., aged thirty-one, healthy, powerful, married, and the father of five children, received a kick in the left groin from a foot inclosed in a wooden shoe, on the 3rd of February. The journal of the Charité Hospital, in which he was medically treated for four weeks, confirmed his statement, that as the result of the injury he had suffered for several weeks from a painful inflammatory swelling of the left cord and testicle, with a hydrocele (not a bubonocele) of the same. After the lapse of four weeks, B. was dismissed cured of all these evils. At the time of my examination, six months after the ill-treatment, a little pain could still be produced by strong pressure upon the left cord, which was no longer swollen, and the left testicle in comparison to the right one was perceptibly wasted and reduced in size. Neither hydrocele nor bubonocele were present. In this case no paragraph of the Penal Code was given for my guidance, but only questions were put to me, which were answered as follows—" It cannot be doubted that the existing phenomena may have been, and have been caused by violence inflicted, which must have violently crushed the part injured, when the nature of the injury and the accurately described illness which was its immediate result are considered. But the power of procreation of the injured party, alleged to have been thus extinguished, cannot be regarded as lost, even although B. asserts that he cannot now perform coition as formerly. But besides that, this assertion is totally devoid of any proof, seeing that B. is young, powerful, in good general health, and with perfectly normal genitals; a certain amount of diminution of sexual power may possibly be the result of the irritation still present in the left cord, but this will certainly disappear in time. As to the smallness of the left testicle, this is of no consequence, for experience teaches us that a man with only one perfectly healthy testicle, as is the case with B., is perfectly capable of procreating. Accordingly I answered the queries put to me as follows :—1, B. is not affected with rupture; 2, he does not labour under the loss of the power of procreating; 3, the phenomena at present existing are to be regarded as the results of the violence inflicted on him."

B. IMPORTANT BODILY INJURY.

1. IMPORTANT INJURY TO THE HEALTH AND LIMBS.

CASE CXXXI.—MANY AILMENTS AND ALLEGED INABILITY TO GAIN A LIVELIHOOD.

Cases are of constant occurrence, in which civil actions are raised—that is, damages are sued for—on account of the previous infliction of injuries or violence. The following one is by no means very difficult of decision, yet it is indubitably instructive. On the 4th of October, 1847, the rag-picker, R., aged sixty-three, quarrelled with a petty proprietor, L., and from words fell to blows. "L.," said R. in his accusation, " seized me with both hands round the neck, and squeezed them together as if he would choke me, and continuing thus to squeeze my throat all the while he thrust me out of his court. When I again returned, he flew upon me once more, punched me in the face till I bled, seized me again round the neck with both hands, squeezed them together, and continued this process so long that I fell senseless and powerless to the earth." On the 8th of November of the same year, he confessed to have been so far restored to health that he was able to go about and carry on his business, and he only declared that he still felt a dull pain in the head and weakness of the left eye.   The day following the infliction of the violence, the forensic physician Dr. R. was summoned to R.   According to his certificate, dated the 20th of October, Dr. R. found him complaining of pain in the head, ringing in the ears, sparks flying before his eyes, giddiness, pain in the region of the liver and stomach, as well as in all his limbs, particularly in the left hip-joint.   There were also present eructations, tendency to vomit, bitter taste in the mouth, and a tongue covered with a thick yellow fur.   Both ears were ecchymosed, there was a large ecchymosis upon the sclerotic of the left eye, and the epidermis was scratched off the face in a few places, and this was also the case in many parts of the neck, both in front and on both sides.   On both sides of the neck a reddish ecchymosis was also visible.   During the night between the 5th and 6th of October, the person injured was again seized with convulsions.   Dr. R. regarded it as a case of gastric bilious fever with congestion of the brain and liver, and that violent emotion and strangulation were the probable causes of the disease.   The physician referred to considered that the violence would leave a permanent injury to the health, and might indeed threaten life itself.   Nevertheless, R. was completely restored on the 20th of October.   On the 25th of January, 1848, on being examined anew by Dr. R., R. complained still of pains in the back and head, and weakness of memory, but the physician could no longer discover any objective morbid phenomena.   After L. had been condemned for his violence to several weeks' imprisonment, R., still not feeling himself satisfied, raised an action of damages against him, asserting that from the bodily injuries resulting from the violence he could no longer prosecute his trade as rag-picker, wherefore he claimed from the defendant the sum of two hundred and sixty-five thalers, ten silver groschen (£39 16s) as damages, and an aliment of ten silver groschen (one shilling) a-day till his complete restoration. Dr. R., once more examined, declared in regard to this, that at his visit on the 5th of October, R. was unfit to carry on his business. How long this incapacity continued he cannot now declare positively, but in any case the evidence of date, January 25th, 1848, proved that at that period the inability to procure a livelihood, at least as the result of this violence, no longer existed. The plaintiff appealed on the 25th of June, 1851, from the decision of the Royal District Court of this city (Berlin), which, considering these facts, had given a verdict for damages much below those claimed, declaring that he had suffered permanent injury from the violence inflicted, L. having "thrice thrown him on the ground,· had kneeled upon his breast, had thrown him so that his left hip had struck the cart pole; had struck him in the face with his fists, and pushed against his genitals with his knee." The injured eye he asserted to be blinded, the left hip still damaged, and that since that time he was almost constantly ailing, and in particular that he suffered from convulsions and from a derangement of his abdominal organs, so that he could not have a movement of his bowels without much inconvenience. In particular he is unable to gain a livelihood, and specially so, because moving about causes his hip-joint to inflame. The defendant disputed all these assertions, and referred in particular to witnesses which should prove that R. on the very day of the receipt of the injury in question worked in the fields, and went about his usual business. On the 5th of this month Dr. R. was again examined. He deposed : that he had visited the plaintiff four times between the 5th and the 11th of October, 1847, but that he did not visit him upon the 7th and 8th of that month, which he would assuredly have done had the severity of the disease required it. His condition at that time was precisely as described in the certificate formerly granted (and quoted above). On the 11th of October his further attendance was unnecessary. R. is rather of a feeble build for his age, he is haggard, walks tottering, and with a stooping gait. In the course of his professional journeys he, Dr. R., has often seen the plaintiff pasturing his horse, or pursuing his trade, or carrying wood, indeed in the course of seventeen other professional visits paid at R.'s house, he was only found at home three or four times, and did not then apply for any medical advice. In the year 1849 the plaintiff was seized with a rheumatic pleuritis, but his assertion that this disease was connected with the injuries received is decidedly rejected by the physician mentioned, who goes on to state it to be his opinion, founded upon repeated examinations, that the phenomena described in his first certificate, dated 20th October, 1847, probably depended upon congestion of the brain, induced by strangulation, especially if convulsions did actually occur, but these had never been personally

observed. The scratches were healed in a short time, and the
gastro-bilious affection removed within seven days. No complaint
was made on the 5th October, 1847, of the alleged affection of the
hip-joint, and no objective symptoms of it were then visible, or are
indeed to be found now. The ecchymoses on the eye and ear
were unimportant, and are falsely regarded by the plaintiff as the
causes of his present blindness and deafness. Dr. R. further goes on
to say, that he had seen R. on the 7th of the present month (November, 1851) riding on his cart in the worst of weather, and
looking quite brisk, that he seldom found him at home, that he had
gone about his business on that very day, and concludes from the
absence of all objective phenomena, from the great discrepancy
in his complaints, which could not have any real foundation, that it
is impossible to ascribe his present incapacity, or lessened capacity for
work, to the violence inflicted on him on the 4th of October, 1847,
but that this depends partly on his advanced age, and partly on his
feeble constitution, coupled with colds which have been brought on
by exposure in the course of his ordinary business, and that this incapacity for work cannot be regarded as the result of the injuries in
question.

I must now add that I *completely agree* with this well-reasoned
opinion of Dr. R. Experience teaches us that complaints, such as
the present, are, almost in every case, more or less completely
unfounded in fact. Rarely, however, has a case come before me in
which the statements of the plaintiff were more evidently a mere
tissue of exaggerations, evident simulations of disease, and untruths,
as in the present one.—On the 4th of October, 1847, the plaintiff
was actually attacked. In his first denunciation, which was emitted
immediately after the attack, he does not, as subsequently, make
mention of any particular weapon, and chiefly declares that he was
injured by the hands of L. The latter, certainly, as was confirmed
by the actual appearances found by the medical man, made use of
his hands to strangle R. by squeezing his throat. No scientific
detail is required in support of the statement that L. might in this
manner have killed R. Since, however, this did not happen, it
must of necessity result physiologically, as is daily observed in
unsuccessful attempts at suicidal hanging, that when the pressure is
removed from the cervical nerves and blood-vessels the congestion of
the chest and head gradually ceases, the circulation and respiration
are again restored, and after a short period of painful sequelæ, which

may last for several days, the state of health, previously normal, is again restored. But violence of this character can never result in consequences lasting for years, such as those which R. declares he now suffers from. It is perfectly agreeable with this that Dr. R. found the plaintiff on the 20th of October, 1847, that is, sixteen days after the injury, ' quite healthy.' But the plaintiff now speaks of other injuries, such as that his breast was kneeled upon, that he was thrown against a cart pole, struck in the face with the fist, and by the knee upon the genital organs. In regard to this, it is in the first place extremely remarkable that these assertions are brought forward for the first time after a lapse of four years, that the memory of this man, now about sixty-seven years of age, not only does not forget former occurrences, but actually recollects fresh ones, which it was of the utmost importance for himself to have brought forward formerly. Further, it is in the highest degree remarkable that not one of the eye-witnesses formerly examined have either seen or testified to violence of the kind last alleged. But there are also professional reasons against the truth of these later assertions of R. These have reference, in part, to the important nature of such injuries as kneeling on the breast, and a blow upon the genitals. Injuries of this character, especially in a man already sixty-three years of age, would have been attended by subjective complaints, as well as objective morbid phenomena, which R. could not have suppressed, and which could not have escaped the notice of Dr. R. when he first saw the plaintiff on the day following the injury, for instance, difficulty of breathing, nervous spasms, swelling and pain in the testicles, &c. There is not a trace of them or any similar affections to be found in the documentary evidence; but, on the other hand, the witness of the greatest professional experience of all those examined, Dr. R., found the plaintiff perfectly healthy sixteen days after the receipt of the injuries, and observed him subsequently for years, and quite recently saw this man, now aged sixty-seven, going about his business quite briskly, carrying wood, &c. From this we may deduce with certainty the untruthfulness of the later assertions of the plaintiff. And this is still further materially supported by two remarkable circumstances; first, by the climax in the pathological description of Dr. R., and second, by the ingenious experiment by which Dr. R. unmasked the simulation of the plaintiff. In relation to the former the documentary evidence contains the following :—

" On the 8th November, 1847, the plaintiff confessed that he was now so far restored as to be able to return to his labour, and asserted that he only now felt a dull pain in the head and a weakness in the left eye.

"On the 25th of January, 1848, he complained of pain in the back and head and of weakness of memory.

"On the 30th of April, 1849, he stated that his left eye was almost blinded, that his hearing was enfeebled, and that he felt pain in the hip-joint and abdomen.

"On the 12th of March, 1850, he said, 'I am feeble from age, and unable to make myself intelligible to others.'

"On the 12th of June, 1851, the left eye is blinded, the left hip-joint damaged, and he is, since that time, almost constantly ailing.

"On the 15th November of the same year, he has finally tearing pains in all the limbs; he cannot make water properly, has often cutting pains in the abdomen, and must therefore sit bent when driving, his 'understanding is gone and he has no more sense:' he is lame in the left hip and deaf in the left ear.

"It would be contrary to all medical experience to assert that such a collection of morbid phenomena ought to be ascribed to a scuffle of no great importance, such as that already described. On the other hand, it is perfectly justifiable to assert that these complaints are partly based upon actual ailments, the consequence of R.'s advanced age; and to this we may refer the weakness of the eyes, the difficulty of making water, and also what R. means when he says 'my understanding is gone,' and that they are partly untrue and fabricated. For instance, of all these complaints Dr. R. found, at the very time when they were brought forward, not a single objective symptom, only a trembling of the hands, 'such as frequently occurs in old people,' a pulse beating one hundred, explicable by the psychical excitement at the time of trial, an occasional lameness, which R. sometimes 'seemed to forget,' and a perfectly healthy left eye, with undiminished power of vision! In investigating the alleged deafness, Dr. R. held a going watch to the left ear, that alleged to be deaf, the plaintiff declared he heard nothing, a standing watch was then held to the right ear, when R. exclaimed, 'Yes! I hear the ticking of the watch quite distinctly with the right ear!'— Accordingly, I gave it as my opinion, 'that the present bodily condition of the plaintiff, the rag-picker R., cannot be regarded as the result of the violence inflicted on him by the defendant L., upon the 4th of October, 1847.'"

CASE CXXXII.—A BENT KNEE, WITH ALLEGED INCAPACITY FOR
GAINING A LIVELIHOOD.

The following case, which occurred fourteen years ago, also belongs to the same category with the foregoing, but it was more difficult to decide.   How I thought it necessary to explain it will appear
from the following relation.   A thief who had been often convicted
was arrested, *flagranti delictu*, and put in prison.   He asserted that
he had been so horribly ill-used by the police officials when cap-
tured that his subsequent illness was the result of it.   As usual in
all such cases, the depositions contained in the documentary evidence
were extremely discrepant.   The following are the most important
results attained by the very long examination.   W., the person ill-
treated, alleged that four policemen had fallen upon him, at his
entrance into his dwelling, bound his hands behind his back, and
carried him off in a drosky to the police court.   In the first place,
it is worthy of remark, that at his first examination he only said
these few foregoing words in regard to the treatment experienced by
him at his capture in his dwelling (in the W. cellar), whilst he subsequently made a perfectly different statement regarding it.   That
his hands had been forcibly torn from his sides in order to take the
money from his purse; that he had been thrown on the floor, knelt
upon, and forcibly pressed down with his face to the earth.   In
regard to this part of the alleged violence, the witnesses present
testified as follows :—Mrs. J. bore evidence to the tying of his arms,
and that he received several blows with the fists upon his head, and
also several kicks; that, nevertheless, he had walked quietly through
the cellar-shop, upstairs, and into the drosky.   Her husband deposed
that he had seen the manacling and the kicks, and that W. had been
treated "like a beast."   But against the trustworthiness of this
witness, the accused criminal officer, Z., alleged that the man J. was
drunk at the time, and this was also confirmed by a journeyman weaver,
M., who was present, and who, moreover, "cannot say that W. was
severely dealt with when arrested;" and Z. has also produced other important testimony against that of the married couple J., which incul-
pates the officials, as that of the woman of the house, Mrs. W.   Finally,
the book-keeper, B., who had been robbed, and who was also in the
cellar, confesses, indeed, that W. "was not gently treated," but he
did not see either blows or kicks given him on this occasion.   " W.

has also made the following further special statements in regard to
his subsequent treatment. When arrived at the police station, he
ought to have acknowledged the theft, but could not. 'They now
bound my feet together, two of the men laid me across the table,
held me firm, and stopped my mouth, while one of the others beat
me over the back and seat with a stick the thickness of two thumbs.
When one was tired, a second took the stick and also beat me, and
when he too was tired I was let loose. Then I was again requested
to say whither I had carried the beds, and when I could not tell I
was again laid upon the table and most unmercifully beaten. This
procedure was repeated some five or six times, till I fell senseless on
the earth. I was abused thus for a whole hour, and received a full
hundred strokes.' The following statement made by him on a subsequent examination was not quite consonant with the preceding ; a
cord was so put round his legs that he fell on the ground with his knees
and the upper part of his body ; then he was placed upon the table and
beaten with the stick, but he did not know whether one only of the men
had beaten him, or whether they had alternated, as he could not see
round, his body being pressed down upon the table. He also does not
know for how long he was thus abused. While, on the one hand, W.
thus materially contradicts himself, on the other, all the four accused
officials, whose credibility I must at least reckon as not less than
that of the culprit, who is known to be both obstinate and
malignant, agree in asserting that all the statements as to his
ill-usage at his apprehension are 'horrible lies,' and that they were
not guilty of inflicting any injury upon W. Further, the deposition
of A., who was arrested along with him, is also opposed to the truth
of his statements ; according to it he begged A. to give evidence in
his favour, when he would not forget him ; his ill-treatment was a
lucky thing for him, he would demand from the officials a monthly
aliment of ten thalers (thirty shillings), which he would spend
quietly in his native place, where he would not be so carefully
watched, to see whether he always went on crutches or not. From
this it is already apparent that there is little probability that W. has
been actually so violently abused as he states, and there are also
medical reasons which considerably lessen this probability. It requires
no very special medical knowledge and experience to understand,
that when a man has been 'beaten for a full hour, and received a
hundred strokes from a stick the thickness of two thumbs,' while
lying bound upon a table, several men having alternated in beating

him, that a man so abused must be reduced to a half-dead condition, that he must be deprived of the free use of his limbs for many days, that his back, &c., must exhibit very important and extensive traces of the violence, and that a violent febrile reaction must occur, &c. How different, however, was W.'s actual condition! The clerk K., who examined him immediately after he was brought in, did not 'in the least regard him as having suffered from violence shortly before he was brought to prison,' neither did he complain of any pain; and what is of more importance, the forensic surgeon L. has expressly stated, upon the certificate of imprisonment dated December the 17th, that 'except the ecchymosis found upon his seat, he is otherwise perfectly sound in body.' Finally A., the witness already referred to, saw his back on the morning following his imprisonment, and indeed 'observed on the loins and seat a few red patches which might have resulted either from falls or blows, but saw no weals;' neither did he hear W. complain of pain, but on the contrary, 'he saw him on this day walk quite well out of the prison.' On comparing these statements and reasons one with the other, I hold myself in the first place, to be justified in stating it as my conviction, that W. *cannot* have been abused in so severe a manner as he alleges. However, the day following his imprisonment, he was removed to the Lazaretto, where the house-surgeon bled him on account of 'inflammatory fever,' and cupped him upon one hip; in doing which the surgeon observed 'blue marks' upon his seat. On the 31st of December, however, he was again put upon ordinary diet, and upon the 8th of January, as we learn from the records of the prison, he was transferred to No. 4 as 'well,' the surgeon remarking upon this, 'that W. was then so far restored, as apparently to require no further treatment, otherwise he would not have been sent back to his number.' On the 13th of January, however, he was again sent back to the Lazaretto on account of 'symptoms of fever,' which the forensic surgeon L., however, 'did not regard as the result of the injuries observed.' On the 1st of February, I myself have for the first time carefully examined W., and in my certificate of that date I have stated as objective and indubitable symptoms, that now only the scars of the cupping glasses and a few red patches are visible, and that the right inferior extremity, from muscular relaxation, is a little longer than the left. At that time, without any knowledge of the previous history of the case, or the personal history of W., I believed that from the totality of the symptoms, I must conclude that W. was exaggerating his ailments, while on the other hand, there was no reason to doubt that external violence had been inflicted upon his back and nates. The disease got worse, and six weeks subsequently, on the fifteenth of March, the surgeon certified, ' that W. labours under a species of *chronic rheumatism*, particularly of the right inferior extremity, which is also somewhat emaciated, and bent and stiff at the knee joint (which was not the case previously), so that he can only move about slowly and with difficulty with the aid of a stick; he is also never quite free from fever.' Surgeon L. therefore advised the removal of the patient to the Charité Hospital, which took place on the 18th of the same month. W. remained in this hospital till the 8th of October, and was then sent back without any apparent amelioration, wherefore he was on the 9th of the same month again sent back to the Lazaretto, ' on account of the crooked state of the right knee joint.' Thence after the lapse of three months he was transferred as ' well,' on the 19th of January, to the police prison No. 12, but returned six days subsequently to the Lazaretto, ' on account of gastro-rheumatic fever,' from which, however, he was, five days afterwards, again dismissed as ' well.' "

" The present condition of W. (May, 1846) is as follows: the right inferior extremity is somewhat longer than the left, and the knee so bent as that it may still be flexed, though not like a healthy joint, while it cannot be extended. There is no apparent emaciation of the right limb compared with the left; but there is an inconsiderable enlargement of the condyles of the right thigh bone. On the knee joint there are also visible the cicatrices of many fly blisters and moxæ. On the right hip joint there is nothing anormal, except the cicatrices of two moxæ, in particular, there is no swelling, and *no dislocation of the head of the thigh bone from the cotyloid cavity;* the contours of both hips are also *quite straight.* The general health of W. is at present quite good, in particular he is quite free of fever; he cannot, however, of course, move about without crutches. From time to time, however, as I myself have had occasion to observe, he is liable to an erysipelatous affection of the skin of the right leg."

" W. himself does not assert that his right knee was specially struck when he was so violently assaulted, and yet his present most important ailment is confined to this joint. His morbid affection is susceptible of a two-fold explanation. Either W. labours

E 2

under chronic rheumatism, or the affection of the knee-joint is the
result of a chronic inflammation of the hip-joint. The latter, however, might be very readily produced by violence inflicted upon that
joint, consequently by violent blows upòn the nates and right hip
joint; and the present permanently incurable condition of the
knee-joint might thus be very explicably proved to be the result
of the blows. There are, however, certain reasons which are
decidedly opposed to such an assumption. I have, in the first place,
already shown why there is no reason to believe that any considerable
violence was ever inflicted on these parts; it is of more importance,
however, that the disease of W. did not, as must have been the
case if the present condition of the knee had originated from the
hip-joint, continue to progress uninterruptedly from the moment of
the infliction of the violence up to the time when the affection of
the knee-joint commenced to declare itself; but rather that there
was an intervening period, of at least one month, during which he
remained perfectly well, after the speedy cure of the slight inflammatory attack which occurred immediately after his arrest. Even on
the 1st of February, that is, six weeks after his apprehension, I found
upon him, as already stated, not a trace of the severe and easily
recognisable disease with which he is now affected; and finally, what
of itself is decisive, even from the present condition of W.'s body,
that is of his thighs, it indubitably follows that he has even now no
disease of the hip-joint, as is proved by the negative signs already
given, and consequently that he could not have suffered from this at
an earlier period; consequently, also, the affection of the knee-joint
cannot have been the result of any such affection, since the possibility of a cure of the hip-joint disease, concomitant with the
persistence of the secondary affection of the knee, is not to be
supposed. On the other hand, everything agrees in making it
probable that the morbid condition of W., from the middle of
January of this year up to the present time, has arisen from internal
causes, and must be called chronic rheumatism, which in this patient
has been most probably complicated with a so-called ill-conditioned
state of the fluids, of which the existing erysipelatous inflammation
affords proof; and not only has this disease already been called
rheumatism by the expert, but precisely such a condition of a knee-
joint as exists in W.'s case is by no means rare, purely as the
result of chronic rheumatism, which therefore, and also because it
may produce a similar affection of other joints, as well as because

of its obstinacy in general, is a disease very justly dreaded by all physicians." Accordingly, I give it as my opinion, " that it is not at all to be assumed that W. at his arrest has been abused in such a manner as he asserts; and that if, upon this occasion, violence has been inflicted upon his person, the ailment from which he now suffers, a bent knee-joint, cannot be regarded as the result of this violence." The complaint was thereupon dismissed.

*CASE CXXXIII.—ALLEGED IMPORTANT INJURY FROM TOOTH EXTRACTION.

The father of a girl now aged fifteen, lodged a complaint against a dental surgeon, that he had a year ago—differences had recently arisen between them—extracted four teeth from his child at one time, and with the aid of chloroform, and that his daughter now suffered constantly from headache and weak eyes. We found the girl approaching puberty, and very plethoric ; all the ocular tissues were healthy. In the upper jaw on the right side there was one molar, and on the left side one eye-tooth wanting, in the lower jaw the right corner incisor was absent. The accused asserted that he had extracted the three teeth referred to upon request, because they were very crooked, and obstructive in speaking; one eye-tooth in the upper jaw was still very crooked, and this supported the statement of the dentist. We further detailed that the alleged ailments were purely subjective, and could not therefore be positively ascertained to exist. But at any rate they could not be regarded as an "important injury," and it was impossible that the extraction of the teeth would have such an effect after the lapse of a full year, and that it was rather to be supposed, presupposing the correctness of the statement, that the headache, &c., was the result of the general plethora and the sexual development. Moreover, it was evident that the gaps between the teeth of this girl had already closed, and her mouth was not in the least disfigured. Consequently this case could neither be referred to § 193 (as a severe injury), nor to § 192 a (as an important one). Because of this opinion, the dentist subsequently raised an action against the father of the child, on account of an accusation scientifically false.

*Case CXXXIV.—Fracture of the Thigh Bone by being
thrown from a Window.

The injury in this case was just as unusual as its occasion. " I
found this powerful man, aged forty-seven, lying in bed, and with
his right thigh still bandaged. On the 10th of March in this year,
while playing at hazard in G. in a half drunk condition, as he
supposes, he was struck from behind with a wine bottle on the left
side of the head, and thus received the injury more fully described in
the certificate of the district physician, Dr. H., dated the 25th of the
same month, and as the result of which there is still a small red patch
visible on the left side of the forehead. This cranial injury is no
longer of any consequence, since there exists a far more important
one. The plaintiff states that after the receipt of the injury above-
mentioned, he was thrown upon the street out of a window fourteen
feet high; this the accused deny. It is certain that R. was found
close to the house and with a broken leg. According to the medical
certificate already mentioned, it was found at the examination on the
25th of March, that 'the external condyle had been broken off, and
the femur itself had been fractured longitudinally; the external
ligaments of the knee-joint, and partly also those of the patella
were torn.' The injury is now so far healed that the bones are
again united, and only require the support of a bandage. In walking,
R. must make use of a pair of crutches, and the knee-joint can be
but very slightly bent. His general health is perfectly satisfactory.
Still R. requires both attention and treatment, and will not, as I
answered the question put to me, probably, be able for three or
four weeks longer to attend a trial at F. As to the future condition of the party examined, it may be almost with certainty
foreseen that in the most favourable case the power of walking
will be very limited, that is, the fractured right leg will always
be more or less stiff. If we apply the phraseology of the statute-
book to these results of such an injury, it cannot be doubted that
it must be regarded as an 'important injury for (the health and)
limbs,' just as the injury itself has already produced 'a long-
continued inability for work.' (§ 192 a). On the other hand, I
cannot regard a stiff joint with its results in regard to locomotion,
even when the stiffness amounts to complete anchylosis, as a 'mutilation' (§ 193), since in ordinary language a mutilation always

presupposes the deficiency of some part of the body, and in this case there is and will be no deficiency. The word 'crippling,' which exactly meets this case, has nowhere been employed by the legislator. And since the other results of an injury included under § 193, neither have nor will occur in this case, I must state conclusively, that the bodily injury inflicted upon R. is to be regarded as an important one in the sense of § 192a of the Penal Code.

* CASE CXXXV.—DID THIS HERNIA OCCUR BEFORE OR AFTER THE 2ND OF OCTOBER?

I quote this case, instances similar to which have three times come before me, only as a fresh proof of the extraordinary questions which come before the medical jurist. A woman, aged forty-three, declared that she had been thrown downstairs on the 2nd of October, and had thus acquired a double inguinal hernia, whilst the accused asserted that she had these herniæ previously. In the right groin there was a small reducible hernia, and in the left one, at least, a dilatation of the inguinal canal. We stated that a hernia which had occurred only two months ago—for our examination was delayed so long—and one of older standing, could not in general be distinguished one from another. The treatment to which the woman had been subjected on the 2nd of October might have produced a hernia, but more probably her's were of an older date. Since, besides that usually when hernia is immediately produced by violence, which is rare, symptoms of great severity at once set in, the woman has given birth to six children, and has a very loose, wrinkled, and pendulous abdomen; it is a well-known fact, that multiparæ are for that very reason liable to ruptures : and it is equally well known, that small ruptures such as the one in question, are often not observed for a long time by those affected with them, because they produce no inconvenience; so that in this respect the statement of the woman, that she had not previously remarked the existence of any rupture, is not incredible. It is possible, that on a more close inspection of the state of her body after the infliction of the violence, she for the first time observed the rupture, and believed it had just been produced. Accordingly, I declared that it cannot be determined with certainty whether the rupture had been produced before or after the 2nd of October; that, however, it is to be assumed as probable that it already existed previous to the day mentioned.

2.—LONG CONTINUED INABILITY FOR WORK.

To this category of the Regulations of the Penal Code belong the greater part of the bodily injuries that come before us. How manifold are the combinations occurring in real life, and how difficult may be their medico-legal estimation will be shown in the few following practical illustrations of § 50, and in selecting them I have restricted myself to those only which presented any striking peculiarity.

CASE CXXXVI.—MANY STABS AND CUTS.

While sitting quietly in her own room and nursing her infant at her breast, a small, but powerful and pretty woman, aged twenty-three, was attacked and thrown on the ground by a man, who had shortly before entered her apartment. She felt herself to be stabbed, but fought with all her strength with the robber till he left her, whereupon she feigned to be dead. When the culprit then fled she hurried after him, and he was discovered and arrested. I found, besides many bloody scratches on the back of the right hand, twelve stabs and cuts upon the right parietal bone, both cheeks, the left ear and both hands, and in particular an incised wound one inch and a-half long upon the right cheek. Her general state was from the first tolerably good, and at the subsequent examination, four weeks afterwards, all the wounds were cicatrized. The woman was, however, much disfigured about the face and complained, as was very credible, that she still felt very much exhausted and unable to perform her work as formerly. Since thus more than "twenty days" had already elapsed since the receipt of the injuries, and "disease and inability to work" were still present, the injuries had to be declared to be "severe," though *now* they would only be reckoned "important" (Penal Code, § 192a). The culprit would nevertheless, even now, not escape the severer punishment, for the jury-court declared the crime to be an attempt at robbery and murder, and condemned him to fifteen years' penal servitude.

CASE CXXXVII.—BLOW WITH A KNIFE UPON THE ARM.

The following case, in which the date of the injury could no longer be ascertained, was also sent to me by a foreign jury court, during the continuance of the former Penal Code, which recognized the " more than twenty days' duration " of the disease or inability to work. A man had struck his wife upon the left arm with a chopping knife. A medical man who had been consulted had certified, that there was an "injury of the elbow-joint which had been laid open." The injury was inflicted about the middle of August, and the woman died of "dropsy" on the 6th of October, and there arose a report connecting the injury with the death, and on the 11th of October the body was exhumed and examined. Those who inspected the body found the joint still open. Accordingly I declared at the time of the trial, that it was of no consequence that the exact date of the injury was unknown, since in any case it had lasted more than twenty days, as was proved by the appearances on the body. With such an injury, however, little was exactly known about it, the wounded person could certainly not have performed her work as usual previous to its receipt ; she must, therefore, have been relatively " unable for work, and that for more than twenty days ;" accordingly the injury must be regarded as a " severe " one (now it can only be reckoned " important").

CASES CXXXVIII. AND CXXXIX.—BLOWS UPON THE HEAD.—
PECULIAR " INABILITY FOR WORK."

CXXXVIII.—Two-and-twenty days before my examination, a Berlin constable received in the discharge of his duty blows upon the head from several men, upon which blood ran from his nose, and he became senseless for a short time. The surgeon found an ecchymosed swelling the size of about a shilling upon the middle of the forehead, with the skin over it injured ; a second similar swelling was on the right prominence of the forehead ; a third upon the right temporal bone ; the nose was very much swollen, and upon the vertex there was a swelling of the scalp the size of a halfpenny. For my part, three weeks subsequently, I only found bluish-red scars upon the parts which had been previously excoriated, but there were no longer any objective morbid phenomena. The injured man also no longer complained of anything except a feeling of weight in the head, and asserted, very credibly, that he could not put on his helmet because its pressure pained him. But it is required in the constable service that the head be covered with a helmet ; and moreover, the constant patrolling of the streets renders this service a very fatiguing one. N., consequently was for "more than twenty days" after the injury unable for duty (work), and the injury (which would now be only reckoned important) had at that time to be declared to be severe.

CXXXIX.—The case of another constable, which occurred subsequently, under the present Penal Code, was precisely similar to the one just related. On the 6th of June he was struck with a pot on the left side of the head, and this was followed by vomiting once or twice, and by loss of consciousness for a short period, while a contused wound of the suboccipital aponeurosis one inch and a-quarter in length was thus produced. He declared that he could not, from pain in the head, perform his duty (with the helmet) for four weeks, and having then attempted it, had again to give it up. His further assertion, that he had become "somewhat deaf" of the left ear could not be reckoned a simulation, since he confessed that this was only the case when he was spoken to amid noise, and during conversation he bent the head forwards and to the left side. This injury was declared to be an "important" one, since it had resulted in producing both of the conditions referred to in § 192a, "important injury to the health," and "a long-continued inability for work."

CASE CXL.—BLOWS UPON THE HEAD.—ALLEGED TEARING OUT
OF THE HAIR.

Fiction and Truth! Three-and-thirty days before my examination, Mrs. P. had received repeated and violent blows with a key upon her head, she had been thrown on the ground, and a quantity of hair was said to have been torn out of her head. I found her still in bed, which she declared she was obliged to keep the greater part of the day, taking physic, complaining of pain and confusion in her head, and visibly quite powerless. The whole central division of the head was quite bare of hair, and a large packet of hair, which was given in by the husband along with his complaint, was said to have been torn out of this part during the quarrel. This must be declared to be untrue. Many hours would not have sufficed to complete such an operation, which would also have produced quite other results from those certified by the medical man in attendance. Upon my representing that this mass of hair must rather have gradually been combed out, as its roots became loose, as hair is thus very easily removed, the married couple not only confessed this, but also produced a second ball of hair, with the assertion that the comb removed a similar packet daily; but that this disease of the hair-roots was first occasioned by the violence. This statement also I was compelled to reject in my report. "The ill-usage had taken place on the evening of the 18th of April, and on the 22nd of the same month the husband lodged his complaint, and with it a large mass of hair. It cannot be supposed that this was removed by the comb during a period of not more than four days, the accumulation of the hair must rather have commenced at an earlier date." Nevertheless, and without any reference to the alleged hair extraction, it was indubitable that Mrs. P. had been "more than twenty days" ill and unable for work, and the injury had to be (at that time) declared a severe one.

CASE CXLI.—BLOW WITH AN AXE UPON THE HEAD.

I received the following case from the district jury court at W. for my opinion. A rural labourer surprised a thief *in flagranti*, and received from him while attempting to apprehend him, three blows on the head with an ordinary kitchen axe, with its edge. On the 17th of June he was taken into hospital, where were found only three contused wounds of the scalp, but no serious symptoms were discovered, so that the wounded man was dismissed after eleven days. In the session of the jury court on the 30th of October, it was, however, proved that M., after a lapse of four months and a-half, could not stoop without feeling pain and giddiness, that therefore *he could not yet mow*, and neither could he be employed in hunting, because he was so irritable that he was terrified at every shot, which was not the case formerly, before the infliction of the injury. Therefore in this case also there was an "inability for work of more than twenty days' duration," and the injury must then be reckoned a severe one. (The alternative accusation was for homicide, and even now, when the injury would only be reckoned an important one, this alternative might, from the concomitant circumstances of the case, be correctly maintained.)

CASE CXLII.—INJURY BY THE BITE OF A DOG.

The overseer of a garden was attacked and bitten by a large watchdog, and raised a civil action against its owner. In the course of

this civil suit, and months subsequent to the injury, the plaintiff was
brought before me. He had upon the right forearm nine, upon the
left five, and upon the right leg four scars of bites, some of them
dark and others pale red, and of various lengths and breadths, from
a few lines to an inch and a-half. The extensor tendons on the
back of the right hand had also been injured, as could be distinctly
felt. The plaintiff declared also that some of these cicatrices re-
opened from time to time, and we found this distinctly confirmed in
one of the cicatrices on the right leg, which was just commencing to
close. The plaintiff was otherwise healthy, and no longer required,
notwithstanding his declaration, any constant medical attendance.
There evidently did not exist any complete uselessness of the limbs
or inability for work ; but in giving a negative answer to this query,
we had to answer the other one put to us by saying, that the
plaintiff was not yet able to work to the same extent as before the
receipt of the injury.

CASE CXLIII.—BITE OF THE RIGHT THUMB BY A MAN.

A private gentleman had the thumb of his right hand bit by a
man upon the 10th of January. Four days subsequently the
medical man in attendance certified that the whole of the right
hand was swollen and erysipelatously inflamed, and that there was
upon the extensor side of the thumb a wound half-an-inch in depth,
corresponding with a similar wound upon the flexor side. Both
wounds were already suppurating. My examination of the bitten
man took place nearly four months subsequently in the course of
the preliminary investigation of the charge against the accused.
Both of the wounds were already healed, and were only visible as
narrow red cicatrices half-an-inch in length. There were also visible
on the thumb the cicatrices of two incisions, the results of the treatment necessary for the former condition of the member. The
thumb was only slightly moveable, and the hand could not be quite
closed. That the plaintiff was not a labouring man could have no
influence on the estimation of the injury (vide p. 23, vol. IV.), which,
as he was certainly, even after the lapse of so long a time, not yet in a
position to exert his ordinary bodily powers in the same degree as
before the receipt of the injury, must (then) be declared to be a
severe one.

*Case CXLIV.—Bite on the Nose from a Man.

The publican A., when in his shop on the 26th of April, was struck with a glass upon the head, and bitten on the nose! According to the certificate of the medical man in attendance, there was on the following day "a circular piece of skin, the size of a farthing, torn off and only connected with the nose by a narrow strip of skin. A smaller wound upon the nasal septum distinctly displayed the form of a tooth." The wounded man had to keep his bed for three or four days, and, on the advice of his physician, as was alleged, his room for three weeks, and to refrain from exercising his trade in the damp cellar in which it was carried on. At my examination, which was made for the first time in June, I found on the right *ala nasi* a semicircular cicatrix, still of a deep red, and the point of the nose still painful to the touch. Considering that this peculiar injury had produced "a long-continued inability for work," it had to be declared an "important" one, in the statutory sense of the word.

Case CXLV.—Stab with a Knife in the Back, with dilatation of the Wound.

This was another example of those cases in which the surgical treatment of an injury was involved in its results, particularly in relation to the question of time; while, according to my opinion, in the present position of our penal code (vide p. 7, vol. iv.), this concurrent circumstance ought not to be included in the estimation of the injury; but the facts of the case ought to be detailed, and their interpretation (according to §§ 192 *a* and 193, compared with § 185 of the Penal Code) left to the Judge. The boy E. was stabbed by the boy M. with a knife in the back. The stab was between the ninth and tenth ribs on the left side, but did not penetrate into the thorax, and the injured boy never displayed, either at first or afterwards, any serious symptoms. Nevertheless, the condition of the wound, which was treated in a public institution by a distinguished and skilful surgeon, rendered its dilatation necessary. I found the boy at my examination on the thirteenth (consequently before the one-and-twentieth) day after the infliction of the injury, perfectly healthy; but the wound, in consequence of the dilatation, exhibited a sore two inches in diameter, and it was evident that a bandage, &c.,

would still be required for some weeks. For just so long, of course, the boy would also be "unable for work," and the injury had to be estimated according to the then § 193, and consequently declared to be severe.

Case CXLVI.—Violent Throwing Down.

This case presented a certain amount of similarity to the foregoing one, inasmuch as in it also § 185 of the Penal Code had to be taken into consideration. The washerwoman, N., aged sixty-three, was pushed out of a room by the accused and thrown down, so that she fell upon the left knee and thigh. The surgeon found her on the third day confined to bed with a reddened and swollen knee, and the whole of the anterior surface of the thigh livid. She declared that she had been five or six weeks confined to bed and unable for work. The witnesses for the defence, however, asserted that she had for years laboured under rheumatism and gout. Summoned as an expert, at the time of trial, I declared that it could not be supposed that the violence which the knee-joint had suffered, though such an injury is always long felt, could have made the woman N. bedridden and unable for work for so long a time as she declared. However, it must be confessed that, from the complication with the rheumatic disposition, the results of the violence might have lasted longer with her than with a healthy person, because experience teaches us that rheumatism is extremely liable to attack joints which have been injured. Nevertheless, I can attach no important value to this complication, since § 185 of the Penal Code expressly excludes all idea of the individual peculiarities of the victim, even in regard to fatal violence or wounds. Still, it must in any case be acknowledged that an injury to the knee-joint, such as has been determined in this case, must have prevented her from prosecuting her business as a washer-woman for from fourteen to twenty days. This is in truth "a long-continued inability for work," according to the present § 192 a, and therefore the injury must be declared to be an "important" one. The deduction was accepted, and the accused condemned to two months' imprisonment.

Case CXLVII.—Throwing Down.—Abdominal Inflammation.

This case occurred while the former § 193 of the Penal Code, with its illness " of more than twenty days' duration," &c., remained in force, and affords proof of the hazardous nature of any such limitation of the time. The boy Rodolph, aged thirteen, was, on the 10th of January, seized by the neck and lifted up as high as the father's own head, and then thrown down with such violence that he could not immediately stand up again, and could only walk in a stooping posture. The day following pain in the abdomen set in, vomiting subsequently occurred, and five days after the infliction of the violence a medical man was called in, who diagnosed an abdominal inflammation. The disease continued up to the 28th of the same month, and the next day the boy could again go to school, and fourteen days afterwards I found him perfectly healthy. The connexion of the illness (and inability for work) with the injury could not be denied, but as the disease had only lasted for *nineteen days*, we could not, at that time, assume the injury to have been severe. It would now-a-days be declared to be an "important" one.

CASE CXLVIII.—CHAINED TO A BLOCK.

This case, so peculiar from the nature of the ill-usage, and so remarkable on account of the individual concomitant circumstances, belonged, along with the following one, in its day, to the *causes célébrés* of our city. The proprietor of an educational institute for boys, was accused of irregular treatment of the children, and on the police making an unexpected examination of his house, the boy D., aged thirteen, was found chained to a block with an iron chain fastened round his belly; the boy was immediately brought before me for examination, and to get my opinion as to the case. He was very far behind in his development, and had only the appearance of a boy aged ten or eleven. He was remarkably pale and thin, the latter particularly on his body and upper extremities. The *nates* were completely covered over with tolerably recent stripes, obviously arising from powerful blows with a rod, and this was confirmed by the child. On both shoulders there were greenish patches, the last traces of ecchymoses, which must have arisen from blows or knocks. The statement of the boy, that he had been punished with a stick, was thus supported. The chain and the wooden block weighed exactly fourteen pounds ten ounces. It was fastened so firmly round the belly, above the navel, that the index finger could with difficulty be pushed below it, and it was fastened above the right hip

with an ordinary padlock. A square wooden block was attached to
this chain. D. stated that he had already dragged this weight
about for eight days, and was to have borne this punishment for
five weeks longer. The chain was ¬not unfastened at night, or
indeed at any other time, not even, for instance, when he had to
go down two stairs from his room to the privy. On the contrary,
the boy declared that he was forced to "walk about" with this·
burden three times a day, for one entire half hour each time. The
abdominal coverings displayed a very evident mark of strangulation,
and the legal and police officials present satisfied themselves of this,
that round the body there was a soft furrow, three or four lines in
depth, in which red stripes were distinctly visible the marks of
the links of the chain. At the place where the padlock had lain
there was a red, round patch, the size of a small bean. There
was also a remarkable distension of the superficial cuticular veins of
both lower extremities—the like was not observed on the upper
ones—evidently the result of the obstruction to the return of the
blood through the veins by the constriction of the chain. The boy
was in a state of nervous depression; he spoke low and timidly, and
was easily made to cry. It was now our task to estimate the results
of this treatment according to the statutes of the (then) Penal Code.
"It is evident, we said, that the whole of the treatment which the boy
has received must have had an injurious influence upon the health of a
delicate boy. In particular, both of the two important functions,
nourishment and sleep, must have been materially interfered with, the
former by the compression of the abdominal organs, the latter by lying
upon a chain, and the impossibility of obtaining an easy posture, of
turning himself, &c. To this must be added the over-exertion of
the nervous system, from the continued dragging about of so
grievous a burden, as well as by the severe chastisement so often repeated, as is proved by the marks upon the body. These disturbances of the functions have already produced an effect upon the
bodily health of the boy as is shown by his appearance already described, and if the boy has not yet fallen into any evident disease, this
only affords a fresh proof of what has been long known to physicians,
viz., that disturbances even of the most important functions may be
tolerated for some time. On the other hand, medical experience teaches
us that it may be assumed as certain, that similar ill-treatment continued for many weeks longer would have made the child decidedly
ill, and that for a long time (that is, for "more than twenty days")

since the continued disturbance of important bodily functions must of necessity produce a 'disease' (in the sense of the then § 193 of the Penal Code), and that most probably some gastric affection." Accordingly we gave it as our opinion, that the ill-usage had an injurious influence upon the health of the boy D., and that a continuance of the same would have resulted in the production of an actual disease of "more than twenty days' duration." The accused was condemned, and his Institution closed for ever.

CASE CXLIX.—BLOWS WITH A KNOUT.—RUBBING WITH SNOW.

This case also was connected with the investigation just referred to. The boy K., aged ten, was admitted into the Institution about Michaelmas. The child was then, according to the statement of his mother, "perfectly healthy," and this was also confirmed by the servant of the Institution, only the boy was afflicted with occasional involuntary escape of urine during the night. According to the statement of the maid-servant during his residence in the Institution, he gradually became "very ailing," and this was attributed to the treatment he received there. "K. was frequently stretched in the *Spanish goat*,* and received both then and also at other times blows upon his clothed nates, upon his back, loins, and feet with a knout of plaited leather made for his special punishment by the accused." Further, in order to wean him from his uncleanly habits, or to punish him for them, K. was frequently deprived of his food, and particularly of his supper, so that he constantly complained of hunger. Finally, it was stated, that for the same reasons he was once, at Christmas, during a severe frost, with his naked feet stuck into wooden shoes, and without trousers, but with his coat, waistcoat, and shirt on, made to stand in the snow, with which he was also rubbed. Chilblains, which already existed, ulcerated after this exposure. After the lapse of half-a-year, his mother found him in a "very lamentable condition, emaciated, with swollen feet and ulcerated chilblains, and affected with an eruption on his head." She therefore removed him from the Institution, and put him under medical treatment. According to her statement, her son then suffered from pains in all his limbs, from a discharge of blood and mucus from the rectum; he was "almost starved, and had blue and

* A framework in which the culprit is placed for the purpose of castigation.—TRANS.

green marks on his extremities from the beatings he had suffered." It must be remarked, that the medical man in attendance was since dead, and therefore no medical history of the case could be procured from him. On the other hand, there is a medical certificate from Dr. L. of this date, according to which the boy was "not emaciated," because he had recruited much during three weeks' residence with his mother. Nevertheless this physician, a few lines further on, says as follows:— "it cannot certainly be denied that he is thin," thus immediately retracting the expression, "not emaciated." Dr. R. also certifies to an ulcerated chilblain and a few small scabs upon the legs. K. himself says, "I have often received as many as ten blows with the knout," and that even whilst he was stretched on the *Spanish goat,* so that thus, as well as by deprivation of food, he was made always weaker and more sickly, particularly in the last two or three weeks, when he was deprived of his supper regularly every evening. He also confirmed the being rubbed with snow. The illness at his mother's, on the other hand, he described as of less consequence, as only a cold, which compelled him to keep his bed only for a few days. The practising physician, Dr. P., had stated in his certificate, dated eighteen days previously, that the boy "at the first glance presented the appearance of a sickly emaciated child, the skin clinging fast to his bones." He also found a few ecchymoses on the left leg, also some redness on the wrist-joints and both ankle-joints not only reddened, but "strongly inflamed" and painful. A frost-bite was also certified to exist upon the toes. Dr. P. assumed that the boy was of a scrofulous constitution, as well as that he must have been extended in the *Spanish goat* within a few days previous to his examination of him; nevertheless, at his subsequent judicial precognition, he, for the first time, stated that it was difficult to say what share the boy's morbid constitution, and what the ill-treatment to which he had been subjected, had in producing the results discoverable on inspection, thus declining to give a decided opinion as to the existence of a "severe injury," in the Penal Code sense of the expression. In this uncertain state of the case it was referred to me for my superarbitrium:—"I must, in the first place, observe, that the disease under which the boy laboured on his dismissal from the Institution, and which was treated by the deceased Dr. M., must be set aside unconsidered, because no facts regarding it can be obtained, and, according to my view of the case, this 'illness' may be very properly disregarded. The only question is (1), did K—

become 'ill' in this Institution? and if so, then (2), was this 'illness' produced by the treatment to which he was subjected? and (3) has this 'illness' been of more than twenty days' duration? (§ 193 of the former Penal Code.) I am constrained to answer all these three questions affirmatively. If 'illness' in the sense of the paragraph quoted, be such a disturbance of the health as either produces a general affection of the body, or a material disturbance of any bodily function, and which, of course, did not exist previous to the infliction of the injury or the ill-usage, then K— did become 'ill' while in this Institution; not because he laboured under involuntary escape of his urine and fæces,—because it is in the highest degree probable that he laboured under this ailment from his earliest childhood,—but because this boy, who, according to the documentary evidence, had been previously perfectly healthy, became sickly and lean,—and this must be regarded as completely made out,—and emaciated even more and more, so that, at last, about Easter, 'his skin clung to his bones.' Unquestionably such a deep depression of the function of nutrition is a 'general affection of the body;' and though Dr. P— leaves it undecided how far this 'emaciation' may be supposed to depend upon the 'scrofulous constitution' of the boy, yet I must remark that this scrofulous constitution is purely a hypothetical and personal opinion of the physician named. The boy's mother knows nothing of any such predisposition; and, besides that, the term 'scrofulous constitution,' when not based upon certain decided and objectively visible bodily symptoms, such as everted lips, fulness of the hypogastrium, tendency to glandular enlargements, &c., is nothing but a very generally prevalent medical idea; moreover, in this general sense, no child exists, particularly among the lower classes, which is free from this scrofulous constitution, and not a single symptom has been observed in this boy, either previously or since, which might have been ascribed to the development of this scrofulous constitution, with the solitary exception of the unimportant eruption upon the head. Least of all could the general emaciation, which was wholly unattended by fever, or the marks of the rod, ecchymoses, &c., be regarded in this light. Since, therefore, the 'illness' which K— acquired in the Institution has not been the development of a predisposition to scrofula, it is evident that its origin is rather to be ascribed to external causes. The ill-usage which the child had to suffer, as described in the documentary evidence, is a sufficient

F 2

explanation of it, even for non-professional people. When a boy ten years' old repeatedly receives such violent castigations, the traces of which were quite recently to be observed on his body, when he has been repeatedly deprived of his meals in succession—leaving entirely out of view the ill-usage during the winter above referred to, since it only happened once, and it cannot be proved that the frostbite was caused by it,—it would be much to be wondered at, if he had not become ill precisely as has happened, that is, become ailing, weak, and emaciated, as if starved, just as Dr. P— found him to be. And, finally, as to the duration of this illness: it has been ascertained that the maidservant who saw him received into the house healthy, saw him also 'gradually,' that is, at least, within the first month of his residence, fall away, and that after Easter, that is, six months after his admission into the Institution, the emaciation of the child, that is, his illness still continued, so that it must have had a 'duration of more than twenty days.'" Having respect to this statement, I declared that the ill-usage and injuries which had been inflicted on the boy K—, were to be regarded as "severe," in the sense of the (then) § 193 of the Penal Code. (Now-a-days they would only be reckoned as "important.")

c.—SLIGHT BODILY INJURY.

Out of the large number of cases of this character which occur, which cannot be included either under the former or the present definition of "severe" injuries, nor even under the present category of "important" ones, and which must therefore be declared to be slight, I will also relate a few selected on account of some interesting peculiarity, even if this be only to present a few examples of simulation, &c.

CASE CL.—ACCIDENTAL POISONING BY CAUSTIC LYE.

This case gave, in a remarkable manner, occasion to the judicial query, whether a (then) severe bodily injury existed? Messrs. Von E. and Von H. had inadvertently drunk a small quantity of the usual washing (caustic soda) lye instead of beer. Both immediately spat out the caustic fluid. The first felt a violent burning in his mouth and difficulty of breathing; he had to keep his room for three days, but was then quite restored. Accordingly, none of the consequences referred to in the then § 193 of the Penal Code had resulted, and the injury could not therefore be declared to be a severe one. The second also felt after drinking the fluid a violent burning in his mouth, his speech and breathing were both difficult; but after "nineteen days" he was so far restored, only, as he said, still perceiving "an alteration of his sense of taste," inasmuch as beer tasted like soapy water; water was bitter, while he did not perceive the taste of salt or vinegar at all. This alteration lasted for eight days longer, inasmuch as "water had not the taste it ought to have." We could not admit that such an alteration of the sense of taste amounted to "disease," that is a general affection of the body or a *material* disturbance of any of its functions, as does not require to be more fully stated here, and we in particular brought forward how often imagination affects the sense of taste in general, and how explicable it was that H., remembering the accident which had befallen him, should for a time taste beer as if it were soapy water (lye). But none of the other results of a severe injury existed, and this, therefore, which did exist, must be declared to be only a slight one.

*Case CLI.—Blows with the Fist and with a Stick.

L., aged forty-six years, declared that five months previously she had been illused by blows with the fist on the left side, several blows with a cane upon the head, and by being thrown down half a stair. The medical certificate testified—a great deal : "an ecchymosis, the size of half-a-crown, upon the right cheek, an inflammation of the right eye, an inflamed patch upon the left cheek and breast, complaints of sharp cutting pains on inspiring, of sharp cutting pains in the right temporal region, which was swollen, and giddiness and faintness." According to the opinion of the physician, a pleurisy "seemed" to be threatening, but this apprehension was not realized. At our examination, which took place at so late a date, we did not find the slightest objective anomaly; but the woman, L., complained that she still suffered from "a certain amount of weakness of her head," and that her memory had failed. Considering, however, that, in spite of the alleged failing of her memory, she could enter into all the particulars of the alleged ill-usage with the minutest detail, that she could give no more definite description of the weakness of her head, and that, after eliminating all the merely subjective statements in the former medical certificate, the objective appearances therein described were only very trifling, we could not announce that the injuries had been important in the sense of § 192 *a* : that is, we could not announce that they had produced "important" injury to the health or limbs, or a long-continued inability for work, and consequently we had to declare these injuries to be only slight.

*CASE CLII.—SOUSING WITH COLD WATER.—A SLAP ON THE EAR.—NERVOUS FEVER.

The girl W., aged twenty-two years, was slapped on the ear and soused with cold water from a stewpan, by the accused, on the 7th of August, while her mother lay seriously ill of a nervous fever, of which she died seven days subsequently. The slapping on the ear was disputed by the witnesses. The following day the young woman became ill, and was sent to the Charité Hospital, and there, according to the hospital journal, there was developed a true abdominal typhus, attended by symptoms dangerous to life. The young woman nevertheless recovered, and four months afterwards, at the time of my examination, she was in perfect health. It was remarkable, however, that in spite of this she, in unison with her father, complained of a host of subjective affections, such as sleeplessness, difficult digestion, giddiness, &c., which were evidently simulated. We could not hesitate, after considering all these circumstances, to declare that the alleged ill-usage, in particular the sousing with cold water in the heat of summer, could have no influence upon the health of the young woman, and that the nervous fever had been the result of infection received from her dying mother.

*CASE CLIII.—KICK UPON THE ABDOMEN.—INFLAMMATION OF THE LIVER.

The accusation in this case was evidently unfounded. The woman S. declared, that on the 2nd of October she had received a kick on the abdomen, and had contracted from it a double inguinal hernia, and an inflammation of the liver. I found her on the tenth of February still in hospital, suffering from chronic inflammation and swelling of the right lobe of the liver, and the medical men in

attendance assured me that they had never heard the patient complain of a double hernia, and had never observed it, any more than
·I was able to find it. On the other hand, they assured me that
the patient seven years ago, had a chronic liver complaint, with
jaundice. Considering all these circumstances, as well as that
which of itself was proof sufficient, viz., that the alleged ill-usage
took place on the 2nd of October, while the present attack of
liver complaint did not commence, as S. herself confirmed, till the
beginning of December, we could not assume that there was any
connexion between the kick and the disease.

*CASE CLIV.—KICK.—INGUINAL HERNIA.

The journeyman G. declared, that on the 21st of May he
was struck with the fist in the face, thrown down and trampled on.
On the same evening, a medical man found on the right side of the
forehead, and on the left side of the neck, several lacerated wounds
of the skin, the size of a lentil. Besides these unimportant appearances, he also certified the existence of " an inguinal hernia on the
left side, in which the portion of intestine (?) lay in the right half (?)
of the scrotum, and that the whole of the right inguinal region, as
well as the said portion of intestine, were extremely painful to
touch." After the results of my subsequent examination, no reliance
was placed upon this perfectly inaccurate description. G. had certainly an inguinal hernia on the *right* side; the abdominal ring was,
however, so unusually dilated that two fingers could be thrust
deeply in without causing pain. This considerable dilatation in a
healthy man, just twenty-four years of age, permitted the deduction
that the origin of the rupture was of a much older date than just
seventeen days. The trifling cuticular injuries were long since completely healed, and the injury must, therefore, be declared to have
been quite unimportant. I may remark that G.'s mother subsequently stated that he had been afflicted with rupture from his
childhood. (*Vide* Case CXXXV.)

*CASES CLV. AND *CLVI.—KNIFE AND DAGGER STABS ON THE
BREAST.

CLV.—The cigar-maker F., eight days before I saw him, had received a stab with a knife two inches below the right collar-bone,

which evidently had not penetrated the chest; because, according
to medical certificate, no serious symptoms had been manifested; I
myself found the man after the lapse of so short a time, already
perfectly healthy, and the wound closed to the size of a pin-head,
so that its direction could be no longer ascertained. The injury
could, consequently, be neither declared to be severe nor important.

CLVI.—Precisely the same was the case with a boy aged fifteen
years, who, seven days previously, had been stabbed in the breast
by a boy, aged fourteen, with a dagger—which the little rascal had
himself manufactured out of a knife-blade, filed to a point, and provided with a hilt! The dagger had gone through the thick winter
top-coat, and all the rest of his clothing, into the soft parts, right
over the most dangerous part, one inch below the nipple, between
the fifth and sixth ribs; but it had not penetrated the thoracic
cavity. After the lapse of only seven days, I found only a scar
one inch in length, and the boy perfectly healthy; having received
but a slight injury, instead of one which might have been instantaneously fatal.

*CASE CLVII.—BLOW ON THE BREAST.—ABDOMINAL INFLAMMATION.

A master furrier was placed at the bar charged with having
inflicted an important bodily injury, by throwing a heavy cap block
against the breast of his journeyman, L. Dr. N. found, next day,
swelling and a slight difficulty of breathing, which induced him to
apply cupping-glasses. Already, on the third day, L. was able to
do light work. Thus the week passed. On the seventh day he
complained of pain in the abdomen, constipation, &c., and was sent
to an hospital, where abdominal inflammation was diagnosed; he
was properly treated, and speedily restored. In this state of matters,
I declared at the time of the trial that I could not assume that any
connexion between the blow on the breast and the abdominal
inflammation *could be proved to exist*, whereupon, of course, in this
case also, the original estimation of the injury was departed from,
and the accused was acquitted.

*CASE CLVIII.—BLOWS WITH THE FIST UPON THE HEAD.— ALLEGED CONCUSSION OF THE BRAIN.

It is specially the case in regard to injuries of the head, which are of such frequent occurrence, that there is no want of the most exaggerated description of their alleged results, as well as of medical certificates painted in the blackest colours. And it is one of the most unpleasant and thankless tasks of the conscientious forensic physician to reduce these statements to their proper value. I have selected the following case as a sample, from out of a host of similar ones, because in it the investigating Judge himself entertained doubts as to the severity of the injury as certified to the plaintiff by Dr. X. R. declared that he received several blows with the fist upon the head and breast from K., on the 20th of October. On the morning of the 21st Dr. X. found "the whole face swollen, especially the right half," and certified further that the injured party had "an excoriated patch over the (?) eye, and complained of giddiness, singing in the ears, and a feeling of stupidity in the head." "The latter phenomena, he went on to say, are symptoms of a concussion of the brain, which might readily have proved fatal;" and it could not be stated "whether or when R. would get the better of these ailments." In our opinion of the case, we stated, "Mere blows with the fist upon the head, such as are given in an ordinary brawl, are a form of ill-usage of such daily occurrence that in general they receive no further attention; showing that experience teaches that they produce no important or permanent injury to the health. Least of all are such blows with the fist fitted to occasion a 'concussion of the brain,' which requires for its production a far more violent injury of the body, such as blows with some blunt weapon, or a fall upon some similar article, &c. How little, however, any such so-called concussion of the brain has been actually produced in this case is already evident from the circumstance, that the injured party, five days after the alleged concussion of the brain, 'which might readily have proved fatal,' was already able to write out with his own hand an accusation of four large folio pages. Under these circumstances no important value can be placed upon his mere subjective complaints, which are embodied in the medical certificate, and the actual appearances testified to in it are so extremely trifling that, according to them, it seems justifiable to declare the injury to be quite unimportant and 'slight.'"

CASE CLIX.—STRANGLING, BURNING, A KICK IN THE FACE, AND YET ONLY A "SLIGHT" INJURY.

The great and important difference in the construction of the

Regulations of the old and of the present universally authorised
Penal Code, in regard to bodily injuries (*vide* p. 2, Vol. IV.), both for
Judge and medical jurist, can hardly be more strikingly illustrated
than by the following case, which came before me under the old
Penal Code.   An artisan apprentice had, on the 23rd of January,
attacked and robbed an old woman in her own house.   He had
put a cord round her neck, thrown her on the floor, kicked her,
piled bedding on the top of her, and set fire to it.   Twelve days
subsequently I found, 1. Both right and left on the neck dark red
stripes, two inches long, and the breadth of a finger; 2. An ecchymosis of the sclerotic of the right eye, and the whole circumference
of this eye bluish-green and swollen; 3. On the back of the head an
excoriation about the size of a shilling, with the hair burned round
about it; 4. Recent venesection wounds at both elbow joints, the
marks of bleedings which were made on account of the state of
unconsciousness in which she was found immediately after the ill-
usage.   Otherwise the injured woman was now, after the lapse of
twelve days, in perfect health!   Seeing that the former Penal Code
directed attention to the *danger* and mere *possibility* of evil results
from injuries and ill-usage, nothing could be easier to assume than
the danger to life from so murderous an assault as this.   Now-a-
days, on the contrary, when the actual *result* is all that is attended
to, it would be a forced assumption to include any such assault
under the head of "important" injuries, since they neither produced
any so-called "important injury to the health or limbs," nor could
an inability to work" for scarcely twelve days be rightly called "long
continued."   In fact, in this individual case strangling, burning,
and a kick in the face could only now-a-days be declared to be
"slight" injuries.   (It was precisely the case of a loaded pistol
clapped to the breast, *but not fired!*)

PART FIFTH.

DISPUTED BODILY DISEASES.

STATUTORY REGULATIONS.

PENAL CODE, § 113. *Whosoever renders himself unfit for military service by self-mutilation, or in any other mode, or permits himself to be rendered unfit for service by another, is to be punished by imprisonment for not less than one year, and by the temporary deprivation of the rights of citizenship. The same punishment is to be inflicted upon any one who, at the request of another, renders him unfit for military service. Whosoever, with the intention of wholly or partially avoiding the duties of military service, employs means calculated to deceive, is to be punished with imprisonment for not less than three months, and by the temporary deprivation of the rights of citizenship. The same punishment is to be inflicted on the abettor of this offence.*

IBIDEM, § 118. *Mendicity is punished as an offence by imprisonment from one week to not more than three months, in the following cases:—1. When any one begs, &c. ; or by pretending to be a sufferer from some misfortune, disease, or ailment ; 2. &c.*

INSTRUCTION FOR MILITARY SURGEONS *in investigating and deciding regarding fitness and unfitness for military service, and the invalidating of soldiers, dated July 14th, 1831.*

§ 53. GENERAL.

I have already (§ 8, p. 190, Vol. III.) pointed out how very often the state of the bodily health of a man comes to be disputed, and becomes the object of medico-legal investigation. A. considers it his interest to be ill, B., from precisely opposite views, disputes this, or B. imputes to A. a disease or ailment which the latter denies. The dispute is sometimes between one private individual and another;

at another between a private individual and some public Board, Judicial, Police, or Life Association ; at one time civil, and at another criminal interests are concerned (§ 54). The simulation of disease is at times carried out by purely mental exertion, lies, adroitness, power of mimicry ; at others with the aid of material means of the most various kinds, caustics, cutting instruments, blood, strongly smelling substances, bandages, spectacles, hernial trusses, crutches, &c. The distinction consequently often made between diseases and ailments merely simulated, and those actually existing, but intentionally produced, is of not the slightest importance in practice, or for the detection of the case. The means which requires the smallest expenditure of mental power, mere lying, is most frequently employed in these simulations. To this category also belongs the exaggeration of those ailments which do actually exist, but in which to one quarter of reality three quarters of lies are added. Experience teaches that by far the larger proportion of all cases of simulated bodily disease belongs to this class. Adroitness and power of mimicry are, on the other hand, not very prevalent qualities, and but few are able to imitate the short-sighted person with his pinched looking, or the photophobic man with his blinking eyes, the deaf man with his head bent forward, a limp, or attack of convulsions, so skilfully as to deceive for any length of time any one really acquainted with the original. So it happens that such cases occur in practice far more rarely than books would have us believe. Moreover, the aid of material means to render actually existing ailments more important and remarkable, or to produce new ones, is employed by but very few persons, even where the ends sought to be attained are important. My own experience at least has shown that such cases are of the rarest occurrence, so that it is to be supposed that the great importance which is usually ascribed to them has been much exaggerated. I have not once been so fortunate as to observe and detect a woman delivered of a piece of a duck (Pyl), or a girl who had introduced stones into the urethra (Klein), or a boy who appeared to pass ink instead of urine (Romeyn Beck), or any vomiter of frogs, or any one afflicted with a marvellous disease, as Rachel Herz (Herold), and yet I have had to examine a very large number of persons condemned to death or to imprisonment for life or for many years, to say nothing of those imprisoned for debt, and it has been a daily part of my official duty to investigate the diseases simulated by these persons. But even our military surgeons in their examinations of recruits, have shared my experience in this matter, as I have learned both from oral and written communications. By this I do not intend to throw doubts upon the truth of the observations made by others, such as the unique and almost incredible cases of a Hutchinson, a Percy, &c., in regard to obstinate and consistent simulations of the most difficult kind, and in regard also to ailments voluntarily produced and maintained, even to the final sufferance of amputation, &c. But the severity of the life of an English man-of-war's man at sea, on the one hand, contrasted with the pleasant life of the invalided seamen in the palaces which stand ready for them, on the other, as well as the frightfully laborious service of the conscript in the armies of Napoleon, which marched through the whole world hurrying from battle to battle, were facts which have no analogy in our life. Such peculiar conditions are well fitted to explain unusually bold attacks upon life or health, in the hope of attaining a great and permanent advantage. But similar attacks have, especially in recent times, become much rarer for other reasons also, particularly because of the great progress made in medical diagnosis, of which the public also have become conscious, and in prisoners probably also, because of the better management and continuous and sharp attention paid them in the penal institutions, which at least render gross deceptions almost impossible.

§ 54. REASONS FOR SIMULATING OR CONCEALING DISEASE.

It is of importance to know the reasons which induce to such obscurations of the truth, because this knowledge alone frequently leads to the mode of clearing up the case. How erroneous it is to speak, as is so often done, as if only prisoners and criminals were the usual objects of examination in regard to this matter, is already apparent from what has been already stated in considerable detail in §§ 8-12 of the General Division (Vol. III., p. 190, &c.) in regard to the object of the medico-legal examination in general. On the contrary, far more of these cases come before the physician for examination in the civil than in the criminal courts. Bodily diseases are simulated, in the widest sense of the word, in order to avoid any troublesome duty, such as to appear before the court as a witness or juryman (or as culprit); in order to repel a declaration of paternity; to obtain a dissolution of an obnoxious marriage; to avoid military or other service; to obtain leave of absence from any service for a summer excursion, or from sordid pecuniary motives; as, for instance, to obtain damages for the infliction of some injury; to obtain public or private compassion; or, in rarer cases, from pure vanity, to get talked about and to excite attention; and, in other cases, to give foundation for an accusation against the perpetrator of some violence; to avoid the infliction of some decreed imprisonment; to avert some disciplinary punishment, as deprivation of food or corporal punishment; to invalidate the accusation of some sexual crime; to get removed from a worse to a better prison, or to an hospital, or to obtain dispensation from penal labour. Diseases are chiefly concealed to avoid the resignation of some office or service; to prevent the dissolution of a marriage; to avoid the refusal of admission into a life assurance association, a fund for the support of widows, or other similar institution; or to conceal the penal cause of the disease, such as certain syphilitic affections, wounds received in a duel, or in the commission of a robbery or murder, &c. The circumstances of each individual case of disputed disease, in which the advice of a physician is required, will at once point out to him to which of the motives in the varied series here unfolded, he has to direct his attention.

§ 55. GENERAL DIAGNOSIS.

Every deceit of this character is essentially a mental process, and mainly to be met in like manner, and the means of doing so will in general be readily found by every good diagnoser. Here we have peculiarly a field in which the medical jurist may make a profitable use of his judgment and practical talents, and just for this reason it is impossible in this matter, as in so many others, to teach the best methods. Practice and experience in such matters make the master. There is no one who can say that he was never deceived during the earlier years of his official experience. Subsequently, it is often sufficient for him merely to glance at the appearance, the whole behaviour, and the manner of speaking of a man, to gain a conviction which is justified by hundreds of similar cases previously observed, while at first he would have hesitated. He has often experienced that men, said to be confined to their room from gouty or rheumatic affections, were not to be found at home during the coarsest weather, when he has surprised them by a visit; he has found others carefully covered up in bed, and on turning down the bedclothes has found them fully clad; he has surprised those said to be ill of fever or severe abdominal disease, &c., when displaying their health by a well-filled trencher at mealtimes; he has often enough experienced that the "patient" did not even know the name of the physician said to be treating him, that when requested to exhibit the remedies said to be employed, bottles were with trouble sought out, the directions on which were dated a long time back, &c. Thus he has learned caution, and thus must all the inexperienced learn it in their turn. It would be a display of great *naïveté*, and still greater want of personal experience in these matters, to say that a simulation ought never to be presupposed. In all cases in which disease is a subject of dispute, or in any manner, even if only alleged to exist, comes to be a subject of inquiry by the medical jurist, he will rather do well to remember that the person to be examined has a reason for saying the very opposite of the truth, whether that be health or disease, and to direct his examination accordingly. And now is the place for general diagnosis to step in with the aid of all the modern scientific appliances, auscultatory, neuro-physical, microscopic, and organo-chemical, which have been of the utmost service to forensic medicine, and have henceforth rendered impossible cases such as Fontana's beggarwoman, who imitated a cancer by sticking a piece of a frog's skin upon her breast; or the beggarman of Paræus, who simulated a prolapsus ani by introducing a piece of an ox's intestine into his rectum, which for hundreds of years went the round of its literature as remarkable "observations." Besides a thorough general diagnostic examination the following rules can be also recommended :—

1. In any doubtful case we ought not to be contented with a single examination, even when this visit has taken the man to be examined by surprise, which ought always to be the case as far as possible; for even though the day and hour of the visit were unknown to him, yet from the very nature of the matter, which is very well known to him, he is prepared for the examination, and has frequently and long previously made preparation for it; but he does not expect a second visit; and this will come upon him most unexpectedly when it follows almost immediately after the first. I have very often succeeded in criminating even skilful malingerers by revisiting them a *very short* time after the first, under any such pretence as a forgotten question, or the like. The bedridden were clad and merry, or no longer at home!

2. Where the case is peculiarly difficult, or where circumstances permit it, as in prisons, hospitals, barracks, and other similar institutions, it is of the greatest consequence to observe the patient when he is quite unaware of being an object of attention. In very dubious cases (particularly in regard to the general behaviour in mental diseases, to which we shall by and by recur) we are as often thus convinced of the actual existence of disease as of the reverse.

3. The general rules for the examination of the sick teach us that we must investigate the origin, exciting cause, and general course of the alleged disease. When the patients' statements in regard to these are not in accordance with general medical experience, a material advance towards the attainment of our end has thus been made.

4. The same is still more true in regard to the alleged symptoms of all internal complaints. In regard to these, it is very often conducive to our end, and easily done, artfully to ask after a series of symptoms, the more extraordinary the better, which have not the slightest connexion with the alleged disease. If the alleged "patient" confesses, besides his pretended pains, &c., also to have, for example, double vision, sleeping of both thumbs, a desire to go to stool regularly every night at midnight, occasional hæmorrhage from the left ear, &c., we know at once how the case stands!

5. It is much to be commended, after listening to all the complaints of the "patient," to put questions of the most opposite kinds to him. He describes, in the liveliest manner, his sufferings from an obstruction of the bowels, which no remedy can cure, or from such constant sleeplessness, that he has become quite emaciated. "Of course you never have a diarrhœa?" "Of course you never have such a thing as a sound sleep?" It is astonishing how often such a "cross-examination" alone makes the lie to falter. The simulator thinks he has given the wrong symptoms, and usually answers such questions affirmatively.

6. I have never yet been deceived in declaring all such "patients" to be malingerers, who come forward with dozens of ailments, and cannot find words to express what they suffer in every part and in every organ. Hysteria need not be brought forward as an instance of the reverse. None but a tyro in medical practice would ever confound an actual hysterical patient, complaining *bonâ fide* of everything, with a healthy malingerer.

7. All local evils alleged to exist on parts of the body usually covered, sores, ruptures, prolapsus, skin diseases, hæmorrhoids, ble-

norrhœa, sweaty feet, &c., require, of course, to be stripped bare for examination, just as they also require to be previously cleansed. On the other hand, I can state for certain that, in ordinary medico-legal practice, it is almost never necessary to examine the *entire* body naked. (The reverse is the case in the medical examination of military recruits.)

8. We must never permit ourselves to be deceived by bandages of every kind, by crutches, rupture trusses, not even by the presence of fly blisters in the very act of blistering, by recent leech-bites, or the marks of cupping; the latter operation in particular, which is popularly believed to be "healthy," is frequently performed for the purpose of deceiving the doctor, and I could relate a number of cases in which the pretence of primary syphilis, which, while it lasts, excludes imprisonment for debt, &c., has been supported, in men, by the application of enormous bandages round the genitals, which, on their removal, were found to be perfectly healthy. One such individual, in whom this proceeding had failed twice already, attempted it a third time, after what he supposed to be a better fashion, inasmuch as he rubbed the whole upper part of the organ sore, so that, after removal of the coarsely-applied bandage, we found an excoriation certainly, but no syphilis!

9. No importance is, in general, to be placed upon the statements of relations, fellow-prisoners, comrades, &c., in cases of the simulation of bodily diseases, and this for evident reasons. (The reverse is the case in regard to mental diseases, §§ 65–67.) The physician must rely solely on his own knowledge, and his bodily and mental tact.

10. I, for my part, have never in one single instance been in a position to make use of anæsthetics as a means of diagnosis. It is evident that these can only be employed when the alleged patient is completely at the command of the physician, as in institutions of every kind. In suitable cases I would not reject their employment.

11. On the other hand, in the case of prisoners, &c., I have obtained good results from the employment of pseudo-physic, homœopathic globules, bread pills, water coloured with tincture of saffron, and the like, and by observing the conduct of the "patient" during the pretended process of cure. In one uncommonly difficult case of simulated insanity (Case CLX.), after long hesitation I got the right clue by this means.

12. One always successful means of unmasking obstinate and con-

sistent malingerers, when everything else has failed, is to threaten the employment of unpleasant, repulsive, or painful remedies or methods of cure; or even their experimental and humane employment. No one can dispute the right of the physician to make use of such means, and experience confirms their efficacy. A woman had wandered from Bohemia to Berlin; scarce arrived in the city, she went into an open kitchen, stole spoons, and was at once sent to prison. On admission she was found apparently cataleptic and lifeless, and was sent to the prison hospital. Next morning we found that she had never taken off her clothes, and she was kneeling on the bed with folded hands looking up to heaven. She had eaten her morning soup, but could not be brought out of this position, and would not answer a single question. Her pulse, general appearance, eye, sensibility, &c., were perfectly normal, and the deceit was a very coarse one. One single emetic put an end to it in a very short time. A pretended deaf mute, a notorious cheat, and a dangerous and often punished thief, who imitated convulsions tolerably successfully, we " cured" by extreme restriction in her diet, which she could not endure for more than two or three days. Cold effusions, threatened surgical operations, most successful when attended by an apparently unintentional display of the necessary instruments, the formation of a small eschar with a red-hot pointed iron on some unimportant part, such as the insertion of the deltoid muscle, the application of a fly-blister, &c., have often enough enabled us and others to attain the desired end. I myself have, however, also seen the firmness of the criminal character, and the stony desire to attain the wished for end, endure even such means with the most resolute disdain. Did not a journeyman potter, apprehended on suspicion, immediately before his apprehension voluntarily cause an acquaintance of his, a barber, to apply *four* moxæ on his back, the better to support his assertion, that he had a constant and unendurable pain in it, and could not, therefore, do without home comforts; and yet he was, and continued in prison to be, perfectly healthy. In such cases, as in all the more difficult ones, which I may repeat are of very rare occurrence, when all other methods have failed—

13. There is nothing else for it, but for the physician to pit his own cunning and skill against that of the malingerer. Success in such a case affords a very explicable satisfaction.

§ 56. SPECIAL DIAGNOSIS.

After what has been already said, it would be perfectly superfluous to enumerate one by one the long list of diseases and ailments which may be simulated. Their diagnosis is taught, not by Forensic Medicine but by Special Pathology. I shall only go over a very few of these conditions which require in doubtful cases something peculiar in the methods employed to discover the truth.

1. If it were still necessary to allay any doubts as to whether *frogs, snakes, &c.* can be (not merely accidentally swallowed and immediately vomited up again, but) *continuously vomited* forth by a man, in whom they are continuously propagated by the laying of eggs, &c., I might point to Berthold's experiments, which have fully proved that none of these animals can exist in a temperature so elevated as that of the stomach; and that they must die very shortly after being introduced into it. The ingenious experiment of Sander's has become trite from repetition ; he took a frog just vomited by one of these frog-vomiters, opened it upon the spot, and found in its stomach half-digested houseflies, which it must, consequently, have swallowed outside of the cheat's stomach not long previously. Complete isolation and unremitting vigilance will put a speedy end to all such simulations in every case.

2. *Incontinence of urine.*—I have seen this often voluntarily pretended, and still more often actually existing. It is in fact not so difficult as it is often alleged to be, to obtain a correct diagnosis in doubtful cases. Hutchinson's advice, to lay the individual in clean sheets after giving him a large dose of opium at night, to examine the sheet next morning, and if it be dry, to conclude that the case has been one of simulation, cannot be regarded as decisive ; Fallot's method, of awaking the alleged patient every hour or half hour during the night to make his water, till he has had enough of it, and becomes healthy, is less deceptive ; but all this trouble is quite superfluous, besides that the aid of a sick tender is required, and neither plan can be carried out, except in certain cases, in prisons or other institutions. On the other hand, the effectual plan of suddenly introducing the catheter can be employed everywhere. But even this is in most cases never required. When actual incontinence has persisted for any length of time, not only is the urinary meatus found to be constantly damp, always becoming so immediately on

G 2

being dried—a symptom which is readily distinguished from any voluntary emission of urine—but also when the complaint has continued for any considerable while, we find the whole neighbourhood of the genital organs down to the thighs irritated, reddened, and even eroded, and the uncovered parts exhale the alkaline odour of putrid urine, phenomena which no amount of voluntary exertion can produce, nor any amount of cleanliness altogether prevent. We have only then to make an unexpected visit to the patient's home, to cause his body and bed linen to be produced, and if these are found to be soiled and stinking with urine, then we are certain not to be in error when we assume the actual existence of the ailment.

3. Extraordinary *hæmorrhages*, especially hæmoptysis and hæmatemesis. Information regarding these will be at once obtained by a glance at the general *habitus* of the patient, and by a thorough diagnostic examination of his condition, both general and local (mouth and fauces, rectum, urethra, &c., according to the alleged source of the bleeding). Whether the fluid actually come from the body is really blood or no, may be at once readily ascertained by means of the microscope, especially when the alleged blood is recent, or when the linen soiled with it has not been too long kept or much rubbed. That even apparent trifles may sometimes prove of importance is proved by the case of an old and dangerous deceiver whom we had for many years to examine repeatedly, sometimes on account of imprisonment for debt, which was to be carried out, at others on account of penal imprisonment, and who, finally, after having in vain attempted to simulate a number of other diseases, was alleged to be seized with a vomiting of blood, just as she ought to have commenced a residence in bridewell, in which she ultimately died. No symptom, no disturbance of any of her bodily functions supported her allegation. But she sent me as proof a linen pocket handkerchief entirely soaked in blood! This handkerchief itself was enough to convict her. For there was not one white spot upon it, as is the case when a pocket handkerchief is used when blood is vomited; but the whole handkerchief had obviously been *dipped* in blood. The microscope, moreover, revealed oval blood corpuscles; the blood was, therefore, that of a bird; and, on my representing this to the "patient," begging me not to make her unhappy, she confessed that she had dipped the handkerchief in pigeon's blood!

4. We read much about *stinking discharges* from the ears, nose, vagina, &c., which are said to be produced by the aid of old cheese, assafœtida, garlic, and the like. I have never seen one single such case, but a syringe full of clean warm water and a good speculum would doubtless make the matter at once plain !

5. I have only rarely observed the simulation of epileptic attacks. It is not so easy, as is alleged even by good authors, to imitate these convulsive attacks in their totality so exactly as to deceive not only casual passers by, but even experts. Certain diagnostic characteristics of the real attacks cannot be simulated ; such as the tonic or clonic spasms of the muscles of the eye ; the insensibility of the iris to light; the constantly irregular action of the heart and arteries; it is very difficult to simulate the insensibility of the skin to even powerful irritants (reflex sensibility continuing perfect) ; it is, moreover, impossible to imitate the very peculiar respiration, or the froth slowly exuding from the mouth—(*soapy froth* artificially produced by means of a piece of soap in the mouth has quite a different appearance, and would at once betray the deceit)—or the entire bodily and mental condition of those really thus affected immediately after the cessation of the fit. If we attend to the time at which the alleged attacks usually occur (in doing which we cannot help thinking of the *epilepsia nocturna*), and particularly the place, on which the alleged patient is in the habit of falling, we shall very soon see whether there is any necessity to attempt to unmask a deceit.* An obstinate deceiver will probably withstand every kind of irritant, though it is credible that Cheyne succeeded in unmasking such a one in the midst of a pretended attack by dropping brandy into his eye, but he is less likely to resist the long continuance of unpleasant restrictions, such as solitary confinement, deprivation of food, &c. It is not opposed to the statements already made to say, that the mere *assertion* of labouring under epilepsy or convulsive affections of a similar character is of frequent occurrence in practice, since the parties in question are, very rightly, convinced that the forensic physician is not in a position sharply to criticise their allegation, apart from an

* In the case of a very cunning deceiver and extremely obstinate simulator of insanity, who also for many years pretended to be epileptic, I repeatedly asserted, with the same decidedness in both cases, that the mental affection was simulated, but the epilepsy real. At her last appearance on trial, which she thus rendered innocuous for long, she fell suddenly from the bar, epileptic, on the floor, striking her head so violently on the wall that some serious cranial injury was dreaded. No malingerer falls thus !

actual attack. For there is not one single, even in any measure certain, symptom, whether of demeanour, physiognomy, features, condition of the teeth, &c., which betokens with any certainty the existence of epilepsy in its ordinary degree, even when several years have elapsed since its first appearance, as is very well known to every medical practitioner, and specially so to the superintendents of institutions for the treatment of these complaints, and all that has been said in opposition to this by recent authors (Esquirol, Cazauvielh, Romberg, &c.) may be true in regard to isolated cases of long standing and inveterate epilepsy, but is certainly not applicable to the large majority of ordinary cases. In those cases also, in which the invesgator has never had the opportunity of observing an attack, he must attend to the evidence of internal truth in the description, by the party examined, of the disease with which he states himself to be afflicted, of its origin, the nature of the attacks, and the methods of treatment which may be alleged to have been employed during by-past years, &c., and must deliver his opinion accordingly. If the matter in dispute be, as it usually is in such cases, the evasion of a penal or debtorship imprisonment, I have very often in suspicious cases, which, for the reasons already given, could not be à priori certainly determined, decided for the carrying out of the imprisonment, which of itself could not be injurious, leaving it to a not far distant future to afford us an opportunity of observing an actual attack, which, in the larger proportion of such cases, failed to arrive.

6. In cases of doubtful *paralyses* of the extremities, even in forensic practice, we must distinguish between paralysis of the sensitive and the motory nerves. Paralysis of the sensitive nerves, when simulated, is in general readily detected by some sudden and painful impression. In paralyses of central origin, particularly when cerebral, certain general symptoms are also ordinarily present, which the malingerer is either unaware of, or does not know how to simulate. Motory paralyses, well simulated, may be very difficult to detect, and in many cases may require the employment of all the usual means of detection (§ 55).

7. *Shortsightedness* is no longer a frequent object of investigation to the Prussian military surgeon,* and comes still less frequently before the medical jurist in any country. Only in regard

---

* According to the cabinet order of June 6th, 1829, otherwise serviceable recruits are not to be rejected merely for myopia, but are to be placed in the second class.

to the capacity of an individual to enter the service of the post or telegraph office, the official medical certificate is specially required to certify as to the possession of normal vision. In these cases of course myopia is not likely to be simulated, but rather concealed where it exists. Should a strongly prominent eye, and a very convex cornea cause a man to be suspected, we have only to hold before him an ordinary printed book more than eight inches from his eye, which he will not be able to read in any measure fluently if he be myopic. In other cases, we may put before the suspected malingerer glasses of from twelve to twenty inches focal distance mingled with others of window glass, and observe his conduct.

8. *Amaurosis*, so extremely difficult to simulate consistently without the deceit being betrayed by a slight turn of the head, or stretch of the hand towards the object actually seen, scarcely ever occurs in medico-legal practice. In such extraordinary men as that obstinate malingerer, whose unique case is related by Mahon (*Med. Leg., I.*), even the sudden presenting of pointed instruments before the eyes, the threat of an operation, &c., will not avail in the attainment of the end desired. I recommend young forensic physicians to make themselves acquainted with the general *habitus* of amaurotic patients, by observing a large number of such cases in extensive institutions for the blind, as this appears to me to be of the greatest service in the diagnosis of its simulation. The perfectly lifeless glance of the eye, the absence of any attempt to fix it upon any object, a certain repose in the whole demeanour, a frequent blinking and shutting of the eyelids—all this would require an adept rendered skilful in the imitation by a careful study of the original, and such an one is not readily found. This assertion is confirmed by the consideration of the demeanour of the most famous actors and actresses when representing some well-known blind characters. This consideration of the general demeanour is all the more valuable that such symptoms as an inactive iris, an angular pupil, a hazy appearance in the posterior chamber, or even squinting, do not occur in all amaurotics. A lengthened observation of the " blind man" readily enables us to recognise artificial dilatation of the pupil, when he is so situated as to be unable to renew it, because it is not permanent.

We must examine with the ophthalmoscope to ascertain whether any alteration in the colour, softening, varices, &c., can be discovered in the retina. In the further examination von Gräefe's ingeuious mode of detecting alleged unilateral amaurosis,* a prism, with its base looking either upwards or downwards, is held before the healthy eye, and the suspected malingerer is asked whether he sces a candle held in front of it single or double ? If he sees two candles placed one above the other, which move towards each other following the turning of the prism, the one of these images certainly originates in the other eye, and the cheat is discovered. In alleged bilateral amaurosis, moreover, the use of the ophthalmoscope with a strong illumination will certainly reveal any sensibility to light which may be actually present. In cerebral amaurosis, which had lasted but for a few months, von Gräefe constantly observed, by means of the ophthalmoscope, the white tendinous-like degeneration of the optic nerve and atrophy of the retina.

9. *Difficulty of hearing and deafness* have frequently come before us as doubtful, and such cases have generally been subsequently proved to be simulations. It seems to the malingerer so very easy to simulate deafness, and he does not dream of any fear of an alarm shot from a pistol close behind him, as in the case of the recruit related by Percy. In the first place, of course, the external meatus of the one or of both ears must be thoroughly cleansed, so as to remove any foreign body accidentally present, any hardened wax, &c.; next, the ear speculum must be employed to search for any ulcerations, stenosis, perforation of the drum, &c.; the examination of the fauces must also not be omitted, specially to see if there be any enlargement of the tonsils present. Great experience is required to catheterize the Eustachian tube, and a negative result from this operation could not in any case be received as proof of the deafness being simulated, since the usual paralytic form of deafness cannot be thus ascertained. Much value is, however, in every case to be placed upon the physiognomical diagnosis. He who is actually dull of hearing, or deaf in one ear, presents in conversation instinctively the hearing ear by means of a slight turn of the head towards the party speaking, keeping his mouth usually more or less open. Farther, it is rare that we find a man actually deaf speaking with a perfectly natural voice; but inasmuch as he does not hear himself speak, a deaf person usually converses either in a very loud or a very low tone. Violent methods are less suited for the detection even of obstinately persistent malingerers than outwitting them by means which must be skilfully

* Archiv. für Ophthalmol. II. 1.

adapted to each individual case. When we speculate on the tendencies, emotions, psychical dispositions of the various individuals according to their social position, their actual situation at the time of the examination, &c., we shall seldom commit a mistake. And the means to be employed is most simple and readily carried out everywhere, as it consists solely and alone in a sudden depression of the tone of the voice at the proper time. A foreign lady, highly educated, and received into the highest circle of society here, in which she moved for a long time, was finally discovered, and arrested as a common thief and swindler. To get out of prison, she simulated, one after the other, a number of ailments, without avail. At last she complained that the dampness of the prison, which she alleged to be so unhealthy (it was perfectly dry and airy), had "affected her ears," and made her deaf. Even when speaking in the loudest tones to her, she answered pleasantly (and apparently unintentionally) quite falsely. At first, apparently agreeing with her complaint, and treating her accordingly, at a subsequent visit I exclaimed, in the middle of the conversation which I carried on in a loud tone, "Good God! are there actually vermin here," and suddenly went on to say, in a low tone, "there is a louse actually creeping on her left sleeve!" The "lady" did not fail to exhibit her horror and disgust in her features, and at once to glance at her left arm, and— thus betray herself most brilliantly! The whole . . . . . district Jury Court were witnesses of the following scene. A wicked old peasant woman had a neighbour called Lemke, whom she had cut upon the left forearm with a sickle during a quarrel while working together in the fields, and she subsequently was placed at the bar accused of having inflicted a severe bodily injury. She was alleged to be feeble from age, afflicted with numerous ailments, and, in particular, to be as deaf as a post. She was allowed to seat herself close in front of the presiding judge, but the trial could not be proceeded with, and was interrupted with the agreement that my attendance should be required, when it was again attempted. On my arrival, the accused, as formerly, sat close in front of the judges, understood no question, &c. With the loudest screaming into her ear on my part, I appeared only to make myself understood with difficulty; and yet the *habitus* of the woman, and all that I had heard of her, gave me the fullest conviction that it was only a malevolent simulation. "You are accused of having severely injured the woman Lemke," I roared into her ear after a long conversation :

"That is not true." "But," roared I again, "the woman Lemke does not think so;" and then I rapidly added in a low tone, "and she is certainly no liar." Her wrath was stronger than her consistence, and she instantly rejoined, to the great amusement of the Court, "Yes, indeed, she is a liar!" and so the trial went on to sentence.

10. Though—except by street beggars, who may indeed deceive children—*deaf muteness* is very rarely simulated in ordinary medico-legal practice, since it requires such a rare amount of self-command to carry on for any length of time, yet two (and only two) cases have occurred to me. One of these was very coarse. The woman W., born a von X., of one of the oldest noble families, was arrested for vagrancy (!), and remained deaf-mute from the morning till after the usual hours of examination late in the afternoon. After that, however, she chatted with her fellow prisoner, who did not betray her, till she fell asleep at night. Nevertheless, she betrayed herself one morning when she was transferred to a worse prison, when she commenced to protest most vociferously, and no longer denied having been feigning. A thief, who had been often punished, in the midst of a fresh examination, during which he had up to that time continued to speak, suddenly ceased to answer, explaining by signs and writing that he had suddenly become deaf and dumb, and that this had happened to him once before some years previously (!!). Having been requested to investigate this case, and give my opinion concerning it, I went, of course, with the conviction that I had to deal with an imposition. In the midst of my conversation carried on with A. in the prison, where he was suffering solitary confinement, at the moment when I was endeavouring, by means of signs, to get him to show me his tongue, the surgeon to the prison, standing behind him, knocked in a preconcerted manner, quite gently with his stick upon the floor. A., as may be conceived, did *not* look round—a more considerable noise would doubtless also have had no effect upon him—and he was precisely for that reason unmasked. Extreme restriction of his diet sufficed to restore him his speech in two days. It is very well known that actual deaf-mutes recognise the vibrations of the sound-wave when a resonant body on which they may happen to be placed is thrown into sonorous vibrations. I have made this experiment in our large institutions for deaf-mutes, and in all those civil cases in which the examination of deaf-mutes was required,

too often, and always with successful results, not to be convinced
of its certainty. Actual deaf-mutes, behind whom a slight noise
is made in the room, as by the tread of a foot, or the fall of a
small bunch of keys, &c., turn *at once* towards the place the sound
comes from, and they willingly acknowledge, by a pathetic smile,
the satisfaction they feel in recognising their connection with the
external world. The absence of any such acknowledgment consequently permits us to conclude that the case is one of simulation.*
Further, another deaf-mute, or better still, a teacher of deaf-mutes,
will easily recognise a malingerer by the way in which he converses
by signs. Though thus convinced that the simulation of deaf-muteness is much more easily and certainly recognised than most other
simulations, yet a case related by Professor Maschka * must excite
some hesitation. In the case of a vagrant supposed to be pretending
to be a deaf-mute, a noise made behind him also produced no effect,
and a teacher of deaf mutes declared that from his unusual pantomimic conversation he must be a cheat. Nevertheless Dr. Maschka
could never elicit a single articulate sound when he caused the man
to be wakened out of sleep, and not even on awaking from chloroform
narcosis, and therefore he declared him to be no cheat. (Which of
these two was right?)†

* Prajer Vierteljahrschrift, 1857, iii. s. 111.

† No illustrative cases are appended to this chapter, because both in it
and in the earlier chapters of this work numerous examples, drawn from
experience, have been already related, and by far the larger proportion of
all the cases that occur are only coarse exaggerations of subjective ailments,
&c., and present therefore nothing instructive.

PART SIXTH.

DISPUTED MENTAL DISEASES.

OF CIVIL AND CRIMINAL RESPONSIBILITY.

STATUTORY REGULATIONS.

GENERAL COMMON LAW, PART I., TIT. 3, § 3. *Where freedom of action is entirely absent, the statutes have no application.*

IBIDEM, § 7. *So far as any action is free the agent of it is always to be held responsible for its immediate results.*

IBIDEM, § 8. *The agent is also responsible for the mediate results so far as he has foreseen them.*

IBIDEM, § 14. *The degree of responsibility attached to the immediate as well as to the mediate results of any action depends upon the freedom of the agent.*

IBIDEM, § 24. *In regard to the responsibility attachable to any free action the statutes have no respect to the peculiar condition or mental power of any given individual.*

IBIDEM, § 25. *Only in regard to crimes and contracts which presuppose a special confidence in the agents, is the amount of responsibility to be measured by the definite personal peculiarities of the agents.*

(RHENISH) CIVIL CODE, Art. 901. *A man must be of sound mind to execute a valid deed of donation or make a will.*

GENERAL COMMON LAW, PART I., TIT. 1, § 27. *Those termed lunatic or insane are all those who are completely deprived of the use of their reason.*

IBIDEM, § 28. *Those who do not possess the power of considering the results of their actions are termed imbeciles.*

IBIDEM, § 29. *Lunatics or insane in regard to the various rights depending upon age, are reckoned as children (under seven years of age—vide Part I., Tit. 4., § 23) but imbeciles as minors.*

IBIDEM, § 31. *Those who from immaturity of years or deficiency of mental power are unable to manage their affairs properly, are under the peculiar care and attention of the State.*

IBIDEM, PART II., TIT. 18, § 12. *Insane or imbeciles, who are not under the care of a father or a husband, must be taken under the guardianship of the State.*

IBIDEM, § 13. *The judge must examine and determine, with the aid of medical experts, whether any one is to be considered lunatic or imbecile.*

IBIDEM, § 34. *Lunatics and imbeciles must be kept under such continual superintendence as to prevent them injuring either themselves or others* (this statute is also in § 346 Ibid., rendered applicable to deaf-mutes).

IBIDEM, PART I., TIT. 12, § 21. *Individuals who are under guardianship on account of insanity or imbecility, are incompetent to make any testamentary disposition of their property so long as the guardianship continues.*

IBIDEM, PART II., TIT. 18, § 815. *The guardianship of lunatics, insane or imbeciles must be annulled whenever they recover the full and free use of their reason.*

IBIDEM, § 816-817. *The Court of Trusteeship must investigate whether this has happened. At this examination there must be consulted, besides the guardian, one expert named by the Court (&c.).*

(RHENISH) CIVIL CODE, ARTICLE 174. *The insanity of an intended spouse may be constituted a legal impediment to marriage.*

CRIMINAL REGULATION, § 279. *The moral character and previous manner of life of the accused in general increases or diminishes the value of the ascertained evidence, or contributes to elicit the amount of responsibility, and must therefore be in so far properly detailed.*

PENAL CODE, § 40. *A crime or offence does not exist when the agent at the time of the commission of the deed was insane or imbecile, or prevented by violence or threats from exercising his own free will.*

IBIDEM, § 42. *When the culprit has not yet completed his sixteenth year, and it has been ascertained that he has acted without discernment, he shall be acquitted, &c.*

IBIDEM, § 43. *If it be ascertained that a culprit, who has not yet completed his sixteenth year, has committed a crime or offence with discernment, the following regulations are applicable to the case, &c.* (Here follow very important mitigations of the statutory punishments in favour of such youthful offenders.) *Vide* further the statutory regulations prefixed to the following §§, 65, 78, 81, 86, and 98.

STATUTE OF 3RD MAY, 1852, ARTICLE 81. *The responsibility of the culprit is one of the facts to be determined by the jury.*

CHAPTER I.

GENERAL PRINCIPLES.

§ 57. Difficulty of the Subject.

Of all the questions which the physician has to treat in medico-legal practice, there is, without exception, no one more difficult to solve than that of the disputed mental condition of any individual. The number of determinate characteristics on which to base an examination and decision is in such questions extremely limited, and even these few are frequently too insufficiently marked to build anything upon them. The modern exact method has also ventured to intermeddle with this problem. The psychical power has been attempted to be explained by the most recent anatomical investigations of the structure of the brain and spinal cord and the elements of the physiology of the nervous system. No one can deny the right of science to make such hypotheses and attempt such conclusions, but no unprejudiced person will doubt that as yet these attempts have not only completely failed, but also that in this way the desired end will never be attained, and this wholly apart from the hazardous results to which these speculations must and have led. But even the method of investigation, solely and alone justifiable and practicable in forensic medicine, the empiric-psychological, meets with the most manifold and serious difficulties. In the first place, apart from the circumstances of the individual case, it is not every one who may be called, not every legally qualified physician whom the judge may appoint for this purpose, that possesses sufficient talent for observation in the abstract, and particularly for psychological observation, a science which more than any other requires thorough general education, practice and experience, a general knowledge of the world and of mankind, powers of reasoning and often enough actual sagacity. Moreover, in the matter itself numerous difficulties are involved, of which I shall only mention a few which daily force themselves upon the attention of the practitioner, especially in criminal cases, though much more rarely in civil ones (§ 58).

To these belong, in the first place, the difficulty, in many cases the actual impossibility, of determining the limits between mental health and mental disease. Even in respect of the individual mental powers the greatest variations are observed. Those fortunate individuals gifted by Providence with a perfectly proportionate and completely harmonious development of all the fundamental powers of the mind, representing as it were the perfect *norm* of mental health, are extremely rarely to be met with. On the other hand, individuals are daily observed who possess for instance along with a most wondrous power of memory, just as feeble a power of judgment; others, with the most extraordinary liveliness of imagination, have a most miserably deficient power of will, &c. In one, an excessive vivacity of character produces despite of all the established customs and forms of society, a behaviour so extraordinary as to cause offence and raise doubts as to his sanity; in another, originality of character, true genius, stamps its peculiarity upon every act, and often enough requires a sharp observation to discover if the limits of sanity be not actually overpassed. In still closer relation to our subject, there are the impulses of the desires, passions, and emotions, which will be entered upon more at large by and by (§§ 86, 88), and in which the query, "where is the limit?" is frequently extremely difficult to answer. Further, to this category belong those cases everywhere of frequent (and in large towns of very frequent) occurrence, which we cannot omit directing attention to, of individuals, men and women, savage and intractable, and wholly deficient in morality, whose bodily functions and nervous system have been wholly destroyed by drunkenness and other excesses, by vagabondism, and defective nourishment, who have long since ceased to be influenced either by conscience or the laws of morality, who have gone on from offence to offence, and from crime to crime, and whose life has been for years a continual alternation between confinement in prisons and workhouses, and a state of vagabondism and savage freedom, in whom at length it is often quite impossible to ascertain distinctly whether they have overstepped the limits of mental disease. Often enough no less difficulty attends the decision of those other cases, in which previous existing insanity is now alleged to be cured, and in which the medical jurist is all the more likely to hesitate in defining the distinction betwixt disease and health, as from the nature of the matter (in pending civil actions) it is generally quite impossible to devote a sufficient length of time (many months) to the examination

of the individual.   It is evident that in all these cases he will have
to state his scruples conscientiously to the judge.

One of the other chief difficulties in regard to the present subject
is the fact that the motives to any action, even the most extraordinary,
and often so strange and peculiar, are often so deeply buried in the
soul of the agent, that great experience and thorough inquiry are
requisite that we may not be deceived and led to•the logical, and
yet erroneous, conclusion that the deed has been without motive.   I
shall return to this subject when considering the subject of the *causa
facinoris* (§ 61).

In other cases it is no less difficult to decide in this matter, when
the object of examination is not placed before the physician clearly
and distinctly, but sophisticated by craft and evil desires for selfish
ends, and thus placed in a false light, perhaps both skilfully and
successfully.   It is readily conceivable that the detection of no other
kind of simulation presents such difficulties as that of an anormal
mental condition only moderately skilfully carried out (§ 67).

A further difficulty in determining the diagnosis of a provisionally
only suspected mental disturbance, is presented by the possibility of
dissembling the actual existence of any such disease, and not a few
of those mentally diseased are truly masters of the art of self-
command and of concealing their disease from the physician.   I
shall by and by (§ 74) return to this subject and show to what
erroneous and dangerous dogmas this art of dissembling in the
insane has given rise.

§ 58.   OBJECT OF THE EXAMINATION—CIVIL AND CRIMINAL
RESPONSIBILITY.

In all cases in which a physician has been requested by a Judicial
Board to examine into the disputed mental condition of any individual,
the object may be two-fold: either it is doubtful and disputed
whether the party to be examined is in such a mental condition as
to enable him, without prejudice to others, to distinguish between
mine and thine, to enter into agreements, to make a will or other
legally binding document, to give evidence or take an oath, to
undertake any office or the duties of any situation, or having long
enjoyed these to continue to exercise them, in one word, in common
law phrase, whether he is able "to manage his own affairs," that is,
possessed of civil responsibility.   Civil responsibility consequently

only comes in question in civil cases (though it is within my own experience that Judicial Boards sometimes, even in criminal cases, make use of the word *Dispositionsfähigkeit*—civil responsibility—as synonymous with *Zurechnungsfähigkeit*—criminal responsibility). In general, it is much more easy to give an opinion in cases of civil than in those of criminal responsibility, because in by far the larger proportion of cases of the former character the party concerned presents himself as he is; we have, consequently, an undisguised condition for examination, and thus for the most part escape at least the difficulties attending the detection of an intentional deception. The special questions, however, which in such cases come to be considered, such as the validity in law of the statements of witnesses labouring under any mental anomaly, or the determination whether the last wills, or other legal contracts of such individuals, should, for psychological reasons, be held to be valid or invalid, the question as to the civil responsibility of deaf mutes, &c., are purely and exclusively questions in law with which forensic medicine has nothing whatever to do, and which are in every case positively regulated by the legislature.

In criminal cases, after the perpetration of illegal and punishable acts, it may be questionable, from the manner in which the deed has been committed, from the conduct and behaviour of the accused, from the evidence of witnesses, &c., whether the culprit at the time of the commission of the deed was in such a mental condition in which "the power to act with freedom was not wanting," so that he "must be responsible for the result of his deeds," or whether, rather, "no crime or offence has been committed,since the culprit at the time of the commission of the deed was insane or imbecile" (p. 92, 93, Vol. IV.). The term Criminal Responsibility is easily defined, since it is based upon unchangeable natural psychological laws—recognised by every man's conscience, and which may be comprehended in the following simple dogmas :—1. Both a good and an evil principle reside and act in man: It appears, in regard to this as well as the following axioms, which, because they are such, require no proof, perfectly superfluous to enlarge upon this point, to point out that the recognition of this fact is coeval with the origin of man, and is expressed in the most ancient religious myths, &c. 2. This double principle itself, as well as the power of recognizing it is born with every man, and is firmly based in his conscience, so long as he remains in his natural condition. Our Penal Code appropriately terms this cog-

ORER

nition the "faculty of discriminating" (Mittermaier's "*libertas judicii*").* 3. Man in his normal condition is at perfect liberty to permit himself to be guided in his actions by either the good or the bad principle, he has *freedom of choice;* according to others, the "power of spontaneous action," "moral freedom," "*libertas consilii*" (Mittermaier), or the "freedom of the will" of our Penal Code. 4. Every man knows, and so long as he continues in his normal condition must be presupposed to know, that, in spite of his freedom of choice, he must allow himself to be guided in his actions by his good principle, and withstand the allurements of his evil one. 5. And that when he acts reversely he exposes himself to the punishment of his internal judge, his conscience, which dwells within him as the natural guardian of the laws of morality. These are eternal inborn truths, the foundation stones of the whole doctrine of criminal responsibility, to which the following may be added, if indeed they do not actually belong to the same category. Every man living in the bonds of society, who has arrived at the normal development of his mental powers, has experienced and knows that society is not, and cannot be satisfied by the punishment inflicted by his internal monitor, but, in accordance with the requirements of the inborn laws of morality, has also laid down and put in execution external punishments for any action opposed to morality. In accordance with these laws, therefore, he must be made responsible for every rash deed, so long as he is in the unclouded possession of his mental powers, since, in such circumstances, he is in a position to foresee all, even the evil results of his actions; he is possessed of criminal responsibility. *Criminal responsibility (imputability) is therefore the psychological possibility of the efficaciousness of the Penal Code.*

The fact that the conditions of criminal responsibility are inborn in man, and are consequently deeply rooted in the whole of his psychical organisation, whilst other psychical processes are the result of education, of mental cultivation, of indoctrination in the ways of life, &c., produces another difference, and a most important one in medico-legal practice, between civil and criminal responsibility. The latter occupies, as it were, a higher position, and in not a few cases must be assumed to exist where the former must be denied. Of this

* *Disquisitio de alienationib. mentis quatenus ad ius crimin. spectant.* Heidelberg, 1825.

a whole class of criminals affords an example; I refer to those very youthful culprits who are in general as yet incapable of exercising the civil rights of life, and consequently possess no civil responsibility, while their illegal evil deeds are carried out with the most perfect consciousness of their punishable character, and with all the unquestionable criteria of freedom of will, and they must therefore be declared to be criminally responsible. Both in the science of law and in the practical administration of justice, so far as I am acquainted with them, this doctrine is assumed as indisputable, and I shall by and by (Case CLXXI.) give the details of an otherwise psychologically very interesting case in law, in which a swindler, who some years previously had been judicially declared to be "imbecile," and at that very time *was still under interdict*, was condemned to the statutory punishment for the crime of which she was accused, on my proving her to be possessed of criminal responsibility.

§ 59. CONTINUATION.

The definition thus given of the term Criminal Responsibility (§ 58), solves at once the much agitated question, whether the decision of it belongs to the medical or the legal forum? Considering that it is always and everywhere, in medico-legal medicine, an unchangeable and established maxim that the physician is exclusively employed with the natural object (for judicial ends), considering further, that the subject of criminal responsibility deals with the applicability of the penal statutes, with something very different, therefore, from the natural object, it cannot be any subject of doubt that the physician has only, in the respective cases, to lay before the judge the scientific material, and to leave it to him alone to decide as to the question of responsibility. The law views such cases from quite a different point of view from medical science. The Prussian physician, for example, declares a man for the most incontestable reasons, to be "lunatic" or "imbecile," and must therefore also declare him to be perfectly irresponsible. In this he is also at one with the legislator, for the latter (§ 40, Penal Code) assumes that such individuals cannot possibly commit a crime or offence. But the legislator of the same country nevertheless condemns such an individual to pay damages, when he states in § 41, Tit. 6, Part I. of the General Common Law, "that lunatics and imbeciles shall, out of their own property, make good the immediate loss they may have caused to any one." A further proof of how the lawyers drag the decision as to criminal responsibility before their own forum is to be found in the positive regulations of all statute books as to the responsibility and punishableness of the illegal actions of children and minors. Finally, the most striking proof of all is given by the legislator, inasmuch as in each individual case of disputed criminal responsibility he submits the decision of this question to a jury; and the apparent paradox of handing over such a question for decision to twelve non-professionals often, as in small country and district courts, belonging to the least educated classes, from this point of view loses its smartness for the physician. Further, I cannot suppress the remark that, in practice, the judicial boards of every character, nevertheless, in drawing up their questions for the physician expressly mention the "criminal responsibility" of the accused, as it were, presupposing that the medical jurist or the medical board consulted will very well know how far they have to deal with this definition and its interpretation and verification. In all such cases it is self-evident that the physician, who must always keep as close as possible to the mode of interrogation and form of expression employed by the judge, is perfectly in the right when, in the tenor of his opinion, he expresses himself as he has been asked in regard to the "criminal responsibility."

After having thus constructed the general theory of criminal responsibility in accordance with certain axioms, it is quite obvious how superfluous, objectless, indeed even erroneous, it would be to treat of any theories of special responsibility, such as the doctrine of the responsibility of pregnant women for criminal actions, or of women in childbed, of deaf mutes, of epileptics, &c., and that this customary and much affected mode of treating the subject ought to be completely discarded from the science. All these conditions, pregnancy, childbirth, epilepsy, &c., may become occasional causes of mental disturbance (§ 71). Should such disturbance be alleged to have occurred, then it must of itself be inquired into and determined. The mania of a puerperal patient, however, does not differ psychologically from that of an epileptic one, the melancholy of a pregnant woman does not vary specifically from that of an individual affected with abdominal disease, or from that of any other person, and must be inquired into and determined according to the same rules. All these cases, therefore, which are apparently specific, and have been so regarded, both by theory and tradition, possess in reality nothing peculiar ; and the much-affected mode of treating these doctrines in separate chapters of manuals, or in actual monographs, is only the objectionable result of generalisation in forensic medicine (*videatur* the ancient doctrine of degrees of le- thality of wounds, &c.), in which *in every matter, and nowhere more than in psychological affairs, individualisation, the critical ex- amination of the individual case, is the only proper method of inquiry*. In the cases here referred to, for instance, the forensic physician has to detail to the Judge, that pregnancy, childbirth, &c., *may* produce such mental disturbance as to destroy the freedom of the will, that according to the evidence in the case in hand, it is to be assumed that the cause referred to has actually produced this result in the case of the accused, &c., &c. In regard to this we have to deal with a much more important matter than the mere objectless writing of books, which frequently has only assisted in increasing the complications of this subject, and leading astray those seeking for instruction, for I shall, by and by, have to show to what important, consequential, and dangerous results this habit of gener- alisation in medico-legal matters has led.

§ 60. CONTINUATION.—DEGREE OF CRIMINAL RESPONSIBILITY.

As the definition of criminal responsibility in itself is no proper object of medico-legal science (§ 59), so it must also leave the decision of the oft-mooted question, whether degrees of criminal responsibility ought to be constituted? to the science of law and to the legislatures, a no small number of which, and among them in particular, the Prussian Common Law (*vide* p. 92, Vol. IV.), the Austrian Penal Regulations, § 95, &c., have decided in favour of there being degrees of criminal responsibility. Certainly, cases con- tinually do occur in which, for example, crimes and offences are committed by children and minors, by drunken persons, by those afflicted with severe bodily disease, &c., in which the forensic physi- cian cannot conscientiously assert that the culprit, at the time of the commission of the deed, has been affected by any delusion or weak- ness of his understanding, and in which accordingly he sees himself forced to lay this negative result of the psychological investigation before the Judge, who will deduce from it the degree of criminal responsibility of the culprit. " But these circumstances are yet actually just as little capable of being explained away, as is their

influence upon the freedom of the will to be denied ? " Certainly
not. But it is the duty of the forensic physician to explain the
latter circumstance to the Judge, who has his Statute Book to teach
him the proper method of dealing with all these, often extremely
difficult cases, by means of its regulations in regard to children,
deaf mutes, emotional diseases, in the comprehensiveness of its penal
regulations, according to his interpretation of them, in the supposition
of mitigating circumstances, &c.   If, however, as so frequently
the Judge has, in drawing up his questions for the physician or
medical board, included the question of " criminal responsibility,"
then I hold that he or they are perfectly justified, in proper cases,
in affirming the responsibility with a diminished amount of imputability, whatever many authors, from a purely theoretical point
of view, may say to the contrary.   This is, at all events, much
better adapted for the requirements of practice, than for the physician to draw up his report ambiguously, saying, " not wholly
irresponsible, but also not fully responsible," which, by an omission,
leaves the matter wholly undecided.   And yet, taught by long
experience, I cannot understand how the physician, when from
theoretical, not even defensible, but only pseudo, reasons—for
bodily and mental health are not arithmetical quantities, from which
not the smallest atom can be removed, without at once annihilating
their existence as such—I say, I cannot understand how a physician can act otherwise than as I have described in those cases in
which, for the reasons already given, he hesitates to declare the party
fully responsible.   Adopting the same view, Ideler* says, " By
accepting as a fact the idea of a diminished criminal responsibility,
all the medical formulæ expressive of ignorance, will be able to be
almost entirely laid aside.   When the physician by such a formula
declares his inability to arrive at any definite opinion in any given
case, what else can the Judge do but decide it according to his own
judgment, wholly irrespective of the medical view ?   Thereby making
the statutory regulations—which direct that the advice of the physician shall be taken in all doubtful mental conditions—perfectly illu-
sory in precisely the most important cases, because the opinion of an
expert is scarcely required when open insanity comes unmistakably
before the healthy understanding of even unprofessional persons," &c.

* Lehrbuch der gerichtl. Psychologie.   Berlin, 1857, s. 49.

§ 61. DIAGNOSIS OF CRIMINAL RESPONSIBILITY.*

The investigation which the physician has to make in regard to the somatic causes existing in the accused, in order to determine the diagnosis of a doubtful mental disturbance, however highly important this investigation may be in itself, is yet but a preliminary step in the decision of the case. For, as is well known, there is not one single (not congenital) somatic cause, the discovery of which in itself permits, with any certainty, the deduction as to the existence of psychical disturbance, as on the other hand, there are hundreds of cases of unquestionable mental alienation, the existence of which is not in the very slightest betrayed by any perceptible bodily alteration. The decisive and most important criterion for the physician in such cases is, therefore, the psychological diagnosis, which results from the consideration of all the circumstances of the former life and actions of the accused, his character, disposition, and demeanour before and after, as well as at the time of the commission of the deed of which he is accused. In regard to this, the consideration of the following circumstances will in every case be found very materially to facilitate the determination of the correct diagnosis, as well as to aid the practical medical jurist in overcoming many difficulties.

1. We must ascertain whether the deed is, as it were, *an isolated act* in the mental life of the culprit or not; whether it suddenly originated in his mind like a flash of lightning in the unclouded sky; or whether it has not rather been the last link in a long chain of sinful, criminal desires, hopes, and acts.† This is precisely what is expressed by the old judicial phrase in these words, Whether the deed might have been expected of the culprit? The investigation of this matter is, however, of the utmost importance; for it does not

* The title of this paragraph has been selected only for shortness' sake. It would be better to run thus:—Determination of the criteria to be employed in determining in doubtful cases whether an action has arisen from mental disturbance or no.

† I have been in the habit of expressing these views, in almost precisely the same words, in my academic lectures, during the last five and thirty years. I state this now, because these views have been published in a recent 'Manual,' almost in my own words, without any acknowledgement of their source, and it is but right that I should at once repel any possible accusation of plagiarism.

readily happen that a man, who, during all his previous life has been guided by his good principle, and has withstood the allurements of his evil one, suddenly acts in the directly opposite manner, unless there be some psychological cause, some mental disturbance from disease, or—we may here assume provisionally—some emotion which has destroyed the freedom of his will at the time of the commission of the deed, and the psychological causes are in general not difficult to ascertain and to estimate. The literature of the subject is full of examples of what I have called the isolated act, and as recent examples of this which have come before myself, I may refer to the two cases to be detailed by-and-by (Cases CLXXIII. and CLXXIV.) in which fathers, from the whole of whose previous life such a deed could not have been in the least "expected," have suddenly killed their children, for whom they had the greatest affection. I shall by-and-by have to show how these cases of isolated acts have been abused, in order to deduce from them the most scientifically untenable, and practically dangerous hypotheses.

2. The inquiry into the *motive for the deed* (*causa facinoris*) has been the occasion of the most contrarious discussions, and there have been quite as many voices in favour of regarding this inquiry as an important, as there have been for looking upon it as an unimportant condition in the psychological diagnosis of the accused. When, in the first place, it has been said that a man's motives for any action are often so deeply buried in his soul that it is impossible for any third party, therefore for any physician or judge to discover them, whereupon we are referred to the Omniscient, such an argument deserves no refutation, since it is nothing else than an unscientific, though convenient, setting aside of a difficulty under cover of mere phraseology. More important, however, is (a) the objection of those who set no peculiar value upon the *causa facinoris*, namely, that the motives to human actions may be of hundreds of different kinds according to the hundreds of different individuals; this apparently includes (b) the objection urged by others, that often enough in the case of indubitably responsible culprits the trifling nature of the *causa facinoris* bears no manner of relation, or may indeed be in the most striking disproportion to the dreadful character of the deed; as well as (c) a third objection, that there are whole classes of offences and crimes which are wholly wanting in any basis of a so-called *causa facinoris*, and which nevertheless, cannot be struck out of the category of responsible crimes, such as all offences against morality (sensual

crimes), offences caused by political enthusiasm, &c. But all these objections are wanting in probative exactness.

a. It is indubitable that there are as many different motives to illegal actions as there are different characters ; hence it follows, that a motive to any deed may exist for A., urged on by which he carries it out, while in from B. to Z. the same motive would never suffice to occasion the perpetration of any such or similar action. Markmann saw in an alehouse that an old woman trafficking there had a clean-washed shirt in her hand-basket. He was seized with a desire to possess it, he followed her, and killed her on the public road. A young, well-bred and educated official of the Woods and Forests, in the course of a sudden altercation with a coarse young man in a public garden, received• from him a blow in the face in the sight of his bride and a numerous public; transported beyond himself, he drew a loaded pistol, which he always carried with him for official purposes, and shot the insulter through the heart, so that he fell dead on the spot. I do not need to say, that H. would not have committed a murder to obtain a shirt ; or, that Markmann was not likely to possess so touchy a sense of honour, as not to be able to cool his wrath, and satisfy himself with his fists alone. And yet both of these two men had his own sufficient motive for his deed. To recognise this, however, the inquirer must in every case *place himself in the position of the culprit,* and divest himself of his own ideas,—a rule that is inviolable and must be strictly kept to, and a proper attention to which would have caused the avoidance of many errors in regard to this matter. Closely connected with this are the remarks to be made on *b,* in relation to the apparently trifling character of the motives. The superficial consideration of this circumstance has been the cause of numerous untenable medico-psychological opinions. For the public certainly no logic is more usual than to say, " How could Markmann commit murder for an old shirt, or how could Brettschneider (who shot his sleeping comrade G., to rob him of a green purse and six silver groschen [7½d.]) murder G. for a few groschen? They must have been mad at the time!" Such logic does honour to the moral feelings of the public, but is a *testimonium paupertatis* when it flows from the pen of an expert. Why should murder, as a responsible crime, not be committed for a few groschen ? The trifling character of the *causa facinoris* may be just as striking a proof of the monstrous criminality, the extraordinary atrocity of the culprit, as it may, under other circumstances, be of his irresponsibility. For we hold the lives of gnats of no account, and kill them without the slightest qualms of conscience only to save ourselves the pain of being bitten by them, and a man so abandoned, and for years fallen so low as Brettschneider was at his examination found to be, would never hesitate to seize sevenpence three farthings, which would carry him on for a day or two longer, even at the expense of the life of another man which was of no value to him, and whose sleeping condition was so very convenient for the deed !

In how many cases has an abusive word, or a box on the ear from her mistress, caused a young peasant servant to set fire to the house. The disproportion between the cause and effect seem too great to permit of so simple an explanation of the act.   There *must* have been some third co-operating circumstance, and this was supposed to be a peculiar morbid propensity to fire-raising (§ 91).   And yet, besides other reasons, it was so evident a reflection that the perpetrator of a deed betraying such a want of education, so silly and yet so malignant, was still little more than a child, and as such might be readily goaded on to cool her wrath in so convenient and secret a manner.   Again, therefore, I have to say, place yourself in the position of the culprit, and the apparent trifling character of the *causa facinoris* will appear in its true light.   The objection urged under *c*, demands a different rejoinder.   It is there stated that there are whole classes of crimes in which no *causa facinoris* can be discovered, such, for example, as sensual crimes.   Certainly this is the case if the idea attached to a *causa facinoris* be of too restricted a character, and confined to property, profit, or earthly advantage.   Such ends are, indeed, only striven after by the swindler, the thief, the receiver of stolen goods, the forger, the burglar, &c., but not by the ravisher or the regicide.   But the true character of the *causa facinoris* is different from that mentioned, it is *the conscious impulse to the illegal gratification of a selfish desire*—a selfish desire of any character, whether it be directed to property, or the gratification of any sensual lust, the desire of revenge, or vanity, the desire of gaining a name in history, indeed, in not a few instances, the mere desire " of wreaking one's anger " (petulancy), which latter is *very frequently* the sole but efficient motive to the slighter offences, or even crimes of both big and little street blackguards, from lamp-breaking up to fire-raising.   To this category belong those constantly recurring cases of a villany apparently so difficult of explanation, as pouring sulphuric acid upon the clothes of perfectly unknown individuals,

the so-called *piqueurs* and the like * (*vide* Cases CCV. to CCXI.).
And I must here direct attention to the fact that there is a positive
enjoyment in wickedness, a pleasure in cruelty, which produces
crimes the unheard-of character of which, for want of attention to
this sad fact, only too frequently has led to the erroneous conclusion
of an irresponsible mental condition of the perpetrator. In a few
wicked boys we see this character already developed in the pleasure
taken in mutilating animals, and the delight found in watching their
sufferings, and thus—from Nero and Tiberius onwards—there have
been, fortunately not above once in a lifetime, bestial natures, to
whom it was a pleasure to cause the ruin, the torment, and most
cruel death of others. The Count von Charleroi (a Bourbon), who,
among other murders of which he was accused, was convicted of
having shot a servant from the roof, like a wild animal (for this life
was worth no more to this degenerate prince); the criminal of whom
Bottex relates that he threw a boy into the water, and gloated over
his endeavours to save himself; the dreadful women Jäger, and
Gesche Gottfried, who poisoned their husbands, children, and neighbours, and delighted in their mortal sufferings ; the unnatural mother
Pohlmann,† who in summer caught a number of wasps and shut
them up with her child, whose death she desired to bring about.
These and similar cases are illustrative examples of the fact just
alluded to. Of course, there have been medical writers who, in their
passion for maniæ and for generalising and cataloguing everything,
have sought to construct from these cases a species of morbid propensity, or desire to murder (not the homicidal monomania of the
French, § 93), though in them there is nothing else but the highest
and most devilish development of the criminal nature, which gives

---

* For instance, " The Girl abuser of Augsburg " (Annalen der Criminal-
rechtspflege, Altenburg, 1840, Bd. XIII.). In the years 1819 and 1820,
fifteen girls were attacked and wounded by night in Augsburg. Karl
Bertle, aged thirty-seven years, confessed that he had wounded all these
girls, and protested that he had taken every pains not to wound them dangerously. (An extraordinarily important statement, as denoting his state
of mind at the time of the commission of the deed!) As his excuse, he
asserted "an irresistible desire." Seven daggers were found in his possession. The Court, " since there was no reason for assuming the irresponsibility of the culprit," condemned him, as a cure for this "morbidly irresistible
desire," to four years' penal servitude, which doubtless would suffice to
cure it.

† *Vide* Case LXI. p. 155, Vol. I.

itself freely up to all even the most eccentric instigations of the evil principle.

Where, now, we can ascertain such a motive for the deed as has been just (p. 106, Vol. IV.) defined, a true *causa facinoris*, in any individual case, and where this motive agrees with the disposition of the culprit, I hold this to be one of the most certain characteristics of the responsibility of the culprit at the time of the commission of the deed, and *vice versâ*. Of course the *causa facinoris* must not be based on a delusion, as, for instance, when B. Hoffmann stabbed his supposed rival, in a fit of altogether groundless and purely delusionary jealousy (Case CLXXXI.), nor must it be in direct opposition to the everlasting natural laws of human feeling, as when beloved children are killed to release them from the sufferings of this world, or murder is committed from the desire of the malefactor to be executed, and the like (*vide* Cases CLXXIII., CLXXIV., and CLXXVI.), cases which are not usually difficult to decide, precisely because of the peculiar circumstances attending them.

§ 62. CONTINUATION.

3. It is always of importance to ascertain whether the culprit, in carrying out the deed of which he is accused, has acted according to a regular plan or not. In most cases, however, this circumstance is of but little diagnostic value, however apparent it seems to be that regular systematic action should prove the undisturbed state of the mind of the malefactor, and *vice versâ*. Because criminal actions are just as often induced by the seduction of circumstances ("opportunity makes thieves"—and other criminals also), the heat of passion, &c., without any previous preparation or contrivance, as in other cases similar deeds are committed in the same unprojected manner by those actually insane. Still more important is the indubitable fact, well known to every expert, that perfectly insane patients, whose irresponsibility could be doubted by no one, often act with the greatest subtlety and well-contrived artfulness, like one possessed of the clearest intellect, and are frequently a long time in contriving and carrying out a plan for committing some forbidden action, as, for instance, what is so frequent, to escape from an asylum. An insane patient, in a private institution near London, with the greatest artfulness, managed to steal a piece of metal from locksmiths who were working in the house; with this he subsequently sawed through the iron bars of his windows, and brought his long-cherished plan of

flight into execution. The circumstances proved the calmest and
most systematic deliberateness. After his escape, he ran straight to
the Duke of Wellington's residence, and announced himself as his
eldest son.* Only in one case can the examination of the systematic
planning of the deed afford any information, and that is when these
plans and preparations themselves evince the stamp of a confused
intellect, and betray the hazy consciousness, the mental darkness in
which the culprit was involved. A certain Baron von X., formerly
an officer at Potsdam, wished to go to Berlin to lay his inadmissible
demands before the Duke of ——. For this purpose he borrowed
a suit of clothes. He accompanied for a third of the way a battalion
marching in the early morning towards Berlin, in which he had not
a single acquaintance, and then, being tired, he seated himself—upon
a waggon laden with furniture, travelling slowly to Berlin! But the
railway trains from Potsdam to Berlin run every two hours from
early in the morning, and X. had more than ten times the amount
of the fare in his purse! The peasant woman Z., who for years had
laboured under mental depression, resolved to cut the throats of her
three children, born in wedlock, with her husband's razor. For this
purpose, six days before the commission of the deed, she secretly
removed the razor from its usual place, and concealed it. But it
happened to be the only razor which the man possessed, and he used
it every other day, he must therefore speedily miss it, and ask his
wife, with whom he lived quite alone in a lonely house near a
country road, about it, and so it happened. And where had she
concealed the razor? In an open cupboard, which always stood
open, because its key had been lost for years! Of course her husband found the razor, and placed it again upon the shelf where it
always lay, and from which the unfortunate woman took it down
immediately before killing the children. Would any person of
sound mind have made such absurd preparations, so likely to betray
him, for a premeditated deed? These examples, which could be
readily multiplied, prove *how erroneous it is to deduce the responsibility of the culprit from the proven premeditation of the deed*, as is
frequently done by nonprofessional people (the public prosecutor,
the Judge, and the jury. *Vide* Case CLXXVI.).

4. We must ascertain whether the accused has taken measures *to*

* *Vide* Knaggs, *Unsoundness of mind considered in relation to the question
of responsibility for criminal acts.* London, 1854, 8vo. p. 14.

avoid the punishment due to his deed. In the first place, however, we must distinguish between preparations made before and those made subsequent to the performance of the deed. In the first place, when, for instance, he has adopted any disguise or other disfigurement of his personal appearance (removed his beard or stuck on a false one, &c.), made attempts to enable him subsequently to plead an *alibi,* made preparations for flight, or waited for night or the absence of witnesses, &c., in all these cases the suspicion will obtrude itself that while perpetrating the deed he has recognised its punishableness. And this suspicion is justified, and constituted an important diagnostic criterion, when there is an evident rational appropriateness in these preparations, when these are not merely the actions of an unsound mind, which very frequently retains a dim notion of the difference between good and evil, and which very often makes such preparations previous to the perpetration of a deed, but will be certain to exhibit, in some way or other, the stamp of insanity, and just for that reason they afford a most excellent criterion in aid of our decision. The peasant woman Z., already referred to, when she prepared to destroy her children, hung her apron over the sole window of the apartment into which she called her children one after the other to kill them, while her husband was working in the fields, obviously to prevent casual passengers along the country road, which ran close to the window, from observing the deed. But how absurd and unsuitable was this step. Measurements subsequently made revealed that the apron scarcely covered one-third of the window, so that any passenger might have very readily seen all her actions! The paperhanger, Schultz, who in a fit of melancholy cut the throats of all his four dearly-loved children, wished to.get rid of the sole witness in his house, his housekeeper, when he proceeded to carry out his premeditated deed. To this end he wrote a letter to a clergyman living some distance off, and sent her with it. But the person to whom it was addressed did not know Schultz, and the letter contained nothing but the words, "Yours sincerely, Shultz." [note]

In a negative point of view this circumstance has no application, and the absence of any proof of preparations of this character made before the commission of the deed is just as little favourable to the conclusion as to the irresponsibility of the culprit as the reverse; because the actually responsible malefactor is often enough not in a position to be able to make any such preparations, as, for instance, when crimes are perpetrated in the heat of passion, or before witnesses and the like; thus resembling in so far the man of unsound mind, who, acting from blind impulse, in most cases does not think of any such preparations, because he is not capable of recognizing the effects of his actions, and their relations to the penal statutes.

Proof of arrangements of this character carried out after the commission of the deed, such as consistent lying, concealment of the malefactor himself, or of the *corpus delicti*, &c., require to be differently estimated. The responsible malefactor will certainly, in most cases, betake himself to some such contrivances; but the criminal made so by desire of vengeance, or by political enthusiasm, or the deeply degenerate common criminal, made bold by his good fortune hitherto, frequently place themselves in fearless opposition to the laws, and scorn to take any steps to avoid punishment; behaving, in this respect, like the irresponsible man of unsound mind, who in his unfortunate *naïveté*, so to speak, feels himself as safe after as before the commission of the deed.

§ 63. CONTINUATION.

5. Though for the reasons already given (page 98, vol. IV.) the existence of *Repentance* might seem to lead to the conclusion of the responsibility of the culprit, and its absence to the reverse, yet every one whose calling has made him accurately acquainted with the criminal world knows that this is utterly worthless as a diagnostic criterion. In the first place, every one knows that whole classes of crimes in most culprits wholly exclude the very idea of repentance, such as theft, frivolous bankruptcy, many sensual crimes, crimes arising from desire of vengeance, &c. The slumbering conscience of such demoralized natures requires for its awaking the ruder shock of a crime from some higher category. Who has ever seen a professional thief, swindler, or cheat repentant? But even after much more horrible crimes we find, alas! that in most cases, even after the lapse of some time, repentance is just as little awakened, as I have elsewhere shown* that most of these crimes have their psychological basis in the ice-cold hard-heartedness of the malefactor. I have seen a very considerable number of murderers, who during their long imprisonment had become very well known to me, either ascend the scaffold or passing their lives in penal servitude, whose

* Mörderphysiognomieen. Berlin, 1854.

unrepentant icy coldness, indifference, and even frivolity up to the
last moment made a most painful impression. Thus, therefore, do
most criminals behave, and thus also, of course, most men of unsound
mind, after the commission of any crime. It is also, however, very
important not to be deceived by the actual existence of repentance;
for the mind is often disordered at the time of the commission of the
deed, and afterwards, for many reasons, recovers its balance, and to
this subject I shall by and by recur. In these cases the accused
subsequently, at the time of the investigation and judicial and
medical examination, exhibits a repentance all the more deep and
unfeigned, as even the deed itself was not actually produced by the
influence of the evil principle. The subject of repentance will,
therefore, in accordance with these facts deduced from experience,
require to be considered with the utmost circumspection in every
case.

6. But as little diagnostic confidence is to be placed in the circumstance, whether the accused has *any remembrance of the cir-*
cumstances attending on the deed or not; whether, for instance, he
can tell the time correctly, describe the locality accurately, or re-
count his movements and actions on the day of the deed, &c.
Irrespective of the fact, that in the opposite case of daily occurrence,
where these facts cannot be recollected, this forgetfulness may be
only simulated, while, on the other hand, it actually occurs in many
cases of irresponsibility from unsoundness of mind, it is very well
known that memory, the lowest of all the mental powers, continues
unimpaired in many of those mentally diseased, and they do not
then hesitate to describe, with the most perfect accuracy, even to the
minutest particulars, all the circumstances of the deed. Thus a
man so perfectly insane as Sörgel,* knew how to describe all the
most trifling circumstances connected with the murder of the old
man " who had horns; " his clothing, position, &c., quite agreeable
to the facts actually ascertained ; thus, also, the insane man, Gnieser
(Case CLXXVI.) during repeated examinations deposed with perfect exactness all the circumstances of the deed, which were not a
little complicated, and which also were confirmed as accurate.

7. A most important circumstance, the want of a thorough consideration of which has been the cause of numerous errors in medico-
legal opinions, is the peculiar *condition of the intelligence* (not of

* Aktenm. Darstellng Merkwürdiger Verbrechen. Von A. von Feuer-
bach. Giessen, 1828, s. 285.

the mind) of the accused, as ascertained on examination to have always existed, or at least for a long time previous to the criminal act in question. For instance, it often happens in criminal cases that witnesses, relations, and acquaintances of the accused agree, and are credible in asserting, that this man has been always "stupid, childish, simple, silly, of no use," &c.; and it is evident how important such a description of the degree of intelligence is, and how readily this might be, rightly or wrongly, employed as a mitigating circumstance in favour of the accused, and his irresponsibility be deduced therefrom; and this, in fact, has only too often happened. But a low degree of intelligence, feebleness of the understanding, absolute stupidity, by no means produce irresponsibility (§ 96); again, therefore, as always, the relative facts ascertained in reference to the case in question, must be examined and applied to the elucidation of this point. This refers chiefly to the consideration of the family history, education, and manner of life of the accused, and also the consideration of the nature of the deed of which he is accused, and its relation to the sphere of his intelligence. A young girl was accused and convicted of child-murder. The witnesses for the defence were unanimous in making statements such as those referred to, and it was in particular asserted, that she did not know the value of money, could not count, &c. But who was the accused? The young daughter of poor cottagers, who for a short time had enjoyed the very meagrest instruction of a village school; and then, as soon as her physical strength permitted, was sent to herd cattle, and so she continued to live up to the time of the commission of the deed. Accordingly, these facts placed the value of the statements of the witnesses in regard to her stupidity in a different, and that the correct light; for that this girl could not count (since she had never learned, still less practised it) was no proof of her want of intelligence. But granting this (in another similar case), the question, does this feebleness of the understanding stand in psychological relation with the crime of child-murder? must still be unconditionally negatived. For the "faculty of discrimination" is not a process of thought, and does not take its origin in the intelligence, but much deeper, in the mind, and even a stupid country girl knows, and must know, that it is not lawful to kill any one, even one's own child. The case would be different were the same person accused of a different crime, such as the forging of a bill, or any swindle, in which she probably might have been made use of by others. Such a matter bears reference to perfect civil responsibility, proper knowledge, clear insight into the value of the respective objects ; in such a case, defective intelligence might— I am far from saying in general that it must—be taken into con: sideration by the Judge as a mitigating circumstance, in regard to the value of which the circumstances of the particular case must decide.   These circumstances also will have to decide how far, in any case similar to the one just related, the physician may feel himself justified in inducing the Judge by means of his psychological representation of the case to assume a "diminished responsibility." Only I must warn against the abuse of the means referred to in all ordinary cases.

8. *Hallucinations,* especially aural, and particularly when voices are alleged to be continually heard whispering *" Thou must do it !"* are often declared to be the proximate, and in not a few, according to the consistent assertion of the accused, and the absence of every other *causa facinoris,* the sole and only cause of the commission of the deed, because the accused has at last become unable to resist the impulse which has become more and more urgent (Case CCVII.). Whoever has no experience of his own in this matter, has only to visit any large criminal prison, in order to convince himself that the larger proportion of all heinous criminals, when the deed has been premeditated for a longer or shorter time, and not the result of some sudden excitement, end their conversation with the following expression : "I myself do not understand how I came to do it, but I felt as if I must do it; I had no rest night nor day for thinking of it ; the thought was for ever recurring that I must do it," &c.   Whenever there is anything in the statement that sounds deceptive and misleading, anything that seems to point to the borderland between conscious voluntary action and the mere blind impulse of insanity and illusion of the senses, such a statement belongs to the same category with the whispering voices " Do it, thou must do it ! " And, in fact, the cases are innumerable in which forensic physicians and medical boards have allowed themselves to be deceived, and have erroneously asserted, and re-asserted, for this reason solely, the blind impulse and irresponsibility of the accused, and for this they have always found a scientific cloak ready provided, inasmuch as they have the one day hung round the blind impulse the mantle of "Hoffbauer's obligatory resolution," and the next of "Plattner's *amentia occulta."*   But these internal voices permit and require a

totally different explanation. In the first place we have to separate
a large class of cases in which the accused, in order obstinately and
for sufficient reasons to conceal the actual motive to the deed which is
not readily perceptible, come forward with the very ready evasion that
they themselves do not know why they did the deed, that they felt as
if they must do it, &c. (in which indeed, as I shall show, they speak
the actual truth). Next, we have to separate another numerous class
of individuals, children and very youthful criminals, and even older
ones of deficient mental power, who cannot give any accurate account
of their mental transactions, especially after the lapse of some time
(at the examination), even when they have the best intentions and
are aided ever so much by the Judge and the physician, particularly
when there has been actually no palpable or generally acknowledged
motive to the deed, such as vengeance, desire of stealing, &c., but
only as it were a more refined one, and one only obscurely felt,
such as petulance. Such culprits, among whom are to be found a
very large proportion of all very youthful fire-raisers (§ 91), do not
then lie when they declare their utter ignorance of any motive, and—
even when it has not been interrogated into them!—continue to
repeat that they must do it, and so they took courage. Nevertheless, all
such internal voices and the like may also be of decisive diagnostic
value, and may prove the actual irresponsibility of the culprit at the
time of the commission of the deed. Especially when (1) it can be
proved that he has already long laboured under hallucinations, particularly aural hallucinations, which are of such frequent occurrence
in the insane, in which case the existence of the insanity itself will
also be capable of being proved; or, even where this is not the case,
(2) when the act of which he is accused has been carried out solely
from a delusionary point of view and from no other. Instructive
examples of this are to be found in the hallucinations of the senses
during sleep, dream phantoms, and the illegal actions of the somnolent (§ 84). When the poor woodcutter Schidmaidzig, whose case
has been so often quoted, killed his beloved wife while somnolent,
thinking that he was defending himself from a white spectre, he
acted illegally from a delusionary point of view and his irresponsibility cannot be doubted. In all other cases, except those two just
referred to, the voices alleged to be heard saying "you must do it,"
are nothing else than the voice of the evil principle itself within the
breast of the malefactor, which, after a longer or shorter struggle of
the moral with the immoral principle, weighing the selfish advan-

tages and disadvantages of the deed, permits the scale to sink in favour of the immoral principle.   The inducements to the deed over-balance its disadvantages and it is done.

The more attractive its advantages appear to the agent on repeat-edly weighing them, the more urgently he feels impelled to carry it out; and thus it is easily conceivable that during such a struggle he has " no rest," and it is also very explicable that in the course of such an internal ratiocination it should appear to him more and more as if " he must do it."   (Compare what has been said on this subject here with what is also said regarding the morbid propensi-ties, § 88.)

From the fulness with which I have treated this subject, the diagnosis of a deed originating in mental disease, comprehending §§ 61 to 63, I shall be able to be so much the shorter in subse-quently explaining the forms of mental disease, for *the application of these diagnostic doctrines to every doubtful case individually consti-tutes the chief task of the physician* in giving a psychological-forensic opinion, and constitutes the most important condition of the accu-racy and power of conviction of that opinion.   Mere sheets full of dry repetitions of the facts of the history in the opinion (such as " we have seen that this and that, and that and this was the case," &c.) with a mere final *quos ego*-conclusion, " from this it follows that N. N. at the time of the deed," &c., such as we so often meet with in official reports and in serial publications, or reports which lose themselves in wide and purely theoretical speculations, and instead of treating the individual case as such, fit it into their gene-ralizing scheme, are both alike unprofitable.

§ 64. JUDICIAL METHOD OF PUTTING THE QUESTIONS.

Since the forensic physician and the medical boards are in general tied down to the mode in which the questions are put by the Judge (or public prosecutor), and these, therefore, form the basis or skeleton, as it were, of the medical opinion, and the physician is, therefore, materially interested in the mode in which the questions are put, I am not likely to be accused of going beyond my subject in making a few remarks on this subject now.   Since, though the competence of the Judge to put the questions in every individual case, as he chooses and as he holds to be most convenient, cannot be disputed, the right of the physician to answer them as he can will also not be controverted. It has frequently been said that it is much more essential for the Judge to determine whether the accused is to be regarded as possessed of "full freedom of will or not?" than that he should ask whether he labours under insanity, imbecility, &c.? I cannot assent to this view. Quite apart from other purely theoretical objections, which have been as often set aside as they have been brought forward, to require from the expert purely his own individual views, placing him at the same time in so unlimited a field, would be to grant him far too large and too dangerous a scope for the development of his own opinion. What, for instance, if he should recognise both the emotions and passions as sources of irresponsibility? Just as little should the Judge—though this often occurs in practice—lay down "criminal responsibility" as the basis of his questions, because this is purely a judicial and not a medical definition (§ 59), and he thus only induces the physician to overstep his own proper domain. This is now and always the natural object to be examined, in this case, therefore, it is the physico-psychico-anthropological object. The Prussian legislator, and many others of the more recent lawgivers, have not left it open to doubt that they regard this view as the correct one; and it is but reasonable to suppose that the Judge will be guided, in the mode in which he puts his questions, by the respective definitions of the statutes. The obsolete Prussian Penal Code says, namely, "No crime is attachable and no punishment can be inflicted on him who possesses *no power of voluntary action;*" but the present Penal Code (§ 40, Penal Code) says, "No crime or offence exists, when the agent, at the time of the commission of the deed, *was insane or imbecile,* or prevented from voluntary action by violence or threats." Accordingly, in the respective cases, the questions must run thus, "Was the accused, at the time of the · commission of the deed, insane or imbecile?" And the physician is thus at once referred to his own peculiar department. Now, however, the Superior Court of Justice* has decided that § 40 of the Penal Code does not include all the circumstances which exclude responsibility, and in any given case some other mode of putting the judicial queries is all the more likely to be employed, that some other cause of exculpation may be brought forward by the witnesses or the advocate for the defence, as I myself, for instance, was once asked in such a case, whether the accused had been in a state of monomania? "When, however," says a decree of one of the superior Prussian

* Archiv für Preuss. Strafrecht, I. s. 50.

public prosecutors to the public prosecutors of his district,—and though this was only in reference to the queries to be put to the jury, yet it is also perfectly applicable to the queries to be put to medical men—" when, however, such circumstances are asserted, they must be specially stated and put in the form of questions, which will then run as in the following example : ' was the accused at the time of the commission of the deed deprived of the power of-voluntary action by drunkenness, somnolence, somnambulism, or the delirium of fever?' As to the mental conditions termed insanity or imbecility, the manuals of forensic medicine furnish a complete series of morbid mental conditions, from mere perversity and folly, from *mania occulta* and *mania partialis,* up to perfect mania, and further, of limitation of the mental faculties, from mere stupidity up to complete fatuity and imbecility. All these conditions, however, have only a judicial signification when they amount to complete insanity or imbecility (partial insanity is also insanity, and fixed ideas are so also, when they govern the whole man). Aberration and distraction of mind can therefore only come in question when they possess the character of insanity or imbecility, and deprive the individual of the power of voluntary action. Any supposed disturbance of the spiritual life, any mental or emotional disease can therefore, in the questions put to the jury " (and to the physicians) " only be termed insanity or imbecility. In morbid bodily conditions, on the other hand, the question requires to be supplemented, as in the instances given above, by the additional query whether the freedom of voluntary action is thereby excluded ? " *   It was most important and satisfactory to find these views expressed by so exalted a legal official, because they exactly agree with my own well considered,
. though apparently unusual ideas on the whole subject, as by and by I shall have to show (§ 70), and which in my opinion are the only views at all suitable for practice, and which at once and for ever solve the perplexities of generalising theorists.

§ 65. NATURE AND MODE OF EXAMINATION.

STATUTORY REGULATIONS.

CRIMINAL REGULATION, § 280. *The Judge must always have his attention directed to the mental condition of the accused, and specially examine whether the criminal at the time of the commission of the deed has*

* Archiv für Preuss. Strafrecht, II. s. 125.

acted consciously. Should traces of confusion or weakness of understanding be found, the Judge must, with the assistance of the forensic physician or any regularly licensed medical man, investigate the mental condition of the accused, and record the means employed, as well as the result in the documentary evidence, and for this purpose the expert must deliver his opinion regarding the presumed cause and probable mode of origin of the mental deficiency discovered.

GENERAL JUDICIAL REGULATION, TIT. 18 (when it is proposed to make a judicial declaration of the insanity or imbecility of any man, and a curator has been appointed for him), § 6. *The court must then cause a more particular examination to be made into the mental condition of the petitioner by means of a deputy appointed for that end, assisted by the curator, the relations, and two medical experts. One of these experts is to be appointed by the curator, and the other by the relatives.*

IBIDEM, § 7. *Should the curator and the relatives not be able to come to an understanding with each other, or with the experts, the unanimous opinion of the latter must be decisive. But should even these not be able to agree with each other, then the Judge must either officially appoint a third expert, and cause the examination to be renewed with his assistance, or he must require the two first experts each to give in his opinion, supported by reasons, in writing, which, with the documentary evidence, is then to be laid before the medical college of the province, and a statement of its opinion on the matter requested.*

(RHENISH-FRENCH) CIVIL PROCESS REGULATION, ART. 302. *Should the opinion of experts* be required, this must be prescribed in a decree which shall distinctly declare the object for which the opinion is required.*

(RHENISH-FRENCH) CIVIL PROCESS REGULATION, ART. 303. *The investigation must take place only before three experts, unless the parties are agreed that it should only take place before one.*

IBIDEM, ART. 317. —— —— *The report is to be drawn up at the disputed place, or at the place, day and hour appointed by the experts. The report is to be written by one of the experts and signed by them all, &c.*

IBIDEM, ART. 318. *The experts have only to draw up one report, they express but one opinion though with different voices. In case*

* Experts of all classes are here referred to, not merely or exclusively medical experts.

their opinions should differ, they must however point out the reasons for such differences, without indicating what have been the precise opinions of each individual expert.

IBIDEM, ART. 322. *Should the Judges not consider the report sufficiently explanatory, they may officially direct a new investigation to be made by one or more experts, whom they may also officially designate, and who may require the former experts to supply them with those explanations which they may consider necessary.*

IBIDEM, ART. 323. *The Judges are not bound to decide according to the opinions of the experts, when their convictions are opposed to it.*

MINISTERIAL ORDER, Dated November 14th, 1841. *The medico-legal investigations and reports as to disputed mental conditions, in consequence of their revision and the regulations issued accordingly, are now generally carried out by the medical men consulted with more of the requisite care and special knowledge than formerly, nevertheless, cases are of constant occurrence in which these investigations are found to be insufficient and unsatisfactory. This insufficiency chiefly originates in a want of that time and leisure by the physicians at the time of the examination, which is necessary for the calm and thorough investigation, so as to form an opinion respecting the mental condition of a person often wholly unknown to them. In order to provide for the future for the most careful and thorough medical examination, and reporting of morbid mental conditions in the course of processes depending upon them, I hereby ordain, after communicating with the Minister of Justice, and in agreement with him, 1. Previous to the time appointed by the judicial boards for investigating, at their request, into the mental condition of any person, the experts must make themselves acquainted with the case by visiting the party as well as by consultation with his relatives and with his medical attendant. 2. At the time of the investigation the physicians, from their own point of view as experts, making use of their previous information as a basis, have to make a special and complete report of the bodily condition of the party examined, his habitual demeanour, &c., as well as of the conversation carried on with the view of ascertaining his mental condition, giving both questions and answers, adding their provisional opinion in regard to the mental condition of the party examined according to the terminology and definitions of the General Common Law, while they are also permitted at the same time to give the scientific denomination of the morbid condition. The protocols of investigations into mental conditions have, in a medico-legal point of view, the same importance and signification as the protocols of dissections, namely, the complete ascertainment, description and determination of the results discovered as a basis for the opinion to be given.   In order to perfect this desirable conformity with the statutory regulations for the examination of dead bodies which have been long in existence.—3. Experts are in general—and the cases in which an exception to this is permitted will be mentioned at the end of this order—at the end of the investigation to give in to the Judicial Board a special reasoned opinion, and in it to state the results of the information previously obtained, the documentary evidence produced and the protocol of the proceedings at the time of the investigation, as well as a complete history of the case, according to the circular order dated 9th of April, 1838, No. 1746, and further by comparing and criticising the morbid phenomena detailed, the proofs, and the facts of the case before them to subject it to a medico-technical review, and thus finally to confirm, according to the best scientific knowledge, their opinion originally given provisionally at the end of the examination, or another which may deviate somewhat from that.   The Royal Ministry of Justice is required to make the Judicial Boards acquainted with the foregoing resolutions, to instruct them and (a) to inform the licensed physicians proposed to be employed as experts sufficiently early of the time appointed for the investigation, so that they may be able previous to that time to obtain information as to the condition of the party to be examined; and (b) to cause the legal official deputed to control the physicians to make a note in the protocol as to whether they have obtained this previous information or not.   Since, on the one hand, it is equitable that the physicians, for a greater expenditure of time and trouble in this matter should be correspondingly re-imbursed, and on the other it is also necessary that the costs which are in general already considerable, and rise considerably when foreign physicians are consulted, should not be increased in a disproportionate degree, and thus burden either the party himself or the funds of the state, the Minister of Justice has ordained (c) that in no case shall the appointed fees be granted for more than three visits made to the party previous to the time of his examination, and (d) the fees for the special reasoned report to be given in after the examination are to be omitted when the result of the examination has been quite indubitable, and the physician has therefore been able to give in the protocol a perfectly definite opinion.   From the physicians consulted as experts it is expected that they shall only make, previous to the examinations, those visits which are indispensable for their information, and where possible, particularly in the case of foreign or poor persons, should be restricted to one single visit. On the other hand the physicians, in agreement with the legal official, may be permitted, in those cases of simple imbecility or insanity in which the result of the examination is indubitable, instead of giving in after the examination a special and reasoned opinion, to dictate at once such an opinion at the end of the protocol, in accordance with the regulations just laid down. The Imperial Government is to bring this order under the notice of physicians and medical men by publication in the official gazette and in any other convenient manner. Berlin, 14th November, 1841. (Signed)* Eichorn.

To the Minister for Ecclesiastical, Educational and Medical affairs.

The passages just quoted from the statutes show in the first place how variously investigations into mental conditions are carried on in Prussia.   In the (French)-Rhenish process the consultation of medical experts on the part of the Judge in civil cases is a purely official act; he may set it aside and actually does so very frequently. Then he may consult one or three experts, and that the legislator may leave no doubt as to the little value he attaches to this consultation, he finally declares that the Judge is not bound by the opinions of the experts.   In the procedure of the old country on the other hand, consequently in by far the larger proportion of the monarchy, the consultation of medical men in both civil and criminal cases in which the mental condition of a man is disputed, is not only positively required by the statutes, but also the legislator does not say in either case that the Judge shall not be bound by the opinions of the physicians.   The modern method of procedure in criminal cases, which always and in every case refers the Judge more to his own convictions, has indeed also made an alteration in this, and that in particular the Judge of the jury court does not even in this (psychological) respect hold himself bound by the opinion of the experts, is one of the most usual occurrences in practice. Yet there is at least not one single instance known to me of a case of disputed civil responsibility, in which a court has not held to the medical opinion (though it may have been that of a medical board subsequently required.)   But there is also another remarkable difference in the procedure of the old monarchy. It requires in criminal matters the consultation of one physician, and in civil matters of two; certainly not because it considers the ascertainment of the criminal responsibility of an individual to be easier or of less importance than that of his civil responsibility, but indubitably because of very evident reasons of general justice, because, namely, in the latter case there are two parties whose rights are opposed to each other, and the law therefore grants to each the aid of an expert.

A slight difference in the verbal construction of the statutory regulations is certainly unintentional, yet I must point it out. The Criminal Regulation requires the consultation (of a physician or) of a " licensed" medical man; while the (civil) Common Law Regulation demands two " expert " medical men. But that every " licensed " medical man is not also an " expert " in medico-legal, and particularly in psychological matters is very well known to each of my readers of either faculty, who is accustomed to tread the halls of justice and who has gathered his—often very remarkable—experience there!

The mode of making the medical examination in cases of disputed mental conditions (by Prussian physicians) has been already partly accurately detailed in the Ministerial Rescript, dated 14th November, 1841, which has just been quoted, and part also of all that has been already detailed (in §§ 5 to 7, p. 184 of Vol. III.) regarding the medico-legal examination of the living subject generally, has reference to this matter, so that but little remains to be added here. Above all things the physician in general must not be induced, as it were *stans pede in uno*, to deliver his opinion, even where he is judicially invited to do so; in civil cases, according to the Rescript just quoted, this cannot happen, but in criminal courts, according to the present public mode of procedure, it is frequently demanded of him. He will be cited to appear at the time of the public trial of N., the case itself and the man being alike unknown to him, and he will require to be present during the whole time, in order to obtain information for a psychological opinion to be given at the end. I have never once found myself, in the course of very many such cases which have come before me, in one single case in a position which I thought enabled me conscientiously to do so, even when I was thus placed in such a position as thereby to stop the whole proceedings, and I can state with certainty that I have not only never had cause to repent of my conduct, but that I have lived to see many important and some even capital cases in which my lucky star has prevented me from giving such a more or less improvised opinion, and which, as subsequent events proved, would have weighed heavily against me ! For all that is unrolled before the physician at the time of trial, is indeed often perfectly sufficient for the decision of a great number of questions, but it is not so, as every experienced alienist or forensic physician knows, for the determination of doubtful mental conditions. Besides the possibility of a well executed deception by the prisoner at the bar, which even the most skilful may not presume to be able to determine at once and within a few hours, the cases are yet more frequent and more important in which the accused betrays scarce a trace of mental disease during the whole proceedings, because for instance he labours under some definite delusion which has not been touched upon, because (as so often happens !) he can restrain himself and appear to be mentally sound, or because he is only occasionally, but not just at present, subject to attacks of insanity, &c. Even the evidence of the witnesses does not always give the physician the most instructive information, for they often say nothing at all about the—for him—most important circumstances, from not knowing them, or from being unaware of their signification, or for other reasons, and the expert has no basis for his opinion. How different is it when he visits the man in the quiet and retirement of his own house (or cell of the prison), when he takes care that writing materials are provided for the use of the prisoner. There is, and this the non-professional public is not aware of, a whole class of insane, actually mentally diseased, who only indulge in their delusions when they are alone and unobserved and brood over them undisturbed, but who know very well how to behave themselves among men so as not to appear remarkable. But in their own house we find in the midst of the apartment some puppet made of straw and clothed, we find all the keyholes stopped with paper without any assignable cause, the doors and windows all barricaded, very often piles of sheets of paper covered with writing of the most obscene character or the most palpable nonsense, string, alleged to be magnetised, stretched through the room, the looking-glass covered with a curtain, &c., and—in half-an-hour we obtain more information than in the course of a whole day's proceedings in court. In other less clear cases the documentary evidence containing the previous history of the case is capable of affording important hints for the further investigation of the case; for this I refer to § 6 of the General Division (p. 187, Vol. III.), and in every case he ought

to beg permission to see this, when the Judge does not of himself send
it to him.

§ 66. CONTINUATION.

In regard to the questions which the physician ought to put to
the party whose mental condition is to be examined, it is impossible
to lay down any rules which may be generally applicable, except this
dogma that those alleged or supposed to be of disturbed intellect ought
not to be asked any questions which the most sensible man would
have a difficulty in answering properly! That this warning is
neither so trivial nor so superfluous as at first sight it appears to be
is proved by the revision of all the transactions in the investigation
of disputed mental conditions throughout the whole monarchy by
the Superior Scientific Medical Board, in which we often find in the
protocol questions such as " What is God ? " and the like ! Apart
from such unsuitable questions every case must in regard to the
investigation as well as the decision, be regarded by itself, and every
separate individual treated as such, and a presumedly insane man of
learning must not be questioned like a weak-minded peasant girl.
It is specially important in the examination of patients, who have
been ascertained to be only subject to periodical or partial delusions,
so to conduct the examination as at once to discover whether the
party responds to the questions like a sane or insane man (§§ 78, 79).
Further, in every case at all doubtful or difficult of course the
necessity of enquiring as to the party's usual conduct and actions at
his relations or connections who reside with him, at his fellow patients
and attendants, should he be in an hospital, or at his fellow prisoners,
and the officials, should he be in prison, will at once occur to the
physician charged with the investigation. Such a procedure is not
only nowhere forbidden, but in Prussia it is even directed to be made
(*vide* Ministerial Rescript, p. 120, Vol. IV.), unquestionably therefore it is not to be regarded as a *quasi* uncalled-for examination of
witnesses on the part of the physician, who, moreover, must in his
report declare the source of his information, so that if necessary the
Judge may have it confirmed ; on the other hand, this procedure is
most judicious and often perfectly indispensable. Only every such
communication must be taken *cum grano salis*, and in each
individual case the physician must consider which of such witnesses
he can put confidence in, and which he must distrust. The first,
for instance, will be the case in regard to the statements of all

unprejudiced persons, such, as a rule, as all classes of institution officials or the sick tenders of public hospitals and in other cases, even the statements of fellow prisoners, always deserving of but little credit, when they describe the accused as mentally unsound, and when their description bears the impress of truth, that is agrees with psychological experience and with what the physician himself has learned by examining the party himself. In regard to this, however, it is not superfluous to point out that the lower classes of prison officials, such as turnkeys, &c., who have the most immediate intercourse with the prisoners, and who from long experience have become acquainted with their tricks and lies, and have thus had their wits sharpened, very, very often fall into the opposite extreme of total disbelief, and even in cases in which the existence of mental disease is indubitably proved by continued observation, they will constantly answer all the questions of the physician with doubts and shakes of the head, so as almost to make him hesitate, if he has not already learned by long personal experience that which I have just been pointing out. Such apparently trifling circumstances increase the difficulties of the medical diagnosis of mental conditions more than could be imagined.

§ 67. CAUTION AGAINST SIMULATION.

The motives for the simulation of mental disease are, in general, all those which in other cases give rise to the feigning of diseased bodily conditions (*vide* § 54, p. 77, Vol. IV.), to which may be specially added the desire to escape the responsibility for by-past criminal actions. I have already had occasion to speak in detail both of the motives and also of the diagnosis, and all that has been already said (§ 55, p. 78, Vol. IV.) is also applicable to the detection of the disputed simulation of mental disease. From the very nature of the subject it follows that this is a much more difficult matter than the detection of somatic ailments; here we require the closest observation, the most accurate consideration of every, even the most apparently trifling circumstance, such as individual answers, indeed even individual words; here we must have the most sagacious combination of all the circumstances of each individual case; finally, to assure the certainty of diagnosis to the physician in this matter, there is required not only a knowledge of the nature of mental diseases and of the behaviour of those of unsound mind, which is only to be obtained by the visiting of large asylums for the insane, a mode of instruction far too often neglected, but also a knowledge of the criminal world. Besides what has been already said upon this subject, the following principles, deduced from experience, may also serve to aid in this matter. There is reason to suspect that the mental affection is simulated, and the traces of this must be further investigated. 1. When the accused himself continually declares that he is insane and does not know what he is doing, or constantly repeats, like the bad boy in Case CLXVII., that he labours under the "mania of persecution" and the like! Nothing more certainly betrays the malingerer than this, far from uncommon behaviour. Those actually of unsound mind are well known to complain frequently of morbid bodily sensations in their head, a feeling of weight, pressure, or deadness, &c., but never that they labour under delusions; of course, because the instant they acquire the knowledge that their delusion is a delusion it ceases to exist as such. When the criminal in such a case as that just referred to knows, as he expresses it, that he labours under the "mania of persecution," he knows that his persecution is only a mania, that therefore he is not persecuted at all! For similar simple psychological reasons those actually of unsound mind are much more apt to assert, often with the utmost feeling to those who doubt it, that they "are perfectly sane and not in the least deranged." A malingerer will seldom make such a confession out of dread lest he should be believed without any further investigation. 2. It must raise a suspicion of dissimulation when an accused person, as very often happens in cases of alleged "weakness of the head," can give no answer to any judicial or medical questions so soon as they in any way concern his crime, whilst his weakness of memory does not prevent him remembering most accurately other facts, such as numbers, dates, &c. In the remarkable Case CLXXI. this circumstance was the first in the course of the long and difficult investigation which gave any foundation for the proper diagnosis. 3. It is also suspicious when the accused has no answer to give to any question whether important or unimportant, parrying all alike with the one unvarying reply, "I don't know, my head is so weak." Where there is no absolute mental nullity, perfect fatuity, or the deepest melancholy, those actually of unsound mind usually answer any simple question that is put to them, as to their names for instance, their dwelling place, &c., facts which have been long deeply impressed on their memory, *bona fide* and often quite correctly, while the malingerer fears to compromise himself by answering thus. 4. Deceit is to be suspected when the man of alleged unsound mind exhibits a variety of symptoms and forms of mental disease, when one day he displays the utmost folly and the next the deepest melancholy, when at one time he sees the devil and all his evil spirits, forgetting that but a few weeks previously his visions had been of a totally different character, &c. This is well known not to be the nature of true delusions, just as little as it is. 5. For the party examined after having, in the course of long and repeated conversations intentionally carried on about indifferent matters, given the most suitable and correct answers, suddenly, when the conversation has taken for him a hazardous turn, to commence the most irrational speeches. But coarse deceits of this kind are by no means rare. An emigration agent who was arrested as an accomplice of the murderer Holland,* simulated insanity precisely in this laughable manner, which could not deceive us for one instant. He expressed himself in the most sensible manner in regard to his business and emigration in general; whenever however, I mentioned the name Holland, he in general stepped a few paces backwards, became apparently enraged, constantly declaring that he knew no Holland, talked foolishly about Holland and England, &c. He was condemned. 6. It is suspicious when the statements as to the origin and course of the alleged mental disease are perfectly inconsistent with experience. Malingerers, when asked, date the origin of their insanity or mental weakness from a certain fever which they had more than ten years ago, from an injury, the mark of which they exhibit as a small, scarcely visible cicatrix on the forehead, &c., whilst we perhaps already know from the documentary evidence, that no witness has ever observed any ˉmental derangement in them, and the case as a whole does not permit the assumption of any such alleged circumstance as an actual etiological cause in the production of any mental disease. 7. There is urgent reason to suspect deceit when the alleged attacks of insanity, &c. only occur at what may be termed convenient times for the accused, for instance, only when he knows that he is observed, always only when he is about to be arrested on account of some fresh offence, &c. That malingerers do not hesitate to put even this coarse procedure in action is proved by Case CLXXI. already referred to.

* Case LXXII. p. 186, Vol. I.

§ 68. CONTINUATION.

On the other hand there are certain phenomena, a certain physiognomical *habitus* in those actually of unsound mind, which are quite incapable of being simulated, or others which it is almost impossible to imitate or feign, and which at once conduct those only moderately acquainted with the subject to a correct diagnosis. Quite apart from the appearance of complete and cretinous fatuity, as well as from that of many years' continuous insanity, which give the patient a brute-like appearance, originals which cannot be copied, I would, in the first place, direct attention to the fact that the natural requisites of nourishment, sleep, and a certain amount of atmospheric warmth, assert claims which cannot be set aside by any person of sound (bodily or) mental health. An actual constant refusal of all nourishment, a constant wakefulness (and raving) throughout the whole night, a continual braving of cold (running about in a shirt in a cold room), are often observed in the insane, as is very well known: but no deceiver can continue to do all these or even only one of them for long, and we cannot err, when we find these phenomena, in assuming the verity of the mental disturbance.—There are not a few insane persons who exhibit an inexplicable inclination to meddle with their own intestinal filth, to rummage about with their hands in it, to spread it on the walls, and even to eat with their dirty fingers! I cannot suppose that any healthy malingerer would so make use of any such naturally disgusting material in order to gain his ends, and I have therefore in all such cases decided against the idea of feigning, of course with a due regard to the other diagnostic circumstances. A generally respected banker was involved in a criminal investigation for detected forgery and arrested. During his imprisonment he suddenly appeared to become insane, and under all the circumstances of the case the reality of this was all the more likely to be suspected, that, in the midst of the most silly speeches, the accused made repeated remarks bearing upon his business and his accusation, such as " Mother must write to X. that he send the thousand thalers (£150)," &c. But in a very few days the maniacal delusions became continuous, day and night, and when he, a man from the educated classes, finally commenced to rummage in his own ordure, I no longer hesitated to assume the reality of the insanity and to send the patient out of prison into an asylum, in which he (quite suddenly and unexpectedly) died of cerebral paralysis.—One peculiarity of a great many imbeciles further deserves to be here pointed out. It is not known to deceivers, and is not very easy to imitate without clumsiness. Such patients, during the conversation, while they are being examined, because they take no interest in it, glance idly round about with their blank and inexpressive looks and delight to repeat each question as it is put to them, before answering it, as if seeking first to impress it more fully on themselves, e. g. "What is your name?"—"What is my name?"—"Have you any children?"—"Have I any children?" &c. If any acquaintance, spouse, brother, or friend is present, such patients are apt, indeed usually do, turn towards them, as if seeking to be put by them in the way to answer a question which to themselves appears so difficult. I may further call to mind the quite peculiar hastiness and restlessness of so many of the insane, particularly women, who while we are conversing with them are observed to be continually wringing their hands, plucking their clothes, springing from their seat, &c. I call to mind also the constant smiling or laughing of other insane people while being examined, without any cause which they themselves can or will assign when asked, and similar trifling physiognomic traits which are so frequently observed in the insane or imbecile, and to which I have directed attention here because they are never alluded to even in good psychiatric descriptions, probably because they are apparently too unimportant, but which are of the utmost importance in the determination of the diagnosis of a doubtful case of simulation, because they are decidedly more characteristic of the patient than the deceiver, and consequently possess a practical value.

In explanation of what has been said, I now proceed to give as various a selection as possible in abstract, from my cases of disputed insanity, preserving, however, all the material points.

§ 69. ILLUSTRATIVE CASES.

CASE CLX.—CARL SCHRABER,* THE MECKLENBURG PRINCE.

The merchant Carl S. was accused of and arrested for very important forgeries. He had pretended to those whom he deceived that he had deposited an inheritance of ten thousand thalers (£1500)

* Pseudonym, as are almost all the proper names given.

with the N. Court, and he produced before them the deed recording this, which, however, both in contents and form (with forged seals of the N. Court), was only a very clever forgery. At the first precognition, on the 7th of September, S. made an open confession, in which he stated that he was the son of the draper S., still living in G. (Mecklenburg), and had an elder brother who was insane (which was confirmed) and a sister. Up to the conclusion of the examination he abode by this confession, and detailed a number of circumstances with the utmost accuracy, but neither these nor his exculpatory reasons are of any further interest at present. In the whole course of the investigation S. had, according to the record of those who examined him, "never betrayed even the remotest trace of mental disease or imbecility, but had rather in everything exhibited himself as a crafty and long-headed swindler, so that there was not the slightest reason to doubt his responsibility for his crime." In such a state of matters no attention was therefore paid to the statement of the culprit's wife, who, after having in her frequent previous examinations and written memorials never said anything in regard to anything remarkable in the behaviour of her husband, stated, for the first time, at her examination on the 11th of November, that "on frequent former occasions her husband must have been completely deranged, because both his words and actions were unconnected and inconsiderate;" and the culprit was, in the first instance, sentenced to two years' penal servitude, and a fine of six thousand two hundred thalers (£930), or, in case of nonpayment of the fine, to six years' penal servitude. This sentence was made known to him on the twentieth of November, upon which he declared that he would adopt legal measures, named an advocate, wished that the managers of the Penal Institute should be desired to prepare work, in writing, for him on the provisional commencement of his punishment; he also begged leave to see his wife in regard to procuring the money necessary. Five days subsequently, writing materials were granted to him at his own request. The writings subsequently given in by S. are dated the thirtieth of November, and consist of two letters, addressed to his Majesty the King and to the Grand Duke of Mecklenburg-Strelitz, and in a so-called written defence. To the King he represented himself as a nearly-connected prince, as a near relation, and begged his Majesty to cause him to be removed to his royal palace, in order that there, "under his exalted relation's royal auspices, he might as speedily as possible dispose of the matters connected with

K 2

his princely birth, as well as the other unpleasant occurrences." To
the Grand Duke he declared, in the other letter, that he was " a
legitimate son of Duke N. of Mecklenburg-Strelitz, now sleeping in
God." It would be all the more easy to make the necessary investigations by means of his "ambassador," Counsellor H. (the advocate
selected to defend himself), as the servants of his "late father" were
still in the grand ducal household. In order, however, "that the
appanage of the house should not be a source of expense to the
country, he was willing to devote himself to the service of the King
of Prussia, and leave undiminished to Prince Z. the inheritance of
his father, to which he had succeeded." In the written defence,
which occupied eleven folio pages, he mentions, in the first place, his
double hernia and ailing body, and opines "that fate and the force
of circumstances must bear a large proportion of the blame which
had been imputed to him, from the circumstances not being accurately known which induced him, for one instant, to leave the path of
rectitude, on which our mighty Saviour Jesus Christ has preceded
us." He further demanded the most thorough, but at the same
time the most indulgent investigation, spoke of his "hereditary
dignity," and expressed a hope that his Majesty would forget his
errors, as he would be of use to the royal house, the state, and the
army. " It is enough," he said, " when I point provisionally to
Russia; we must make ourselves strong to resist this Colossus; we
must strengthen the land forces, but it is also high time to create a
naval force also for our Prussia. I will, therefore, add a naval force
to the land one," &c. He then proceeded to detail, in a more comprehensible fashion, that according to the wish of his "*pseudo-
father*," or his "*ita dictu* father," the draper, he had learned the
trade in Hanover, relating, by the way, the actually correct details
of the fate of his insane brother. Then he suddenly flew off, remarked that in Göttingen he used to "beat the drum," that to-mor-
row would be fine and there would be a parade, and he begged
Counsellor H. that he " would be pleased to mention to his privy
councillor" to notify to the King that he would appear there in the
uniform of an officer of the guards, and that, therefore, horses,
weapons, and uniforms for his retinue would require to be placed at
his disposal.—From Hanover he went into a trading house in Bremen,
where he remained about two years, and then returned to his native
place in 1827, when his sufferings commenced. His "foster-mother"

—and I may remark, by the way, that the syllable "foster" was not written at first, but had been subsequently entered—was dead, &c. Finally, he concludes with the statement that only "the most incomprehensible blindness and stupidity" could have prevented him from sooner asserting his rights. In consequence of this, an investigation into the state of his mind was ordered, and many witnesses were examined. His wife deposed that "he often got up at night and sat writing for hours; at such times he spoke much of his large property, and next morning often knew nothing about it. Further, it not infrequently happened that he took me for another woman, and said to me at night, for instance, what would his own wife say if she knew that I was lying beside him. He has also often told me that he was the illegitimate son of the deceased Duke of Mecklenburg-Strelitz." Dr. S., who two years ago had S. for a short time under his medical care, was convinced that S. was "an eccentric man of a somewhat limited and confused mind," who communicated to him the "most minute details of his family relations" (witness says nothing about the alleged prince), in whom, however, he did not remark any particular symptoms of mental derangement. The wool-broker S. and his uncle, advocate R., who are well acquainted with the accused, call him frivolous, eccentric, cracked, but have never observed in him "the slightest trace of mental derangement." But there is a peculiar mental condition in all the S. family except the father, and, besides the brother, the sister also is said to have occasionally exhibited symptoms of unsoundness of mind. To the master carpenter V. (who was swindled by S.), and to E., he always appeared to be a "perfectly rational man." The statement of his father, the merchant S., is important; he details the frequent excesses of his son towards friends, his own family and himself, which he ascribes to his "unbounded pride." He never, however, remarked that these amounted to mental unsoundness. "He knew very well what he did, and when it was time to become more yielding, both to himself and others." As to his alleged princely descent, his father does not know whether this statement "ought to be regarded as a sign of derangement, or as a mere fabrication." The court of justice of the town in which he was born declares that S. had frequently robbed and cheated his father, and had frequent quarrels with him, which sometimes ended in blows. In an investigation against N. and partners, it has even been recently ascertained that it was extremely probable that he had, in the summer of 1842, made an attempt to poison his father with arsenic, so as to send him out of the world, and get speedier possession of the shop.

"My own repeated examinations of the accused have yielded the following results: the merchant S., of middle stature and dark, plentiful hair, has a somewhat yellowish complexion, a piercing, unpleasant glance, but nothing odious in his physiognomy, a well-developed forehead, something open and resolute in his features, with a corresponding assurance in his manner. He does not appear to place any particular value on his external appearance. His conversation is fluent, but select, coherent, and manifests the certainly only half-educated but skilful man, and his ceremonies and forms of civility, &c., as a rule, evince the same. S. is not of sound bodily health; various so-called liver-spots upon his chest and abdomen show even more certainly than his complexion that there is something anomalous in the discharge of the functions of his abdominal organs, and the sudden bursts of passion to which he has been proved to have been liable, may be thus accounted for, though not justified. His double hernia unquestionably increases the tendency to constipation under which S. frequently labours. Finally, from the abdominal cause referred to undoubtedly comes a symptom which S. has complained of for several months, namely, an incessant hunger and particularly thirst, which he can scarcely assuage by the most copious draughts of cold water, and which causes an excessive flow of urine, which is otherwise normal. No alterations of the spleen or liver are to be felt. At my first visit, on the 9th of December of the same year, I commenced the conversation by referring to his double hernia, and then I went on to speak of his residence in Hanover. For a long time he spoke of this in a clever, rapid, and fluent manner, and then referred to his present position, which "was doubly horrible to him, as he was a born prince," bringing out this *quite suddenly*. He spoke of his (pregnant) wife—he always speaks of her as his "consort"—and said, as he often subsequently repeated, he was anxious to know whether she would bring a prince or princess into the world. Apparently quite unintentionally, I endeavoured always to bring him back to his family history, and I succeeded in getting him to say, in answer to a question, that ' *his insane brother resembled him.*' Taking advantage of a statement as to the violence of his father's temper as well as his own, I said that it seemed to be a family failing; to which he replied, ' Yes, it *runs through our whole family.'* On the 18th of the same month I found that S. had the day previously taken an emetic because of bilious symptoms, and was lying on his sack of straw. His appearance was open and natural, his tongue still somewhat coated. He very soon commenced to speak again about his Majesty and the Grand Duke, the latter of whom had visited him just eight days previously, but to this I did not further refer. I rather went on to state that I knew one of his countrymen, a physician in this city, and I enquired whether he had any other countrymen here? He mentioned, as I expected, that the advocate K. was related to him. Inquiring into the nearness of this connection, he replied that K.'s father had married *his father's* sister. His two fellow-prisoners both complained to-day that S. made a noise at night, speaking of the King, &c. On the 23rd of the same month I saw him again. During the time for exercise in the court to-day he had attacked one of the prisoners, and he had made so much noise during the night, about one o'clock, that the overseer of the prison had to be called. To-day I entered more upon the ideas of the accused, and exhorted him to reflect upon the improbability of his assertion. I said to him that he did not in the least resemble the duke, and to this he had nothing to say but ' No? that is remarkable—he cannot, however, imagine how else he should always have loved him so much ?' Why then did he not sooner bring forward these statements? because he lived happily, and had no need to alter his position, now, however, he thought he must assert his rights. His appearance to-day was more restless, more excited, and he complained of a bitter taste; his tongue was slightly coated. On the 27th of the same month, S. for the first time complained to me of his great hunger and thirst, and frequent micturition, of which I have already spoken, and said that he felt weak. Besides those already mentioned, there were no other objective morbid symptoms to be discovered. I asked him if he was now convinced of the erroneousness of his statements? He answered quietly, 'I have thought much over it, but it always comes back to me the same. I have never trusted you from the first, because I believed you to be sent by the opposing party,' by which he meant the Grand Ducal House. I again referred to his bodily condition, and stated, that his excessive micturition was a symptom of a very rare disease called Diabetes. How did he come to be attacked by so remarkable a disease—which was always hereditary (a statement without foundation)—did *his father* also suffer from it? S. *thought for an instant,* and then replied, ' he no longer lives,' &c. On the 31st of this month I found the accused chained, because he had a few days previously attacked a fellow prisoner during the night. He was, or appeared, to-day to be quite altered, made use of obscene language and was very coarse. ' What matters it to you?' he replied to my interrogation, 'only get me something more to eat and drink,' but he subsequently spoke of his 'consort,' the expected son, ' the prince,' and his ' hereditary rights,' &c. This behaviour of S. gave me the desired excuse for giving the examination a different direction at my next interview with him (on the eighteenth of January). *Without any inducement he again* began to speak about the duke. I replied to him in an unfriendly tone, referring to his recent unseemly behaviour, that I had listened patiently to his absurdities long enough, only bearing with him because he was labouring under bodily illness. Now, as he was better, as he confessed, he must dispense with such senseless talk to me, as I had long since seen through it, &c. The impression made by this speech was remarkable. S. was *evidently surprised*, and replied, fixing his eyes upon me, *speaking after a little consideration and in a dejected manner*, 'I am, however, firmly convinced of it, and I am sorry that you won't believe it.' Fourteen days subsequently he told me, in answer to a question, that his 'consort' was born a C., and he described in a lively and connected manner her accomplishments, and the history of his marriage," &c.

"These are the chief results of my frequent examinations of S. upon which I have to base the required opinion as to ' his present mental condition, and as to his responsibility at the time of the commission of the deed;' in this opinion which follows, I shall endeavour to show, *that Schraber has only simulated insanity.* I do not mean to deny that in his case there was a concurrence of circumstances which predispose to mental derangement, and which in a greater degree might even produce it. A man, such as S. is, affected with abdominal disease, whose ' pride is unbounded' (a character which experience teaches us is more liable than most to insanity), who, moreover, is generally ' eccentric, cracked, high,' and, besides, given to drink, such a man *may* readily become insane, as well as thousands of others. Seeing, moreover, that it is also ascertained, that his brother labours under incurable mental derangement, and it is said that his sister also is not free from occasional derangement, it must be confessed that the possibility of actual mental disease existing in him is made yet more probable. In the judicial confirmation of any matter, however, even of a man's mental condition, bare possibilities are not sufficient, but as far as practicable, certain proof of its actual existence is required. And now it is evident, that from the circumstances already mentioned, it cannot by any means be concluded that the accused is actually deranged in his mind. However prone any fundamental character, such as his, is to mental derangement, daily experience teaches us how much more frequently this does not occur in men of this description. Further, in regard to his insane brother, the documentary evidence gives no more accurate information; we do not learn what purely individual psychical or bodily causes may have probably occasioned his mental affliction, &c., and, moreover, as to the alleged derangement of his sister, we have only the deposition of the advocate K., who speaks in regard to it only from hearsay. It would be, therefore, all the more hazardous to draw any conclusions as to the accused from the alleged state of health of his brother and sister, as even if that were much more accurately ascertained, even if it were positively determined that both his brother and his sister had from a similar fundamental character gradually become insane, even then, it could not be deduced from that alone that the apparently insane behaviour of S. must be attributed to actual hereditary insanity. But also, even from his peculiar behaviour—which shall be by and by more carefully criticised—no deduction can be drawn, since the obvious suspicion that the words and deeds of the accused are alike voluntary has been quite recently brought prominently forward. On the other hand, the statement of his wife at her precognition on the eleventh of November, which has been already detailed, cannot be disregarded; according to it, as well as her subsequent oral deposition, if these are to be assumed as incontrovertible, it must be regarded as certain that the accused laboured previous to the commission of the deed under evident and precisely the same delusions with those with which he is now affected. For my part, I cannot assume that the statements of Mrs. S., which, moreover, were not made upon oath, are so incontrovertibly true, and in stating this, I by no means overlook the fact that her written declaration is dated the eleventh of November, while the sentence of her husband was not made known till the twentieth of the same month. Nevertheless, there is reason for entertaining suspicions as to the truth of this statement, and that all the more that it is not easy to perceive what could have induced Mrs. S. to come forward on the eleventh of November, after her husband had been two months and a half in prison, with a communication of so much importance both for him and herself, not a trace of which was to be found in any of her previous communications, and which, moreover, was not elucidated at her first oral precognition as it were, but was made perfectly voluntarily and unrequested. These suspicions are also considerably strengthened, when we consider that Mrs. S., in her petition to the Directors of the Criminal Courts, dated the twenty-ninth of October, states 'that her husband was daily expecting to receive his sentence from the court of first resort; and after its promulgation would be at once removed.' It is not the province of the medical jurist to ask whence she obtained this information, and whether in the same manner an agreement regarding their subsequent procedure might not have been entered into between the spouses. It is, however, psychologically of greater importance to point out that *even then*, in the immediate expectation of his sentence (on the twenty-ninth of October), she still kept silence in her petition as to her very important knowledge of the mental condition of her husband, and brought it forwards for the first time *fourteen days subsequently*. Under such circumstances the isolated statements of Mrs. S. are deprived of all value in the determination of the case; and I have spent so much time in their examination only, because they apparently formed the chief argument against my view of the mental condition of the prisoner. Therefore I feel myself necessitated also to point out an internal reason against the allegation of Mrs. S., that her husband had entertained for a whole year (that is since her marriage) the idea of his princely birth, and that is this circumstance, namely, that in concluding his bargain with the merchant E., who was swindled by him, he disclosed to him 'that *his father* was a very rich merchant in Mecklenburg. When on the one hand it is evident, that in any money transactions such a statement must be more advantageous for him than that he was the illegitimate son of a deceased prince, so on the other hand, it contradicts all medical experience in regard to men actually labouring under a fixed idea, that this should be not only occasionally forgotten or denied, but that the *direct opposite* should be believed or declared. Granted, however, that, according to what has been already stated, S., both previous to and also at the time when he committed the forgeries, laboured under a fixed idea, it will not be difficult to prove that even in that case he is not deprived of responsibility for his crime. It would be to stretch the domain of irresponsibility far beyond the bounds and limits set to it by the unprejudiced observation of nature, if it were to be made to include the mere fixed idea of itself. This is the fettering of a mind, in itself and in general free, to one particular delusion, and experience has taught us by innumerable cases that men affected with such a monomania display beyond its influence not a trace of mental derangement in their speech or actions, so long, namely, as the mind still rules the fixed idea, regards it objectively, recognises it as a delusion, from which, however, it cannot free itself. Only then, when the fixed idea on its part gains a greater power and dominion over the understanding, which then can no longer recognise it as delusion, when the man then becomes driven to perverted actions, undertaken from his monomaniacal point of view, only then can such actions, when illegal, no longer be judged by the ordinary standard, and in such cases, as experience teaches us, the patient will always progress more and more from mere monomania to general insanity.   Accordingly, had S. for instance, during the bygone summer, possessed with the fixed idea of being a prince of Mecklenburg, molested or personally addressed the reigning prince of that or of our own country, &c., he would have acted in accordance with his delusion, whilst the forging of false documents—not such as had any reference to his princely birth, but—to get money to be employed in furnishing the means of trade, had not the slightest connection with his (alleged) delusion.   On the other hand, everything is in favour of the assumption, that both previous to and at the time of the commission of the deed the accused was in the free and undisturbed possession of his intellectual powers.   Neither his father, nor any of the other witnesses examined, had ever previously observed any trace of derangement in him; the manner, already described, in which he carried out his forgery proved incontestably the most systematical, carefully considered, cunning, and judicious deliberateness, and the criminal act stands in the closest psychical relation with the ' unlimited indiscretion' of S., a man who hesitates at no means, probably not even at his father's murder, in order to attain his end, an independent existence as a citizen.   Accordingly, I dare not hesitate to give it as my opinion, in answer to one of the questions which have been put to me, *that Schraber was criminally responsible at the time of the commission of the deed.*   When, however, I think I have proved that at that time he was not affected with any fixed idea, I may almost assume it as proved that his present alleged delusion has no real existence, and that he only simulates it for his own ends, in so far, namely, as the present delusion is said to be merely a continuation of the former. But there are also more evident reasons for thinking so. In the first place, I am not disinclined to refer for this end to his external demeanour and manner of clothing himself, to which I directed my attention not without an object. However restricted he may be as a prisoner in regard to this matter, yet the slovenly manner in which he dresses himself is remarkable for a man who imagines himself a prince. We have only to glance at the supposititious kings and princes, &c., in any lunatic asylum, and compare their carriage, the pride that speaks from every feature, with the external appearance and demeanour of S. ! It is true he also speaks of his "consort," and of the prince which she is about to bring into the world, but he does so in a manner which could only deceive those unacquainted with the subject, and the intentional character of which only too clearly appears. The numerous external and internal contradictions to be found in his oral precognitions, his writings, and his statements to myself are, however, of more importance. There is an internal psychological untruth involved in his assertion that he had not chosen to assert his alleged title to exalted birth previously, because he felt himself happy in his position in life. Had he actually imagined himself to be Prince of Mecklenburg, as he now writes and calls himself, he could not have felt himself happy in his paltry and confined position, a position which the forgeries he committed were undertaken for the purpose of bettering. His behaviour is also contradictory, when at one time late at night he makes a disturbance in the prison, at another time attacks another prisoner in the court, and a third time behaved uncivilly to myself, and made use of obscene language, none of these things being at all in accordance with his usual conduct. In so doing—like so many other deceivers—he has confounded the symptoms of fixed delusion, mania, &c., together." — — &c.

"Of great importance further for the decision of the case are those statements elicited from him during his conversations with me, in which, as I must assume, S. has plainly fallen through his character, and betrayed himself as a cheat. When he confesses that his insane *brother* resembles him, that a violent temper is peculiar to all the *Schraber* family, that the father of the advocate K. had married *his father's* sister, he confesses that *he is a Schraber*, and forgets that he comes from a princely parentage. But unless he be cured, no patient affected with a fixed delusion ever forgets his part, because, unlike S., he does not act one. Finally, for the proper determination of his present responsibility, I cannot omit to direct attention to the diligent manner in which the accused has sought both orally and in writing to *exculpate* himself. When he ‘has for a moment left the path of rectitude, fate and the force of circumstances must bear a large proportion of the blame’—he has, if we may believe him, not sought to deceive, but only as it were to procure a loan upon property inherited from his mother, or grandmother—(*and not princely!*), &c. By such statements, however, he only shows too plainly that he is even now very well able to understand that he has left the ‘ path of rectitude,’ he recognizes the punishableness of his actions, inasmuch as he seeks to clear himself of it, and thereby unwittingly confesses that even now he possesses the power of distinguishing between right and wrong, in spite of his would-be insane writings and actions. All these contradictions just pointed out can only be explained by supposing that Schraber is not affected by any actual general or partial insanity. When to all this we add the consideration, that the prisoner is just a man from whom such actions as he is accused of could be very well expected, and that the alleged delusion first made its appearance after he had been sentenced to a severe punishment, it seems justifiable for me, with reference to the details already given, to sum up my opinion finally in the following words: *that the criminal, Carl Schraber, must be regarded as having been a responsible agent at the time of the commission of the deed, that he is so now, and that his alleged delusion is only simulated.*"

In consequence of this opinion the many years' penal servitude to which S. had been sentenced was adjudged to be carried out. Previous to his removal to the Penal Institute, however, he declared, that as it was now of no consequence to him he would confess that he only desired to deceive us all, and that he was "satisfied" with my opinion !

CASE CLXI.—ATTACK OF MANIA AT THE TIME OF TRIAL.

The journeyman joiner, Claus, who had been repeatedly convicted, had a severe punishment to expect in a new matter under investigation. About the end of November, after a short imprisonment, he commenced to make use of insane language, and to behave himself in an unruly manner; but so very awkwardly did he act, that not one of those who saw him ever doubted its being an attempt to deceive.

He got a douche-bath daily, which made him quieter, he ceased to
lay about him, &c., but now he appeared to become more imbecile,
answered the simplest question with " I don't, know," had to be
forced to work, &c., so that I was asked whether, now that the precognition was closed, he could be brought to a public trial? After
due consideration of all the former, as well as the present behaviour
of C., I answered this question affirmatively. On the seventh of
February he was brought to the bar, but immediately on entering
the hall of the jury court he commenced to behave in a frantic
manner, crying out and striking round about him, so that two
policemen could scarcely manage him. The jury court requested my
opinion of the case. Even then the intentional character of the behaviour of the accused was quite indubitable. I explained in detail,
and in a manner comprehensible also by C., the reasons why I must
decidedly declare that he was of sound mind, and I did not hesitate in
conclusion to address myself in a loud tone of voice to C. himself,
and to declare with the utmost decidedness that his behaviour was
not that of a man actually " insane," but only a bad caricature. The
effect was that desired. From that instant C. became quiet, the trial
commenced, he answered every question like a sane man, and was
finally condemned.

CASE CLXII.—ALLEGED WEAKNESS OF MEMORY.—PERJURY.

The wife of a merchant S., had ordered clothes for her son in
January, 1849, and in a subsequent action for non-payment she
declared on oath that she had no remembrance of having given any
such order. The circumstances occasioned a suspicion of perjury,
but in the course of the investigation Mrs. S. alleged that she .
laboured under a great weakness of memory. Charged with the task
of investigating this matter, the following question was put before
me to answer, "Is the memory of the accused in such a state, that it
is with probability to be assumed that an order given by herself
during the year 1849 could be completely forgotten by the
twentieth of November 1850, the day on which she took an oath to
that effect?" Mrs. S. very soon betrayed herself, inasmuch as in the
course of an apparently indifferent conversation with me about matters
wholly unconnected with the investigation, she gave the readiest an-
wers to questions such as, How old her husband was? the ages of her
children, the diseases they had gone through, &c. Then I recurred to

"the unfortunate circumstance of this investigation, and its possible lamentable results," and Mrs. S. did not fail to fall into this strain and to describe most feelingly her unfortunate position, her poverty, and the embarrassment in which she was placed by the giving of this oath, evidently to procure a favourable opinion from me. Consequently, she herself involuntarily let drop the pretext of a weakness of memory, and brought forward evidently the true *causa facinoris*, she forgot herself, and ceased to keep up her character! Of course the foregoing question was negatived.

CASE CLXIII.—ALLEGED LOSS OF CONSCIOUSNESS AFTER THE BIRTH OF A CHILD.—DEATH OF THE CHILD.

On the fourteenth of September, the unmarried woman, E., aged thirty-six years, and who had already given birth to a child fifteen years previously, produced a boy, which was proved by the medico-legal examination of its body to have lived during and after its birth. The accused had concealed her pregnancy, although its existence was "not doubtful" to her. About noon on the day men-tioned, she says she felt a rigor, and gave birth to the child quite unexpectedly and without pain. At this time she was lying in bed closely covered with the bedclothes, and, as she alleges that she did not perceive any sign of life in the child, she allowed it to lie with-out further examination for three or four hours between her thighs and under the bedclothes. Then for the first time she "tore away the cord close to its body," and laid the child by her side. Next morning she dressed herself, swaddled the child, whose birth she had also concealed, in old linen, and carried it to a public building, where she laid it down. At a subsequent examination she confessed " that she felt that something passed out of her body, although the rigor had made her almost completely unconscious." Only after the lapse of three hours, during which she "lay in an ague fit, which prevented the performance of any act," she felt her loins "perceptibly cold," and then, for the first time, it "became evident" to her that she had given birth to a child, since up to that time, she "had not enough of consciousness" to think what had escaped from her. When she then took the child into her hands, it was " cold and stiff," without the smallest sign of life ; and with this statement she corrects her former declaration, adding, that she had been scarcely a quarter of an hour in bed when the child was born. The cook, after returning home, and before bedtime, brought her, at her request, as she complained of headache, some camomile tea to the bedroom, where the accused already lay in bed closely covered to the throat; and she answered affirmatively the question, had she a violent headache? and to the cook, who after this time was going backwards and forwards, she also stated, " that she still had a violent headache, but that it was getting better." The cook did not observe anything remarkable about her as she lay in bed, except a heightened complexion; in particular she did not remark that she had been unconscious, as she answered all her questions correctly. She complained of a violent headache also to her mother, who returned home about noon, that is, very soon after the accused, who came home between eleven and twelve, and who at once went to see her in bed. After three o'clock, her mother and another daughter both went to her bedside, where they remained till dark. It is not a little important to add, that both her mother and sister were continually putting questions to E., which she answered clearly and distinctly, so that there is no doubt as to the state of her mind at this period of the day. At a subsequent examination of the accused she declared that it was not till later, when the tea already stood before her, that she recovered sufficiently from her bodily and mental weakness to take a little tea out of the cup, but that she was immediately after so shaken by a feverish rigor, that she fell into a state of unconsciousness and perfect powerlessness. The medico-legal dissection of the child proved that it had certainly been alive even after its birth, and that it had died from pulmonary and cerebral apoplexy. We, however, did not discover that there had been any active cause in the production of this form of death, but, rather that there was the very highest probability that it had been produced by leaving the live-born child lying for three or four hours beneath the close covering bedclothes of its mother. " When I am now asked as to the credibility of the mother's statements, I must in the first place point out, that delivery during a state of partial or total unconsciousness of the mother is nothing very unusual. Moreover, a usually only momentary unconsciousness during the final and concluding period of the birth, produced by the violence of the pain, and the tendency of the blood to the central organs, is of very frequent occurrence. But, should the congestions referred to, particularly that of the brain, last for a longer period, they may render the patient wholly unconscious, and produce in her the most dangerous condition. A middle state betwixt momentary and long

continued unconsciousness, such as that alleged in this case to have
lasted from three to four hours, especially when no scientific means
were adopted to shorten this condition, would, according to experience, be a rather unusual occurrence, though its possibility could not
be denied. There are, nevertheless, important reasons for considering
that the statements of the accused in regard to the existence of such
a condition are not founded in truth, apart from the contradictions
into which she has fallen. She deposes that she gave birth to a child
during the dinner of her family, and it is certain that the dinner was
over, and the cook had already finished hers when she brought her the
tea. At this time, therefore, she had already given birth to the
child, and that not long before. But at that time she was by no
means in a "state of unconsciousness, weakness, or apathy," such as
is mentioned in the question put before me, because she spoke sensibly
to the person who brought her the tea, who also saw nothing remarkable
about her. It is not unimportant to remark, that the birth in question was not the first which the accused had, since she had given birth
to a child fifteen years previously, and that she is now thirty-six years
of age. Under such circumstances, no such ignorance of the events
of childbirth as may be attributed to a youthful primipara can at any
time be supposed. In particular, the accused must know that before or
at the birth—even if her statement that the child was born rapidly and
without pain is to be believed—under all circumstances, a considerable
quantity of water and much blood escapes from the sexual parts; and
it cannot be assumed that a person who was capable of expressing
herself sensibly in regard to her headache did not know that she was
lying in warm fluid. At all events, she herself has confessed that
"she felt something pass out of her body." And she seeks to solve
the contradiction by subsequently stating, that at first she recovered
somewhat, but subsequently fell back into her former condition.
This statement is devoid of credibility, in so far as such a fluctuation
in the psychical condition of parturient women is not usually observed, especially, as I may repeat, when this condition has been left,
as in this case, to nature alone, and no temporary and not persistent
improvement has been artificially produced by the use of blood-letting,
&c. Now, since we must assume that E. was perfectly conscious
when the cook brought her the tea, and must have felt the child,
and the fluids of childbirth beneath her, then we must also concede,
that she was also in a condition to bestow upon her child those necessary attentions which were unquestionably well known to her; in

particular it must be assumed, that at least it was well known to her that there was a possibility of her child being suffocated, if it continued to lie between her thighs and beneath the bed clothes, and did not come in contact with the external air. On the other hand, the statement of the accused, that she perceived no signs of life in the child, and after the lapse of only three hours found it to be cold and stiff, is also undeserving of belief. Because, from the very considerable distention of the lungs with air, which was proved by the dissection of the child, it is to be assumed as certain, that the child had cried beneath the bed clothes, as every living and healthy child does immediately on being born; and even if it had been suffocated very shortly thereafter, it would not have been either cold or stiff after the lapse of only three hours, because neither cadaveric coldness nor cadaveric rigidity come on so soon after death, especially under circumstances like those of the present case. In the foregoing statements, I think I have shown sufficient reason for giving my opinion in regard to the question put before me, as follows: "that, according to the documentary evidence, it is not to be assumed that the accused during or immediately after the birth of the child, was in a state of unconsciousness, weakness, or apathy, by which she was prevented from taking the child to herself while it was still alive, and the statement of the accused, that she never observed any sign of life in the child appears not to be credible."

Case CLXIV.—Disputed Civil Responsibility.

The widow Z., who was deep in debt, had been long known to me as an arrant simulator of bodily diseases, from the repeated examinations made by me in regard to her arrestments, was required to declare her means, but asserted that she was unable to take an oath on account of "mental derangement," and this was required to be determined. I found her this time as usual, when she was aware of my visit, in bed, but—with all her clothes on—as was proved by throwing down the bedclothes. The not unskilful deceiver knew how to avoid with some cunning this examination of her mental condition, and answered, for instance, my first question as to her age, in the following words: "eighty, seventy, sixty, ah, very old!" On being sharply spoken to, however, she ceased to make these statements, and not a single trace of mental weakness or confusion could be discovered, however hard Z. laboured to make such appear credible. Upon my asking her as to her means, she replied, that she had nothing, nothing at all; upon my further asking her, however, why, if this were true, she should refuse to confirm it upon oath? she made no reply, but drew the bedclothes over her head! My opinion could not be doubtful, and must of course be given against her.

CASE CLXV.—DISPUTED CIVIL RESPONSIBILITY.

In itself this case, in the form it assumed at the period of our examination, was one of daily occurrence, for it was unquestionable that the old lady, at that time sixty-eight years of age, a sensible, educated woman of pure morals, who a long time previously had been judicially declared to be "imbecile," that is, "incapable of considering the effects of her actions," and was still under guardianship, and the effects of her interdiction which she was endeavouring to get removed, was perfectly restored, and, as the statute expresses it, had re-attained "the full and free use of her understanding." I relate this case, here, however, because it gives a most instructive peep into the interior of a mind under the trammels of disease, and gives from a credible witness a confirmation of the well-known psychological experience, that those mentally diseased may not only have a dim cognizance of the difference between good and evil (*vide* p.110, Vol. IV.), but also, that even in respect of this obscure consciousness they can command themselves *up to a certain point.* The discipline of every lunatic asylum is based upon the very proper recognition of this fact. Our convalescent had gone deranged fifteen years previously, and had remained so many years. With the greatest composure she communicated to me many particulars in regard to her derangement and its excesses, describing with the utmost distinctness her then process of ratiocination. For a time, she was impelled to break panes of glass with stones. But she knew how improper this was, *therefore she threw the stones carefully,* so as not to break the glass, but if this happened *then she rejoiced at it!* She set about tearing her paper bed-screen; but, as she found it abundantly replaced, she came to the conclusion, that this was done intentionally to keep her from destroying more valuable articles, *therefore she left off* tearing the screen. Similar tendencies and similar logic were exhibited in many other proceedings, and she could not sufficiently describe to me how puzzling to her now seemed the reasoning, which then seemed clear to her. Similar statements are made every day by insane people, who have become truly restored; but similar confessions to these in respect of individual minor traits, are not so frequent.

CASE CLXVI.—DISPUTED CIVIL RESPONSIBILITY.

In a civil process the accused merchant, W., asserted that, from his ailing condition, he was unable to prepare a statement of his affairs, and to confirm it by an oath. I had to satisfy myself in regard to this, and at the same time to give an opinion, whether he could be arrested personally, if necessary? The investigation proved, that W. certainly laboured under the well-known common disease called hypochondria, which in itself could be regarded as a mere simulation, though it could not be denied that the manifold ailments alleged to exist, were either intentionally or unintentionally exaggerated. "Granting, however," I said, "that W. is ill, nevertheless, since he is not feverish, not confined to bed, and is of clear intellect, it is not easy to see why such an employment as the one in question, the preparation of a statement of his affairs in his own apartment, should be impossible for him or likely to be injurious. When he alleges, that the mere addition of sums causes him much anguish, such a statement is to be rejected as inconsistent with medical experience. Only if he were to be forced and hurried in the performance of such work could there be a possibility of injury resulting." Accordingly, I declared that W. was in a fit condition to prepare a statement of his affairs, and to confirm it by oath, provided a few weeks were granted to him for this purpose, and that if necessary, he might be personally arrested. This opinion was communicated to W., and a statement of his affairs was very speedily thereafter handed in.

CASE CLXVII.—ALLEGED "MANIA OF PERSECUTION" IN A DANGEROUS CRIMINAL.

This case is very instructive in many points; the subject of it was a most obstinate and malevolent individual who had been repeatedly punished as a criminal, and whose deceit I succeeded in unmasking in opposition to the opinions of a physician to a lunatic asylum and of a prison surgeon, both of whom regarded the insanity as real. I first became acquainted with the barber, Teck, aged twenty-six years, by attending his examination at the request of one of the officials of the criminal police, as an accidental listener, unknown to Teck. In my opinion, given in subsequently, after I had also had sufficient opportunity of observing T. in prison, I stated, " Already in course of this long examination I began to suspect that T. was simulating, and that tolerably coarsely, though consistently. This suspicion has in the course of subsequent observation been raised to certainty. In the first place, T. is in perfect bodily health. Every objective symptom points to the normal discharge of all the functions, and consequently, from this point of view there is not the slightest reason for supposing the existence of any mental derangement. IIis look is certainly striking and unusual. Every one, however, who has had much intercourse with criminals will not recognise in T.'s deep-set and piercing eyes the look of one mentally deranged, but rather that saucy insolence, which T. has constantly displayed in his actions and statements—as, for instance, in constantly refusing to perform the work of straw-plaiting, given him in the prison, undeterred by threats of punishment with the rod. In the next place, it is remarkable that in the course of conversation T. sometimes gives himself out to be of sound mind, and at others to be mentally deranged, according as the turn of the discourse seems to make it his interest to appear to be now the one, and now the other. I considered it requisite in the course of my observation of the behaviour of the culprit, to place him on a species of starvation diet, and to remove him from his sociable employment 'with the hat-makers, which he liked very much,' back to his solitary cell. Against these restrictions he protested repeatedly in the strongest manner as perfectly unnecessary and superfluous, ' because he was quite well both in body and mind.' On the other hand, he complained that he had been 'illegally sent back from the Charité Hospital to the town's prison, since he was dismissed uncured from the Hospital, and the town's prison was no proper place for detaining those mentally deranged.' In any case, he said at another time, the consent of his father should be obtained to this treatment of one 'mentally diseased,' and he petitioned to be sent back to the Charité IIospital. I may here point out that T., who was very well acquainted with this Hospital, from having served his apprenticeship as a barber in connexion with it, had previously absconded from it. Such a concatenation of facts must appear remarkable even to a nonprofessional person, even though he may be unaware that an insane person never declares himself at one time to be of sound, and at

another of unsound mind.   To this there is another important circumstance to be added.   When we investigate the statements of T.,
and ask him as to his delusions, he always replies that 'he labours
under the delusion of persecution.'   This expression, derived from
recent French scientific terminology, has quite recently been very frequently employed by the physicians of the lunatic department of
the Charité Hospital, and the source of T.'s science is therefore very
easily discovered.   The alleged persecution consists, in his statement,
both in the Charity Hospital and in prison, that he was 'electrified'
by invisible agencies; the physicians in both of these institutions considered that he really believed this.   Such hallucinations are unquestionably of very frequent occurrence in the insane, and mental
delusions of this character have received the name of *Manie de
persecution*.   But never, and at no time does such a patient say, nor
does he believe, that he labours under the delusion of persecution;
indeed, he protests against the physician who explains this to him,
because it is, as may be readily comprehended, precisely the character
of a morbid delusion that its false conceptions are believed to be
truths by those labouring under them. Therefore, he regards the whole
world to be insane, because it does not believe in his ideas, but he
looks upon himself as mentally sound.   When T. knows that his
'delusion of persecution' is a delusion, then he knows that he is not
electrified, &c., and he must himself recognize that all his demands
and ailments founded upon that idea are unjustifiable.   But T. is
alleged to suffer not from this hallucination alone, but also from
other manifestations of mental derangement.   He has, it seems, discovered a carriage more rapid than a locomotive, and he makes the
most confused speeches as to how he will with this discovery lessen
the price of provisions, &c., indeed in the course of to-day's conversation he brought forward a hitherto unmentioned matter, namely,
that he had also discovered the secret of perpetual motion.   With
such statements, T. only gives, without knowing it, fresh specimens of his folly.   The actual so-called *manie de persecution* (if we
choose to regard it as a distinct species) is a fixed delusion, which
is a somewhat different form of mental disease from general insanity,
that form which alone can be discovered in the latter statements just
quoted.   It is most unusual for one and the same patient to exhibit
both forms of the disease.   Further, it is quite contrary to all experience of such patients as fancy they have made such extraordinary
discoveries as that of the *perpetuum mobile,* &c., that T. in his employ-

ment is perfectly passive in regard to it. Whilst actual patients with this delusion, who are of no infrequent occurrence, are continually racking their brains about their discovery, and writing out whole pages of descriptions or drawings of it, T., who has writing materials in his cell at his command, T., who since he refuses to work, has for weeks passed the whole of each day in his cell alone and unemployed, has never for one instant thought of making use of these writing materials for sketching, &c., his invention. This apparently trifling circumstance is of great value in estimating the psychological condition of the culprit. We are forced to inquire what possible motive T. could have for simulating? The documentary evidence, the character, and the behaviour of this individual, give a most conclusive answer to this. T., who has been described to me as one of the most obstinate of criminals, has been condemned to seven years' penal servitude for repeated violent robberies; three years of this he has yet to serve. I have already spoken of his—for him well founded—desire to be sent back to the Charité Hospital. When it is represented to him that, according to his own statements, he is perfectly healthy, he always replies, that at least he would like to be sent to Spandau or Brandenburg.* When we also consider, that, according to the certificate of the prison surgeon, the first traces of 'mental disease' observed immediately disappeared when T. was released from solitary confinement and sent to work in the open air in company with others, and that the ' mental disease' recurred whenever he was sent back to solitary confinement—then we find the question as to a motive for simulation not so difficult to answer, especially by any expert who knows how often criminals do everything possible, even commit new crimes, only to get transferred to some other prison from the one they are in. When I add, finally, that T., who, as cannot be denied, acts his part with that insolent consistency which is his characteristic, and may thus readily deceive, still often forgets, and by a peculiar smile seems to laugh at his jokes, for at such moments he indubitably regards his statements as such, then, I think, I have shown sufficient reason for stating it to be my opinion in regard to the question put before me, that T. is not mentally deranged; but that his insanity is only simulated, that consequently he may be allowed to continue in his present situation, and that neither in regard to discipline nor in any other respect, is it necessary to pay any further attention to his alleged mental derangement." On the subsequent

* Ordinary bridewells, while T. was in a cell-prison.

emptying of the cell prison, T. was conveyed to a penal institute out of town, without any further attention being paid to his "mental disease."

CASE CLXVIII.—FRAUD COMMITTED IN A STATE OF ALLEGED
IMBECILITY.

On the 1st of July, the shoemaker, F., brought Samuel Walter, aged twenty-two, and the son of a Jewish watchmaker, to the police-office and informed against him; that one evening, about six weeks previously, he had come to him and asked if he had any old gold or silver; some of this being put before him, he desired to take it with him to have it valued, and when F. would not consent to this, the accused ordered a pair of boots, and told him to call upon his family, whom he named Abramson, where five other pairs of boots would be ordered from him. Upon this, F. gave the accused articles of gold to the value of about three shillings. The accused, however, never returned to him, and F. was all the more convinced that he was intentionally deceived, that no family of the name of Abramson could be found at the address given him. On the 1st of July, F. met Walter accidentally upon the street, and immediately, as I have already stated, seized hold of him, and brought him to the police-office, where the accused denied having ever purchased the gold, but this was sworn to by eye-witnesses. The accused, moreover, stated, that after he had ordered a pair of boots, the married couple F. had shown him articles of gold, demanding three shillings for them, which he paid; he immediately, however, perceived that the metal was brass, and when he complained of this, he was attacked by the journeymen and beaten senseless. All the other statements of the couple F. he disputed, especially at the precognition of the seventeenth of November, with much dexterity. I shall presently detail how differently Walter behaved at his medical examination. His father had asserted that his son was mentally deranged, and laboured under frequent attacks of convulsions, and in testimony of this he appealed to a certificate of Dr. D., which forms part of the documentary evidence, and is dated the second of July, and to the acts of the district commission for indemnification, which was alleged to have discharged the accused as an invalid " from imbecility." The certificate of Dr. D., however, only certified, that the accused " had for some time suffered from rheumatic ailments, and was of weak mind," making no mention of any convulsions or actual mental derangement. " In order to make the necessary examination, I went on the 24th of this month to Walter's house, where, however, I found only the father at home, and of him I have recorded that I suspected that he had induced his son to simulate mental disease. This suspicion was at once con-firmed by a somewhat long conversation with W. He gave evidently such an exaggerated description of his son, and such cunning and evasive answers to all my questions, as showed most distinctly his desire, not only to embarrass me, but as far as possible wholly to prevent my investigation. Thus, he stated that his son was actually never at home, because he ran about the whole day; and when I asked when they dined, he answered, 'when dinner is ready, sometimes at twelve, at one, or at three,' &c. Upon this I required that the son should visit me, which he did to-day in company with his father, though the father declared that he allowed him to run about everywhere alone. W. is about twenty years of age, and apparently in perfect bodily health. I permitted the father to retire, but could not succeed in extracting any connected speech from the accused. He slunk into my room like one half paralysed, with his arms hanging loosely at his sides, and his head shrunk on his breast; out of this position I could not bring him, and to all my questions he answered nothing but 'I don't know.' Subsequently, becoming more urgent, and letting him know that I knew all about the matter in question, he still kept to his posture and his answer; and there was finally nothing for it, but to break off the interview! Nevertheless, I have not the slightest doubt but that Samuel Walter was dissembling in the coarsest possible manner. The voluntary and intentional character of his posture as described could not be mistaken even at the first glance, and this all the more, that it had no resemblance to that of an actual imbecile. To this I may add, that if this slinking gait, this apparent semi-paralysis of the whole body, was not simulated, the accused could not go about the streets the whole day alone, as he was alleged to do, because such a man, who might lose his way, or go astray at any moment, would be certain to be stopped and laid hold of every day. Nevertheless, as already stated, his father alleged that he allowed him to go out alone every day, yet he deemed it necessary to accompany him to my house, hoping, perhaps, that his influence and his communications might have been of importance in relation to the examination. Further, the accused must have conducted himself quite differently during his precognition from what he did to-day, otherwise it would have been impossible to deal

with him at all.   But for such a radical change in his mental condition in the space of only two months, there is not the slightest
apparent occasion or explanation, and the cunning and loquacious
father would certainly not have withheld an explanation, were it only
hypothetical, had he suspected it would have been of consequence.
Finally, the same statement may be made in regard to a comparison
of the present behaviour of the accused with his conduct at the time
of the commission of the deed, that is, eight months ago.   He, who
now appears as a bad imitation of a true cretin, and can say nothing
more than a half stuttering 'I don't know,' then, as the documentary
evidence proves, and as I have already shortly stated, acted with
much cunning, not only in perpetrating the swindle under a fictitious
name and residence, but also in the defence which he set up.   It is
contrary to all medical experience that such a material deterioration
of the mental faculties should take place in such a short space of
time, without there being some very important reason for it, such as
severe bodily injury, and it is the less probable in this case, inasmuch
as the father declares, that his son has laboured under this mental
deficiency from his earliest childhood, as the result of a fall from his
nurse's arms.   All this is most probably nothing but a lying invention; medical experience, however, has certainly and indisputably
ascertained, that when mental deficiency has arisen from a cranial
injury received during early infancy, it does not run its course as
that of W. is alleged to have done, developing itself almost suddenly
to so great an extent, and for the first time in his twentieth year."
Accordingly, I answered the judicial question put to me as follows:
that Samuel Walter only simulated imbecility, and that he was a
perfectly responsible agent, both at the time when the deed was committed and also at present.   He was condemned.

CASE CLXIX. — SIMULATION OF INSANITY AND DEAFNESS.

The behaviour of the weaver S., aged fifty-two years, who was
accused of having repeatedly committed incest with his daughter, was
such as to necessitate an investigation into his bodily and mental
condition.   His disjointed and insane babbling made him appear
to be mentally deranged, while the difficulty with which he could be
made to hear the questions during his precognition seemed to show
that he was deaf.   It was very speedily discovered, however, that
both of these conditions were only simulated, and that coarsely

enough, in order to escape the severe punishment which awaited him. Even before me he continued to carry on this crazy talk. When, however, I led the conversation into indifferent matters, without giving the slightest idea of my object, he spoke like a perfectly sane man, while it only required the least possible approach to his own affairs again to excite his frantic declamation! So also was it in regard to his deafness, as in the first case, it was not apparent when I talked to him for a long time in an ordinary tone about his trade, &c., but it at once returned when any question was put to him, even in a loud tone of voice, touching upon the deed of which he was accused! My opinion could not be doubtful.

CASE CLXX.—ALLEGED MORBID SEXUAL FURY.

A vigorous married man, a chemist, aged fifty-three years, was accused of having committed the most lascivious deeds with three little girls of from nine to eleven years of age on one afternoon. He had taken advantage of the absence of his wife, and called the children under various pretexts into the room one after the other. In the first place, he requested the children to be as jolly as possible; "to dance, and jump about, and play the fool." And then he — — (this was one of those horrible cases coming under § 23, and over which, as has been already stated at p. 337, Vol. III., a veil must be thrown!). At his precognition the accused had declared that as the girl Marie related to him that the girl Augusta allowed herself to be gripped beneath the frock by boys, he was seized with such a "paroxysm" that he had kissed the children. He did not remember any unchaste actions. It seemed to him as if "after a time his consciousness returned, the sweat standing on his forehead. For a quarter of a year he has been sensible of a morbid condition, so that at the sight of little girls he was suddenly seized with a kind of fury, and he felt as if he must seize the children and bite them." He ascribed this morbid condition to the influence of arsenical and cyanic vapours, to which he had been exposed as a chemist in a chemical manufactory during the years 1845-48, and by which his "nervous system had been completely destroyed." The accused also made the same statement to me, but he also added thereto many more details. He said, that he ascribed this effect to the influence of a negative electricity upon his body, and he had for long racked his brains as to discover how he might restore himself by means of a positive electrical current. At this time an angel in the form of a female child with wings appeared once to him in a dream, she pointed to her sexual organ with one finger, and then laid her finger on her tongue. This he took for a hint which he must make use of, &c. "It is," I stated, "perfectly sufficient to justify the supposition that all this is pure fiction and pretence, in order to give him the appearance of acting from a blind impulse, and therefore, irresponsibly, when I certify that D. displays not the most remote trace of any mental derangement either in appearance, demeanour, statements, nor manner of speech, &c. It is true that he is afflicted with bodily disease, disease of the chest, but this of course is of no importance in relation to the present question. His defence is also proved, by the circumstances attending the deed, to be perfectly untenable. That he took advantage of the absence of his wife; that he forbade the children to speak of what happened to them; that he promised them cakes to keep them quiet, proves not only that he did not, as he pretends, lose his consciousness, but also that the punishable character of his actions was very distinctly known to him. Upon my representation that the impression conveyed by his alleged dream had not the slightest connexion with the fact that he had acted (so and so) with the children, and rather tended to prove that in the whole matter he had been actuated by gross lasciviousness, he could give no other answer, save that he did not remember this occurrence at all. Irresponsibility dare not be presupposed, but must be proved to exist. In the case before us, there is not one single fact discoverable in the bodily or mental condition of the accused which could be serviceable in proving this. I must, therefore, state it to be my opinion that D. was of sound mind, and perfectly responsible, both at the time of the commission of the deed and also at present." Whereupon condemnation ensued.

CASE CLXXI.—THE DEVILSEER, CHARLOTTE LOUISA GLASER.

The case of Glaser has justly excited the greatest attention in Berlin. It is, indeed, one of the most remarkable psychological criminal cases. It has certainly seldom happened that any one has, for *ten years*, succeeded in befooling so many and so skilful judicial boards, and deceiving them as to his responsibility; seldom that a woman, such as this Glaser, could have, in the course of years, kept in error, as to her mental condition, six physicians, one after the

other, and among them three forensic physicians and one master of
the science; seldom that such a deception has not ceased during a
residence of *one entire year in a lunatic asylum;* seldom, I may
say, that a *judicial declaration of imbecility* was promulgated, based
upon the opinion of the two medical men consulted, who were both
deceived; seldom, finally—and this makes the case of great interest
for jurists also—that at last, after I had succeeded in unmasking her,
who for years had been regarded as a " lunatic," and exhibiting
her in her true character, as an insolent cheat, a penal sentence was
carried out upon a culprit who still stands under *the interdict of the
civil law on account of imbecility !* In the report given in by me,
from which I can only give an extract here, I gave, in the first place,
a *curriculum vitæ,* extracted from the seventh volume of the documentary evidence :—" According to the evidence before me, the
woman Glaser first turns up in the year 1847, as a homeless vagrant,
wherefore she was taken to the workhouse. At that time, she said
that she was a married woman of the name of Kuffner, that she was
twenty-three years of age, and that she did not know her parents;
whilst, in later years, she always said that her father was a labourer,
or lodging-house keeper, whose Christian name was Martin ; she
further declared that she was one of the company and foster-daughter
of a circus rider, called Wohlbrück, with whom she had hitherto
travelled about, but having been seized with an attack of fever by
the way, she was left lying senseless in a ditch by the roadside. All
these statements have been proved to be untrue, inasmuch as there
never was a circus rider of the name of Wohlbrück, &c. On the
eleventh of August she was dismissed from the workhouse, but on
the twenty-sixth of the same month she was again arrested for disorderly vagrancy, and the suspicion of hedge-whoredom, and she
was found to be affected with syphilis. On the twenty-second of
October, she was sent back cured from the Charité Hospital to the
workhouse, where she denied that she had ever been syphilitic, and
again brought forward her former falsehoods about Wohlbrück, &c.
On the thirty-first of March, 1848, she was dismissed from the workhouse, and on the second of April she was again arrested as a vagrant,
and brought to the police prison. Here she told the house-surgeon
' that up to the day before yesterday she had been almost a whole
year uninterruptedly in the Charité Hospital on account of *mental
disease.*' ' At present,' adds the surgeon, ' she still labours under
so much mental weakness that she is unable to give the name or

place of abode of her aunt, with whom she says she resides.' This fact, extracted from the documentary evidence, and *now for the first time become known to me*, is very important, and of itself justifies the opinion which I have held and expressed for years, that the woman Glaser has only simulated mental disease, and continues to simulate it whenever it suits her purpose. For it was a lie, as I have already shown, for her to state that she had been for a whole year previous to the day before yesterday in the Charité Hospital, and a pure fiction that she had been there on account of mental disease at all, and it seems as if she had then for the first time taken it into her head to simulate insanity, in order to remain free from punishment, or to obtain other advantages. After she had once brought forward this untruth (as to the mental disease she alleged she had laboured under), *and that it was such cannot possibly be disputed from any point of view*, she held fast to it, and only two days after this arrest we find the house-surgeon certifying (on the fourth of April, 1848) that the woman Glaser, in the prison, was 'rather ill-natured and dangerous, that she swore, called names, raged, tore the clothes off her back, and sought to break everything,' &c. From this conduct, evidently *purely voluntary*, and easily carried out by any one at any moment, the surgeon, who I may remark by the way did not then know Glaser, and who has long since adopted a different opinion, concluded that she was mentally diseased, and applied for her speedy removal to the Charité Hospital, 'on account of partial insanity,' and she was admitted there on the sixth of the same month. The physician of the lunatic wards, Dr. R., delivered his first opinion in regard to her case, upon the eighteenth of May. In it he states that after the restlessness and depression apparent in the woman Glaser on her admission had abated, she now states that she belonged to a company of circus riders, and had led a dissolute life, for which she now bitterly upbraids herself. As the result of the constant stinging of her conscience she soon became tormented during the night by Satanic appearances; she saw herself surrounded by black figures without heads, and by the flames of hell, &c. This, however, she concealed, and therefore when her syphilis was cured she could be dismissed from the Charité Hospital. Scarcely had she returned to a private dwelling, when the devilish figures commenced to follow her by day also, and she therefore took refuge in flight. 'As she,' continues Dr. R., 'thought she saw at a show-window a picture of Christ, she threw herself on her knees before it, beseeching deliverance from the devils. By this conduct she attracted attention, and as in her confusion she could not declare who she was, she was carried off to the town prison.' Here she continued to see the devilish figures, she broke the windows and was sent to the Charité Hospital. 'She must, therefore,' concludes the report, ' be declared to be imbecile in the statutory sense of the word.' It grieves me very much to say that I do not share this opinion in the least. It is evident that this opinion also is *solely* based upon the *personal* statements of the woman Glaser herself, and their trustworthiness has been already pointed out, and will in the course of this report be still further shaken. In the police prison, as I have shown, the idea occurred to her of alleging herself to be insane, and it must have been much easier for her to deceive by the mere narration of visions than by deeds of violence! But here, also, we can bring home to Glaser another manifest lie, though certainly not known to be such by Dr. R.; I refer to the scene before the picture of Christ as described by her. It is not for a moment to be supposed that so remarkable a scene upon the public street could have remained unknown to the police officials who arrested her, nor that a relation of it would be omitted in the report of her arrest ; there is not, however, one single word about it in it! When, however, Dr. R., with his wonted power of psychological intuition places a high value, and in itself justly so, upon the pangs of conscience as a cause of mental disease in evil-doers, and therefore assumes the alleged tormenting of Glaser's conscience to be a fact, then appealing to all those who during a long course of years have become acquainted with this individual in prisons, and during her various trials, I say that I have the best founded and strongest doubts as to the existence of any qualms of conscience in such a person, whose doings I shall presently still further describe. Meanwhile, the woman Glaser continued a long time in the lunatic asylum, and as the epileptic convulsions under which she laboured, and which ' rendered her recovery more difficult,' were here observed, according to existing regulations a fiscal application to have her declared an imbecile was lodged on the sixteenth of February, 1849, and on the third of April of the same year the inquiry in regard to this was commenced. During the course of this she again brought forward the same untruths as formerly in regard to Wohlbrück and the ditch by the roadside, and then detailed very fully and quite correctly, giving *the exact dates,* her repeated admissions and dismissals from the workhouse and the Charité Hospital, and it did not

seem to strike the medical men as anything remarkable, that, in the
course of the same inquiry, when asked how much is left after
taking twenty-six from forty-nine? she answered, 'I cannot
answer such a question, my head is too weak!' She then related her
Satanic visions in considerable detail. The opinion of the medical
men in the first place distinctly testified from their own observations
to the existence of tubercular disease in the lungs of the woman
Glaser; from this, however, they very properly drew no conclusions
as to any mental derangement, and then assumed from *her own
statements* that she laboured under convulsions, and that these
attacks, *as she herself states*, commence with a preliminary stage of
furor; the report then went on to say, that 'at present Glaser no
longer presents any symptoms of mental disease, and is apparently
quite restored to health.' A new inquiry to decide definitely as to
her state was therefore moved for. On the first of May, the woman
Glaser was sent back to the workhouse as 'uncured,' and on the
twentieth of June discharged on leave to go to the woman Dauter.
And now, for the first time, we have an occurrence recorded which
is subsequently repeated so often in a precisely similar manner,
namely, that so soon as she had appropriated a complete suit of
clothing from the woman Dauter, she disappeared secretly from her
dwelling. At her arrest next day she did not know her own name,
she said she was Christ, &c.; two days subsequently, she knew how
to reclaim her purse 'containing one thaler, and from ten to fifteen
silver groschen (four shillings to four and sixpence) and two gold
rings, with my name inside of it,' which had been taken from her at
her arrest! The deputy physician found her to be diseased bodily,
and 'also to display decided symptoms of mental disturbance,' and
she was sent to the Charité Hospital, from which she was very
speedily, on the ninth of July, sent back as 'improved' to prison. Any
improvement in one 'decidedly deranged' in fourteen days is something remarkable, and of itself sufficient to excite the supposition that
there must have been some deception!—After two days, upon the
eleventh of July, she was dismissed, and on the eighteenth of the same
month, she was again arrested for 'disorderly vagrancy during the
night,' and sent, this time for bodily ailments, to the Charité Hospital,
from which she was once more dismissed at the end of four weeks. On
the thirtieth of August, information was given, that she had gone
to the widow Kliche, and having represented herself as the sick-nurse
of her daughter, had received several articles of clothing for her,

with which she (for the second time) secretly made off. That she, at this time, did not say that she was Christ, nor display any remarkable symptom of mental disease, nor speak of her satanic visions, &c., is self-evident, because if she had, the woman Kliche would certainly not have trusted her with the clothes! She was eleven days in prison.—In the commencement of October of this year there arose that remarkable episode in the trial of Schall, which is only too well known to those acquainted with this extraordinary criminal case. It was extremely difficult to identify the body of this man who was found murdered. The woman Glaser came forward of her own accord, stated that she was *the wife of the commissionaire Fröhlich,* from Driesen, and asserted that the murdered man was her husband. The name itself was a fiction, as well as the whole array of facts with which she supported her assertion, which she maintained, as the Judge said, - 'with such well considered consistence' that the body had to be exhumed, but was not recognised as that of her husband by the woman Glaser. From the documentary evidence connected with the case of Schall, which I became acquainted with officially, I am very well aware of the careful and continuous researches which were made even in the most remote places on account of the statements of the widow Fröhlich, till it was finally proved to be a mere piece of mystification. Thus, for instance, no such man as that named by her had ever existed, &c. From this it follows incontestably that *no Judge, or any other person could have had the slightest suspicion that they had an insane woman before them,* because if they had, these distant and costly inquiries would certainly not have been made. If it should be asked, what could possibly have induced the woman Glaser, if sane, to perpetrate such an apparently aimless mystification? I may reply, that it is not the object of the present opinion to answer this question; I will only hint, that in a person who, in an official report proceeding from the Royal police magistrates of this city, has been termed 'the most arrant swindler that possibly can exist,' many motives may combine to originate such a deceit, and that a probable connection of the woman Glaser with Schall, who subsequently confessed to the murder, or some of his companions, was openly spoken of during the public proceedings connected with this trial. On the 9th of November she was sent from Spandau to the workhouse, where, on the 8th of April 1850, the inquiry into her mental condition, which had been postponed, was carried out. Now she is a very different person from what she was, either when engaged in her embezzlements, or when before the precognoscing Judge at Spandau. She has 'thousands of fathers and mothers; her father is called Pappendeckel, one gets nothing here to eat,' her birthday is every day, when she ' gets anything to eat,' &c. She vented the lowest curses and abusive expressions upon those present ; she did not need to answer, she was more than Judge, she wasEmperor and King—(not Christ this time !) and poured forth the most non-sensical babble. Here I must remark, that she indeed alleged that she did not know all her names, and could not otherwise define her age than that she was ' either thirty or forty,' but, nevertheless, she knew to give the names of Lejars and Kolter (actually existing and well known) circus riders, with whom she said she had been a professional horsewoman ! Further, it is to be remarked, that she knew very well that she had been in the lunatic department of the Charité Hospital, and added, without any incitation, ' Professor Dr. R. ought to come to me '—a trifling, but very distinctive trait, when we know that at subsequent trials in following years *she repeatedly appealed to the opinion of Dr. R.*, whose view of her state was very well known to her. Further, it is exceedingly worthy of remark, that in the whole course of this examination she did not give expression to one single word in the least indicative of religious insanity, indeed, that on being questioned, she repeatedly declared ' I am of no religion, I am a Turk, a heathen,' and at last, when asked—and not before this—if figures did not appear to her at night, she answered ' Devils' ! ! The physicians then recapitulated all that had taken place in the Charité Hospital, and at that day's examination—though, as I have already stated, all this was purely voluntary both in word and deed on the part of the party in question—they referred to her unsettled life and epileptic attacks, and then declared in conclusion, that there was 'no doubt but that the woman Glaser was imbecile in the statutory sense of the word.' Upon this, a declaration of imbecility was judicially promulgated upon the 15th of May. It is much to be lamented, that the physicians before giving their opinion had not at least thought of the possibility of simulation, although from their unacquaintance with the personal history and antecedents of the woman Glaser, they certainly had no peculiar occasion for any such supposition. She remained till the 13th of October in the workhouse, was then dismissed, to the care of a relative, upon the declaration of the police physician, that though she was of a morbidly violent character, abusive, blustering, and refractory, yet she was sufficiently mistress of her reason to permit of her discharge. In regard to the year 1851, all that we find in the evidence is, that on the 24th of February she was found lying on the streets in a fit of epilepsy and sent to the police prison, where she—once more—gave herself out as ' Mrs. Kuffner.' In the year 1852, on the 13th of January she was brought to the guard-house, as it was ascertained that she had lodged in her home the dangerous thief Kuffner, her paramour, without acquainting the police with the fact. When this man was arrested and carried off, she made a most obstreporous noise, and abused the arresting policeman in the vulgarest manner, which resulted in her own arrest. On the 16th of August, 1853, she was again arrested for 'night-strolling unannounced to the police,' she signed the protocol ' Kuffner,' and was discharged on the 13th of September. Again on the 25th of October she was found ' wandering about during night.' She accurately described various dwellings in which she said that she lived ; but her statements were all untrue, nevertheless, she was speedily discharged. On the 23rd of October, information was lodged (for the third time !) that she had made off secretly with articles of clothing and money from the house of Mrs. Reinike, who had taken her in out of compassion. Her person was ascertained, but she herself managed to keep out of the hands of the police till the 7th of November, when she was arrested upon ' urgent suspicion of hedgewhoredom,' and sent to the Charité Hospital; from it she was discharged on the 6th of January, 1854 ; but she was again arrested ' for drunkenness' on the 23rd of this month. During the time intervening (on the 14th of January) she had lodged with the widow Lange, and finding herself alone in the house had (for the fourth time !) purloined five articles of clothing and gone off secretly with them. On the 20th of January she perpetrated (for the fifth time !) a precisely similar deed in a precisely similar manner, inasmuch as she secretly went off with clothing and money from the house of the woman Bendler, with whom she lodged one night. To the widow Lange she had lied, in telling her that a certain director Balk had her alms certificate, and she begged her to go to him and get her an alms. *During her absence* she committed the theft. On the other hand, she begged the woman Bendler to lend her clothes and money. She disappeared with both. At her precognition upon these charges, she immediately brought forward as her exculpatory plea—mental weakness !—She also did not know whether she was married to Kuffner, and justified in bearing his name, &c.

M 2

In the opinion then required from me upon the 9th of February, I detailed how in the course of conversation with me, she had endeavoured in an apparently plausible manner, and not without a certain amount of clever shrewdness, to excuse her crime and exculpate herself, thereby proving that she possessed the power of distinguishing between good and evil; I proved by facts which occurred and by her demeanour during these conversations, that she possessed 'an evil, passionate and vengeful disposition;' that, however, she did not exhibit any traces of any special affection of the mind, and I therefore declared her to be *a responsible agent.* At her trial, which took place in consequence of this, upon the 27th of March, the woman Glaser again appeared as if mentally deranged. She digressed evidently quite intentionally from the questions put to her, &c. I will, however, abstain from giving my own observations at this trial, and will only quote from the official protocol in which it is stated, ' The accused made no rational statement, constantly evaded the questions put to her in regard to the accusation against her, and instead of answering, abused the Judge and all about her, but *whenever questions not bearing on the matter in hand were put to her, she answered quite correctly.*' She displayed, however, such low blackguardism, that she had to be removed. The three witnesses examined testified that *she had never exhibited any trace* of any mental derangement; I held firm to my opinion, which, in the opinion of every unprejudiced person, must have received the strongest support from her behaviour that day, and the woman Glaser was condemned to six months' imprisonment ' for theft, repeated acts of swindling, the use of false names, and insulting a witness,' this punishment was completed on the 7th of October. Already by the 27th of November, she was again arrested for swindling; but had to be discharged for want of proof. On the 21st of April she was again arrested for theft and swindling. On the 11th of that month she had gone to the married woman Fürstenberg, and, after having on the previous day carried out the same manœuvre by the widow Schön (for the sixth time!), under pretence of having been invited to a marriage, she borrowed several articles of clothing and ornament from the woman Fürstenberg, and (for the seventh time!) went off with them secretly. At her examination she contested every statement, declared that she had returned those articles of clothing taken with her from the woman Fürstenberg, which she enumerated with the utmost correctness, giving the exact day and date upon which this had happened;

nevertheless she impressed the legal deputies with the idea that she was insane; wherefore my opinion was again requested, and this was given in on the 12th of May. If the slightest doubt had ever arisen in my mind, in the course of my previous observations, as to whether the woman Glaser had only intentionally endeavoured to delude us with her pretended visions and other disturbances, these doubts would have been completely removed by the results of a very long conversation which I now had with her. I endeavoured this time to win her confidence, and I perfectly succeeded in doing this, as well as in gaining the end in view. I communicated to her, as it were in confidence, that it was now distinctly known that she had actually taken the money belonging to the woman Fürstenberg—which she obstinately denied having done—and that she must regulate her subsequent statements accordingly. She entered upon this subject with the utmost composure, and assured me that no one could have observed her, that there was only a little boy in the room (!), and that she was innocent, &c. Upon my apparently perfectly friendly representation, that it was impossible to be of any use to her, because it was known from experience that she was no sooner discharged from prison or punishment than she immediately commenced afresh to swindle, cheat, or steal, she replied with the utmost composure, begging me to stand her friend, *that if it could only be passed over for this once, this should certainly be the last time.* After such a remarkable declaration, which no one who had lost all knowledge of the results of his actions in consequence of mental derangement would have made, and which rather tended to prove in the most incontestable manner that the woman Glaser knew that she had done wrong and must expect to be punished, I could do nothing else than again declare her to be a responsible agent, and her trial was proceeded with. This was probably unexpected by her. In June she again behaved so outrageously, that in order to subdue what the official report terms 'her irrational conduct,' a straight jacket had to be put on her. On the 30th of June she desired an interview with her advocate, and requested that, '1. The soap-manufacturer, Adolph Kuffner, No. 29, Wassmann-street, residing in the court with the widow Felgenhauer; 2. The wife of the whalebone manufacturer Herrmann, No. 5, Stein-lane; 3. The soap-manufacturer, Burmeister, No. 9, Blumenstreet, residing with the widow Gerlach; 4. The journeyman soapmaker, Becker, No. 26, Prenzlauer-street,' should be cited to attend her trial as witnesses for the defence. In providing for her defence, that is, in the consciousness that she was again threatened with punishment, she forgets herself once more, and the same person who eleven days subsequently could not declare her Christian names or place of birth '*because her head was so weak,*' now—eleven days previously—proved herself to possess a very exact memory, by giving seven names, Christian and surnames, of strange persons, and four perfectly accurate addresses! ! Such a concatenation of documentary facts speaks for itself! I shall presently consider the possible objection that her 'weakness of head' was only occasional—her derangement periodic. The trial just mentioned affords also another highly remarkable illustration of Glaser's behaviour. For after she had made the accurate statements mentioned, and the motion for her defence had been put, it is recorded, 'when the affair had been so far transacted, the accused violently seized the written records of the case, tore them, and commenced to abuse the court in a loud voice, and had to be removed with force'! If we were to ascribe this behaviour not to that peculiar wickedness, rudeness, and violence of character in the woman Glaser, which does not remain long unknown to those who become acquainted with her, but to the irrational excitement of a mental disease, such a sudden change of mental expression in the course of one short examination would be contrary to all medical experience. At the time of the trial she again behaved herself in a most unseemly manner, abused the court in the lowest manner, making use of the most vulgar expressions, evidently with the intention of giving grounds for suspecting her to be mentally deranged; however, after hearing my iterated and perfectly decided declaration as to the opposite nature of her disposition and character, the proceedings were gone on with in spite of her raging. The royal criminal commissioner Pick was examined as a witness, and deposed—' I have examined the woman Glaser when first arrested, and have also repeatedly observed her during other examinations. She has always appeared to me to be *perfectly rational. I know that when at liberty she is quite rational, but acts the imbecile whenever she is brought to prison.* To myself she has made a full confession.' The trial was put an end to for formal reasons, and the matter adjourned. Meanwhile, on the part of the advocate for the defence, the aid of Professor Dr. R., as a second expert, was requested, and he appeared at the new trial on the 28th of November. At this time the woman Glaser did not appear wild and violent, but like a sick person moving slowly and tremulously. She answered the questions put to her in a composed manner ; suddenly, however, she was seized with an epileptic attack, which, nevertheless, passed so speedily off that the trial could be proceeded with. *Nine* witnesses who were examined, had not one word to say in regard to having *ever observed any symptom of mental disease in the accused.* I once more declared my own conviction that the woman Glaser only simulated insanity, and that in a very coarse manner, and that she was a responsible agent.   Dr. R., for his part, declared that, during her residence in the Charité Hospital, he had no reason to regard her as a dissembler.   Fourteen days ago he saw her again.   He asked her, according to his oral statement not included in the minute of the visit, whether she still saw the devil ? and ' she complained that she was still persecuted by the devil in the form of a he goat.'   These observations, however, he continued, could not be held as decisive in reference to the time when the accused perpetrated the act in question.   To-day she seemed to him quite rational.   He could give no decisive opinion as to what state she was in at the time of the act in question.   She might ' possibly have done the deed while in a state of mental weakness, induced by epileptic convulsions.'   In the sentence *the responsibility of the woman Glaser was acknowledged,* but for other and purely judicial reasons she was acquitted."

"The remarkable occurrence for which the woman Glaser is at present under examination took place in the year 1856.   She was found lying in convulsions on the last night of the year upon the streets of M—ch—g and taken to the hospital.   This time she was speechless, and continued so for some time !   On the following day she could not make herself understood by speech, but by her ' symbolical description ' it was understood that a violent robbery had been committed upon her.   This recent fact in the life of Glaser is of the utmost significance.   Even a temporary loss of speech, to say nothing of one lasting so long as an entire day, was never observed to follow an epileptic attack in this woman, and, moreover, according to all medical experience, such a long-continued speechlessness consequent on an attack of that nature would be a most unheard-of occurrence.   If, therefore, without further enquiry, we must in this matter again pre-suppose intention, it comes to be asked what could possibly be the object of such a very easily carried out simulation ?   The answer to this, as well as to the whole of the extraordinary mystification of the judicial boards about to be referred to, is not far to seek.   The woman Glaser knew very well that she was wholly unprovided with a passport or other authorised permission to travel, and that she

would be treated as a wandering vagabond; and this treatment she
had all the more reason to expect should her person be known. She
had evidently endeavoured to devise a plan to prevent another imprisonment and punishment, and she simulated speechlessness, and consequent inability to give any explanations in order to gain time to
fabricate a falsehood. The result proved how admirably she succeeded in this. On the third day, for the first time, she wrote upon
a slate the word Krüger, and on the seventh day she made the following declaration in the course of her examination. She said she
was Charlotte Luisa Emilia, wife of a dealer in fancy goods, named
Krüger, her maiden name being Kroschel, that she was forty-one
years old, ' evangelic ' (this time, therefore not as formerly, ' a Turk
or a heathen '), and she then gave a description of a nocturnal attack
made by thieves upon her supposititious husband and herself during
an imaginary journey from one market to another, during which her
husband and the waggon were lost. This same individual, who had
so often exculpated herself on account of the ' weakness of her head,'
now fabricated a complete romance in regard to this attempt at robbery
and murder, with the most astonishing minuteness of detail. This
weak-headed woman described her husband, the waggon, the horse,
the contents of the chests and boxes on the waggon, the linen
marked with their initials, the persons and clothing (from head to
foot) of both robbers, the dog which accompanied her, &c., in repeated
and long examinations, with such descriptiveness of detail, and such
wonderful consistence, and *without ever once saying that she was in the
habit of seeing the devil or speaking one single word out of joint* that
might have led to the suspicion of mental disease, so that the examining Judge held himself bound to issue an official proclamation for the
more speedy discovery of the perpetrators of this murderous attack.*

---

* The official proclamation is sufficiently characteristic of the woman
Glaser to be quoted entire. It is as follows:—
" Proclamation, regarding an attempt at robbery and murder.—On the
evening of the thirty-first of December last the fancy goods dealer, Charles
Henry Emilia Krüger, with his wife, Charlotte Luisa Emilia, maiden name
Kroschel, came in a one-horse tilt waggon from Klein-Posemuckel into the
district of Bomst, along the road from Cüstrin to Berlin, and about six
o'clock in the evening, as they were between Jahnsfelde and Müncheberg,
two men with a dog came out of a small pinewood and seized the horse by
the reins, they struck the man Krüger, who had got off his cart, with a
cudgel on the legs, and took from his wife, who had also got down to run
away, a purse containing from forty-six to forty-eight thalers in specie

It is self-evident that *all these statements were pure fabrications,* and amongst them we find it stated that her own and her husband's (£6 18*s.* to £7 4*s.*). The woman Krüger arrived the same evening on foot in M—ch—g, and remained speechless for several days before she could detail the circumstances which had occurred. Neither her husband, nor the waggon, nor the perpetrators of the deed have as yet been discovered. According to the statement of Mrs. Krüger, the waggon was an ordinary four-wheeled fork-beamed, one-horse waggon provided with a boot made of deals, all painted blue, with the name C. Krüger painted black in Roman letters upon the hinder axletree. The four wheels had been recently painted green. The name Carl Krüger was also painted in black upon a wooden drag attached to a chain. Beneath the waggon there was a board to which a small black and white Pomeranian dog, answering to the name of ' Lady (*sic in orig.*), was attached by means of a chain. A small black mare, with a white spot upon her forehead, was yoked to the waggon, and the name C. Krüger was cut upon her harness. The waggon was not painted inside, and above it a tilt was extended, marked behind with the initials C. K. in red. On the waggon there were two boxes with sliding lids and padlocks, the name C. Krüger being burned into the lids : in the one box there were the small remains of a stock of children's toys, the other was still tolerably full of fancy goods and hardware, such as knives, forks, spoons, scissors, &c. ; 2. a skin-covered trunk, in which there was both clean and dirty under-clothing, the man's underclothing marked with the initials C. K., the woman's with E. K., and also the numbers one to six, those marked from four upwards probably clean, while numbers one and two were dirty. The underclothing consisted of shirts, stockings, pocket-handkerchiefs, aprons, &c. There was also in the trunk a red-leather pocket-book containing Krüger's legitimacy papers : 3. a down quilt and two large bolsters or pillows, the ticking being blue and white striped home-made linen, the covering similar linen, with a small red and white check, all marked E. K. in red ; 4. two sacks, marked C. K. Posemuckel—in one of them there were still some oats left ; 5. an oat sieve, marked C. Krüger ; 6. a pail with iron hoops and handle, and the name C. Krüger burned into it ; 7. a black sheepskin coat with the wool outside ; 8. in front of the waggon a tin lantern with an oil lamp was suspended from a wooden bar. The man Krüger, who got off the waggon when it was attacked, and left behind him upon it a grey cloth mantle with three capes, is forty-one years of age, of a large, powerful build, has a moustache, and was clad in brown and grey striped buckskin breeches, large water-boots, a brown cloth paletot with velvet collar and large covered buttons, a red and white striped plush waistcoat buttoned to the throat with mother-of-pearl buttons, a blue, green and white crôchet shawl, six ells long, a new grey fur cap to fold over, covered outside with green cloth, a linen shirt, marked in red, C. K. 3, and a brown woollen knitted under-jacket. He carried a whip covered with green leather, and took with him when he got off the waggon a loaded pistol, with a blue barrel and brown handle and percussion lock ; he also took with him a portmonnaie or purse of brown leather, closed with a green elastic band, in legitimacy papers, which were in a trunk, had been carried off with it and were lost! In the meantime, therefore, she was no vagabond, and, till further enquiries could be made, she must be kept in the hospital and provided for! The result of the official proclamation referred to, and the enquiries of the police here was that the royal police magistracy of this city began to suspect that the so-called woman Krüger was no other than our friend Glaser, and they requested, in the words of the official document, that 'this most mischievous swindler and deceiver, who can only be supposed to have simulated insanity,' be sent to Berlin. When brought here on the 31st of January and recognised, she again made use of the mode of defence she had so often found good. She declared that she had wandered from this city during a fit of lunacy, and that she knew nothing at all of all that had happened to her at M—ch—g. She was placed provisionally in hospital, but escaped from it secretly, taking with her (for the eighth time) the clothes of her fellow patient, the unmarried woman Kühlstein, and a pawn ticket. Again arrested

which he had his hawker's license, a common pocket watch in a tortoiseshell case, blue steel chain and watch key, and a German silver snuffbox gilt inside. Mrs. Krüger describes one of the robbers as a tolerably large and stout man, of good appearance, with dark hair, strong, reddish whiskers, moustache, and beard, a pointed nose, healthy colour, and apparently about thirty years of age. He wore a light-coloured pilot coat buttoned to the throat, with a collar of different cloth, with thigh boots over his trousers and a stout cudgel in his hand. She cannot describe what was on his head. His dog was the size of a shepherd's dog, of a yellowish colour, with a long tail and long smooth hair, and answered to the name of 'Karo.' His companion was a less man, of slim figure and miserable appearance, with reddish, close-cut hair, and similar whiskers, which were continued beneath his chin; his face was thin and pale. He wore a torn coat of dark cloth, dark trousers over his boots, a green shawl wound closely round his throat, and a grey cloth cap with a shade and small tassel in front. He is said to have been scratched in the face during the struggle with Krüger, and to have been called 'Julius' by his companion; apparently about thirty-eight or thirty-nine years of age. As Mrs. Krüger attempted to make off with the money out of the trunk, which was contained in a grey crôchet long purse, with steel tassels and rings, the man in the pilot coat took this from her, and stuffed into her mouth a piece of muslin apparently belonging to a small window curtain, such as is fastened up with strings. All those who can give any information as to what has come over Krüger, his gear, waggon, or anything which was on it, or in regard to the probable culprits, are earnestly requested at once to communicate with the nearest magistrate or police board. No expense will be incurred in any case. M—ch—g, January 7th, 1856. Royal District Court Commission."

on the 4th of May, she once more brought forward the excuse of lunacy, which the criminal commissary, Bock, who committed her, declared to be 'manifestly simulated.' At her first examination she declared that she laboured under 'spasms of the brain,' at the same time excusing herself in every accusation of theft by fictitious statements. She was brought forward for trial on the 27th of June. She, however, applied for the appointment of another day and for the citation of Professor R., 'who would prove that she was insane.' The trial then took place upon the 23rd of July with the assistance of Dr. R. and the subscriber. At the commencement of the proceedings she behaved herself quite quietly, and deposed to the particulars of the accusation. Suddenly, to the great astonishment of the court and of the assembled spectators, she began to nod continually against the wall with her head, and to speak in whispers to it. Then she called out aloud, 'you must first drive the devil out,' and then she commenced anew to rage in the most furious manner. When asked where she saw the devil? she exclaimed, 'there, there he stands,' seized a stool, and let drive with it at the public prosecutor who sat near her, so that she had to be removed. Of the two experts who were summoned to the trial, Dr. R. and myself, the former was first examined. He declared that the woman Glaser was 'actually insane and wholly devoid of criminal responsibility, indeed, altogether unfit to be tried,' basing his opinion upon the 'visions of devils,' which she had seen in the asylum, and upon her epilepsy of many years' standing. On the other hand, my opinion was that the present occurrence only strengthened my former view that the accused was a subtle deceiver who only simulated insanity; that it was very remarkable that insanity had recently only broken out in her when she stood under any accusation, and when it might be employed as a judicious means of defence, while it at once left her whenever she was dismissed from prison. That the confession made to me by the accused while in prison, during a former investigation (and already detailed), was of great importance. Further, I must dispute the alleged fact that epilepsy was *usually* attended by insanity : there are many epileptics whose mind has not in the least suffered from their complaint. I must, I declared, regard the accused as a perfectly responsible agent, and I considered myself justified by my medico-legal experience in declaring this with the utmost confidence. I concluded with, 'should the accused be successful in inducing the court to regard her as insane, and should she thus remain unpunished, then she will leave the bar in triumph; I prophesy, however, that her freedom will very speedily be employed in the commission of fresh crimes.' This prophecy was only too speedily fulfilled. The trial was put an end to and the experts were requested to give in their opinions, which were so diametrically opposed to one another, in writing, so that they might pass through the statutory sequence of the courts, and, in the meantime, the woman Glaser was set at liberty. On the 21st of September, however, she was again arrested, because she (for the ninth time) had gone off secretly with articles of clothing belonging to the divorced woman Hahn, maiden name Schmid, a dealer in clothes, with whom she lodged."

"The woman Glaser is at present about thirty years of age, of a slender form, tolerably thin, and of pale complexion. No material disease can be discovered about her, except a slight degree of pulmonary tuberculosis, which is of no consequence in regard to the present enquiry. She has, however, a semiparalysis of the right upper extremity, the source of which cannot be ascertained. Her look· is piercing, unpleasant, and decidedly malicious. I cannot, however, say that it is staring, restless, or that in any other way it is indicative of insanity. It cannot be denied, and I have satisfied myself of this by personal observation of her seizures during one of her trials, that she actually suffers from epilepsy, and has probably done so for many years. Dr. R. has repeatedly and very properly insisted upon the importance of this fact in regard to the determination of her mental condition, because it is well known that the long continuance of epilepsy frequently exerts a morbific agency upon the mental faculties. Still, it is just as well known that, fortunately, this is much more frequently not the case, and that a very great number of such patients even during a long life, are never thus affected even in the slightest degree ; and I have already, in giving my opinion orally, referred to the example of several famous men who are known to have been epileptics. The forensic physician must, therefore, in any given doubtful case, first of all, investigate the preliminary question, that is, he must determine that mental disturbance actually exists, and only after that, when giving his reasons for his opinion and developing the history of the origin of this insanity, is it allowable to refer to the long continued existence of this serious nervous disease ; then only does this disease acquire importance in relation to the opinion, while, in the opposite case, it is not of the slightest consequence, because, as I have already remarked, epilepsy is not necessarily and in every case

followed by mental disease. The question, therefore, is, is this epileptic woman Glaser also insane, or not? Her words, expressions,
and actions have led Dr. R. to the—presupposing their internal
truthfulness perfectly correct—supposition that she labours under
that peculiar form of religious insanity which has been termed Demonomania. In regard to this it is, in the first place, very remarkable
and not at all consonant with medical experience, that epilepsy should
have produced this peculiar form of mental disease, since it is, on the
contrary, known that this nervous disease, when it does produce
injury of the mental powers, acts by gradually enfeebling them, and
thus producing finally complete stupidity or idiocy. Dr. R., however,
further adduces in support of his idea of a ' Demonomania' another
important circumstance, the tormenting of the woman Glaser's evil
conscience, the result of her course of life, which she herself (! !) had
mentioned to him. Besides what I have already stated upon this
subject, I must further direct attention to the circumstance that no
lively feeling in regard to right or wrong could be expected of the
woman Glaser, who always presented an example of uncommon bar-
barity and vulgarity of character, who constantly mystified the judicial
boards, and swindled and stole from poor people, even from those who
took her into their houses out of pity. Moreover, it is decidedly
against this view that the torments of an evil conscience are not
usually periodic, whilst the alleged ' Demonomania' of the woman
Glaser is of a periodic (intermittent) character; but to this I will
return presently. Nevertheless, it is apparently a fact that Dr. R.
' has treated the woman Glaser for a whole year in the lunatic asylum
for a severe mental disease (Demonomania).' I regret to have to
state that the official journal of the Charité Hospital, which I thought
it necessary to inspect, by no means confirms this to be a fact. The
woman Glaser was sent to the lunatic asylum on the 6th of April,
1848, as I have already stated. She detailed all the untruths which
have already been related. ' On her reception,' this journal states,
' she displayed great anxiety of mind. Tormented by the thought
of her great sinfulness, she saw black figures, the preacher and the
devil.' She was seized with an inflammatory affection of her wind-
pipe, but still saw ' the preacher ' on the 30th of April. Already,
in the following month it is recorded, ' In May *no trace of any
mental disturbance could be any longer remarked,*' and during this
month there are only recorded several attacks of purely corporeal
diseases. catarrh, rheumatism, and phthisical phenomena. In June,

'there occurred several very severe fits of excitement, during which
she raged violently, and was abusive, and required to be put under
restraint; yet these attacks were of no long duration.' Here there
is no mention of any 'Demonomania,' while the 'violent raging and
abusive language' are, as is only too well known to the prison officials of
this city quite a daily occurrence with the woman Glaser, who can only
be subdued by measures of restraint, and are only the natural result of
her character as it has been already described. In July, 'she had more
bodily than mental ailments to complain of.' Epilepsy, an attack of
diarrhœa, and her pulmonary complaint are mentioned, and again
this month it is recorded, ' *she no longer appears to have any mental
affection.*' In August, ' voices called to her that she was a great
sinner and ought not to eat, so that much persuasion was necessary
to make her take anything.' This phenomenon of voices alleged to
be heard was not previously observed and has never since occurred.
The woman Glaser had evidently heard in her sick ward this very
frequent hallucination made mention of, and made use of it in a
manner convenient for her own ends. This supposition is confirmed
by the other phenomenon observed, which also frequently occurs in
those actually insane, I refer to the refusal of nourishment, which she
had probably also observed in the patients around her. This also
was never observed either before or since. And though this of itself
must appear remarkable, yet it is not less so that she was induced
' by persuasion ' to take nourishment, while it is well known that
those actually insane, when they commence to refuse nourishment,
are not in general induced 'by persuasion' to discontinue their refusal, but very frequently require force to overcome their objections.
A long abstinence from all nourishment would be, however, indeed
much more difficult for the not insane woman Glaser intentionally to
simulate, than the mere declaration that she saw devils, &c., or than
the mere employment of abusive language and violence, and it is very ·
significative that this result of insanity, as well as another which is
also both of frequent occurrence and impossible to be simulated for
any length of time, I refer to continuous and complete sleeplessness
for many nights together, has never once been observed in the woman
Glaser, either in the Charité Hospital or anywhere else. The journal
further records, during September, besides epilepsy and chest complaints, ' her numerous self-accusations, as well as her wicked and
stubborn character were just as prominent as in the earlier months,'
and in October, again excepting the convulsive attacks, ' in the fre-

quent disputes which she had with the other patients, her coarse
and shameless disposition were evinced in a most glaring manner.
There were no self-accusations.' Further, in November, 'she was
quite well mentally.' In December, it is recorded, 'she refused
her food for a few days, because she declared that she did not
deserve it.' We learn nothing further as to her mental condition in
this month; in particular, we do not learn whether 'persuasion'
this time also induced her to cease refusing. In January, 1840,
excepting epilepsy and hysterical complaints, 'her mental condition
was the same' (as when ?). In February she had a slight attack of
diarrhœa; 'no other change was observed.' In March, 'no trace
of mental disturbance' was exhibited; and in April the woman
Glaser was dismissed. According to this description of her condition during her one year's residence in the Charité Hospital, exactly
copied from the official records, I think I do not err in asserting that
it is true indeed that the woman Glaser resided for a whole year in
the lunatic wards of that Hospital, but tha the assertion that 'she
suffered for a whole year from severe mental disease' is not supported by the records of the Hospital journal. Excepting just at
first on her admission into the asylum, even according to her own
statements, there is not a word more about the appearance of devils
for a whole year. It must seem, however, à priori, most remarkable
to any one possessed of any considerable experience of criminals and
simulators, and who knows the culprit and her character and manner
of life, that insanity by her should assume a religious character.
But in truth it did not do so. This particular form of insanity im-
presses such a peculiar character even on the outward expression of
those affected by it, that during a whole year of observation it could
not have failed to attract attention. And yet not one word regarding
it is to be found in the Hospital journal. Further, it is perfectly
contrary to general experience that one actually affected by religious
insanity should declare, as the woman Glaser did—though that is
probably still unknown to Dr. R.—that he 'is of no religion, that
he is a Turk, a heathen.' She betrayed herself in this matter, however,
from her ignorance of the inner nature of mental disease precisely as
those deceivers do who jumble together all the most important symptoms of different mental diseases. Thus, at one time she said ' I
am emperor,' ' I am king,' &c., expressions which one affected with
Demonomania would never make use of, as he moves within quite a
different circle of ideas. Moreover, religious insanity (Demonomania)

is precisely a form of mental disease which does not usually assume a periodic or intermitting character. Since, nevertheless, the fact that the woman Glaser for a period of ten years only appeared to be insane at times, might very readily be interpreted to arise from a periodicity in her alleged mental disease, this periodicity itself being so well known a phenomenon in mental affections, it cannot be too impressively pointed out, for the refutation of any such interpretation, that, according to indisputable documentary evidence, the *woman Glaser never appeared to be insane when at liberty, but immediately exhibited her ' Demonomania,' raged and raved, &c.*, whenever she was arrested for some fresh crime, or foresaw some punishment when in prison, or found herself *vis-à-vis* with medical men whose opinions she thus desired to influence, and in this her great cunning made her only too successful. None, I say *not one single one* of the many witnesses brought forward in the many charges against her has ever testified to having observed the slightest symptom of insanity about her; and if we must confess that non-professional people are not always capable of making such observations correctly, yet this truth is not applicable precisely to the 'Demonomania' of the woman Glaser, since her raging and employment of abusive language, her crying out, ' there stands the devil!' throwing herself towards him, &c., &c., are certainly circumstances which must have excited the greatest astonishment in even the most illiterate observer. Her cunningly devised mystifications and deceits would also certainly never have succeeded had she exhibited any similar behaviour. That she, however, made use of this, as well as of her entire allegation of insanity, solely as a means of defence, invented purely for this end, is proved afresh by the extremely remarkable lie told at one of her earlier arrests on the second of April, 1848, on which occasion she declared that she had been, up to the day before yesterday, almost constantly for a whole year an insane patient in the Charité Hospital. The documentary evidence proves that this was a pure fabrication. She now commenced, as already remarked, to give further effect to her device and again, as it were, to relapse (apparently) into insanity! This object the ' weak-headed ' culprit never for one instant forgot, and at her examinations, precognitions, and trials, in which she was so ' weak-headed ' as to be unable to state her nation or Christian name, her memory was still capable of remembering the name and title of Dr. R., whose assistance she continually demanded for reasons which have been already sufficiently alluded to. Thus her whole conduct and behaviour during the last ten years, as comprised in the documentary evidence, and observed by myself during repeated examinations, displays indeed a rare energy of vulgarity, a shamelessness unusual even in those lost to all sense of morality, and also an equally unusual amount of shrewdness and clever reasoning, but not a single one of those symptoms which experience teaches us to be characteristic of actual disturbance of the mental faculties, either during its progressive development or when subsequently come to maturity. This unusual combination of peculiarities of mind and character might well prove sufficient to deceive even the best qualified judges, physicians, and boards, but not men who, versed in intercourse with criminals and characters similar to the woman Glaser, and probably after similar previous deceptions by similar characters, have been excited to greater carefulness in forming their opinion."

In connection with the details just given, I once more gave it as my opinion, "That Charlotte Glaser has hitherto only simulated mental disease, and that she is to be regarded as criminally responsible, both generally and also particularly in relation to the punishment due to the crime of which she now stands accused."

The trial was continued on the 31st of December 1856, my respected colleague Dr. R. having meanwhile given in a reasoned report of his opinion. The matter, however, was not protracted by transmission through other courts; it did not require it. For after Dr. R. had accurately informed himself from the documentary evidence as to the life and actions of this remarkable personage, and had again submitted her to further examinations, he also arrived at the conviction which he expressed in his report, drawn up with all his wonted perspicuity, that he now perceived that he had been "grossly deceived" by the woman Glaser, and that he was now perfectly convinced of the correctness of my opinion of her character, which he now fully shared. In this state of matters Glaser's trial was commenced, and she was requested to keep herself quiet to-day, as the court were now fully convinced that her insane behaviour was merely simulated, and it would be of no further use to her. She was not, however, in the least led astray by this, for though she said nothing about the appearance of devils, yet she answered all the questions put to her either with such defiant insolence or so perversely, that she had speedily to be removed, and was sentenced in absence to penal servitude. Thus this person, so peculiarly and extremely dangerous to the public, is now, I hope, made harmless for a long time.*

CASE CLXXII.—ALLEGED INSANITY OF A THIEF.

A swindler, well known in the criminal courts of Berlin, of low origin, though now a Madame von W., had stolen a fur coat, worth two hundred thalers (£30), from an open lobby. At her precognition her remarkable behaviour raised doubts as to her sanity. And these appeared all the more well founded, that it was well known that W. had been a few years ago treated for some time for insanity in a public asylum. She was sent thither at the request of the surgeon to the prison, who stated that she was "of a passionate and malicious character, also very hysteric and of weak nerves," for which reasons she had been sent to the Lazaret immediately after her admission, and there she, "particularly when she thought herself observed, had never left her bed," and at the time of the signing of the certificate she laboured under "mental disease, namely, partial insanity." While in the asylum the accused behaved herself in the most unseemly manner. She was noisy, raved, broke utensils, &c.; declared that she was the Queen of Prussia, and again, that Christ had appeared to her as a white lamb; she fell at the physician's feet, her urine and fæces were passed beneath her, she used the most obscene and vulgar language, then suddenly became dumb, and only lisped out perfectly unintelligible words, &c. "This condition," it is stated in the sick report, "nevertheless disappeared very speedily when the means of restraint were placed in her view." It must appear very remarkable that, while it is recorded in the Journal quoted on the 12th of November, "many reasons are in favour of the supposition of the case being one of simulation," and this opinion is reasoned out, yet the physician in attendance on the 17th of December of the same year makes no mention of any such suspicion of simulation, but on the other hand, declares the accused

---

* Since the above was written, the woman Glaser has nevertheless again (in 1858) turned up. Very soon after her discharge from prison, she again swindled a servant-girl, under some pretext or other, of all her clothing. When placed at the bar she now conducted herself perfectly quietly, almost modestly, and behaved like any other person of healthy mind. She received her sentence peacefully, and only begged for delay in carrying it out, for pertinent reasons. This was not, however, granted.

to be "insane" in the statutory sense of the word, that is, wholly deprived of the use of her reason.

In my report I stated that "my view of the case does not permit me to share this opinion, and my conviction is rather that W. has hitherto only simulated insanity. She is a woman of thirty-six years of age, without any apparent objective symptoms of disease, as was also remarked by the Hospital physicians, so that her complaints of rending pains in the head and giddiness must be taken for what they are worth, as purely subjective statements. The unusual shameless-ness and extreme moral depravity of this woman, who has been already punished seventeen times for police offences, are well known to those who are in any measure acquainted with her. Thus she, in plain terms, boasted to myself and others that she was one of the registered (that is a prostitute), and spoke with the most impudent openness and almost vauntingly of her unchastity, and that in the most vulgar language which is peculiar to her. Besides this extraordi-nary conduct, I have, during repeated examinations, never observed anything which could specially justify the assumption of mental de-rangement. After she had at first during her present imprisonment been very noisy in the prison, raving, talking nonsense, and attempting to make unchaste assaults on men, &c., she has been for a long time perfectly quiet and collected without the employment of any means of restraint except solitary confinement. Further, it is remarkable, that in regard to the theft with which she is charged, she exculpates herself by asserting that the fur coat is her property, &c., and does not even attempt to state that it was committed when she was an irresponsible agent. Nevertheless, it might be assumed that W., formerly at least, suffered under mental disease, namely, from periodic (intermitting) insanity, and that at times, and particularly at pre-sent, she has a lucid interval. But important facts are opposed to this supposition. The surgeon to the prison has certified, that the 'partial mania' of the accused first appeared after she had made acquaintance with a fellow-prisoner, the unmarried woman B., who lay near her, and who was also sent to the asylum on account of 'partial mania.' This woman Bath has in her day been declared by me to be a deceiver, and, as I now perceive from the Journal of the house, was also regarded as such in the Hospital. Further, the physicians there have very properly brought forward the following facts as stated in the Journal. Moreover, there is no natural cohe-rence in the delusions expressed by the patient W., on the contrary, the craziest and most absurd seem to be selected, as it were, to make her insanity appear the more remarkable. The same is the case with the mania which has never again appeared since she was apprised of the unpleasant personal effects which result from it. There cannot be a doubt but that her speechlessness is simulated,' &c. I fully agree with the reasoning just quoted, and since her behaviour at the time of her apparent insanity did not present the true character of actual derangement, while her present look, carriage, expressions, demeanour, and performance of her functions are quite as little indicative of insanity, I must state it to be my opinion, that the woman Von W. does not labour under mental derangement, and is to be regarded as criminally responsible." At her public trial, four weeks subsequently, she presented us with a copy of Charlotte Glaser. She came in clad in a shameless manner, naked to the belly, would not stand at the bar, chattered and raved, murder had been attempted on her, she had ten thousand thalers (£1500) lying in London, &c. She had to be repeatedly removed. When she ought to have recognised the stolen fur coat, she put it on and danced about in it, *repeating meanwhile several times, I am insane*, and, finally, the trial had to be concluded in her absence. When brought in at last, she spat in the face of the witness she had robbed. Of course, in the face of such coarsely intentional behaviour I maintained the opinion I had already given, whereupon she was sentenced to imprisonment for four months for the theft, and to three months more for the spitting.

CHAPTER II.

§ 70. GENERAL.

FROM the earliest times up to our own day the legislatures have, with great unanimity in regard to important principles, arranged the various forms of mental derangement in a very few, only two or three, classes. The Roman law speaks only of *dementibus*, among whom those *mente capti* and *furiosi* were distinguished as subdivisions. In accordance with this great precursor, all subsequent statute books, in particular almost all the German statute books and the *Code civil*, have laid down insanity, mania, and idiocy as, so to speak, legal forms of mental disease,* and these forms alone have been acknowledged (with a few unimportant modifications in certain of the German statute books). The Prussian law also only takes cognizance of, in its Civil Code (Gen. Common Law), insanity, mania, and idiocy, of which besides, in its definitions, it identifies insanity and mania; the Penal Code acknowledges only - insanity and idiocy (*vide* p. 93, Vol. IV); the (Rhenish) civil statute book also acknowledges only idiocy, insanity, and mania (*imbecillité, demence* and *fureur*) ; and the French *Code penal* only the single category *demence*. It is remarkable that whilst in the course of this long period the legislators, who must have known best what was most advantageous for their ends, have contented themselves with such a simple classification, the medical faculty have pursued a perfectly opposite course, and not only for scientific noso-logical, or practical psychiatrical purposes, neither of which concerns us here, but specially also for *medico-legal* ends have laid down numerous divisions and subdivisions of forms of mental disease,

---

* The Gen. Common Law in § 815, Tit. 18., Part II., also makes use of the word " *Wahnwitzige*"—lunatic, but in this paragraph lays down the same regulations in regard to the abolition of guardianship, as it had done in regard to idiocy, mania, and insanity ; so that it is indubitable that the legislator in employing this word did not intend to constitute a new form of mental disease.

classes, species, and subspecies, which we find extending in recent
authors to sixty, eighty, or more.   Of course, those teachers who
employed such a mode of proceeding could not be expected to
have any unanimity among themselves, and new classifications, and
new subdivisions were constantly being evolved, so that the confusion
in psychological matters, which has wrought so much mischief in
forensic practice, has been continually increased.   In the first place, it
may be asked whether, contrary to the views of the legislators in all
times, there was or is any actual necessity for the special classification of mental disturbances for forensic purposes?   This I decidedly
deny.   As the three chief properties of the soul, imagination, sensation and will, do not act separately, but by their continual harmonious
co-operation cause and constitute the healthy mental action, so also is
there a similar co-operation in morbid mental action.   *Mental health
is, therefore, as a unity to be placed in opposition to mental disease.*
From this, of course, it by no means follows that one man mentally
diseased behaves himself like all other mentally diseased men ; just
as little as this is true of mental health.   For as in every healthy
man one faculty often preponderates relatively over the others, as
there are men of a lively and others of a dull imagination, men of
great energy of will, and others of precisely the opposite psychical nature, so what we thus find to be the case within the limits of
mental health we also find within those of mental disease.   In a
hundred insane persons the faculty of imagination is chiefly affected,
in a hundred others, who have less disturbance of the faculties of
imagination and perception, the energy of the will is chiefly weakened.
This indubitable fact certainly justifies the recognition of a certain
limited number of principal forms of mental disease, which, as they
actually occur in nature, have also their value in forensic practice ;
but it by no means justifies, especially for the object referred to, the
division of mental diseases into an unmentionable number of species
and varieties, amidst the confusion of which the forensic physician is
only too apt to lose sight of *that which alone is necessary, the thorough
examination of the individual case in itself.*   This view is also that
of Mittermaier.   He requires, with the greatest propriety, in a
medico-legal report no such opinions in regard to "general ideas,"
which, from their uncertainty, can only give rise to arbitrary decisions, but he requires that the report should duly consider the physical and psychical sphere of the party concerned, and should then
refer the case " to one of the categories recognised by the law" (that

is, " insanity or idiocy"). Too much cannot be said against generalising in forensic psychological matters, or against the construction of
innumerable species, varieties, and subvarieties, such as psychiatric
authors have laid down, and which have introduced a much greater
confusion into forensic literature and practice than the similar generalising, but now, fortunately, exploded theory of lethality in regard
to the fatality of any injury in any given case. Many of these supposititious species are, however, in themselves perfectly unscientific,
inasmuch as they elevate individual cases to the rank of species or
degrees, such as, for example, Hoffbauer's manifold " degrees " of
idiocy, or because they include under the head of new species one or
more defectively observed cases, in which all that appears peculiar
disappears whenever the defects in the observation are filled up
(vide *Mania occulta*, &c., §§ 74 to 76). Another difficulty presented by all too fine-drawn classifications, and which of itself would
make them untenable, is the many transitions which experience
teaches us there arc from one " species " to another. Thus the
mania occulta often becomes but too manifest, melancholy passes into
" silliness," and silliness into melancholy ; insanity passes into mania,
mania into idiocy, &c. However, therefore, psychical therapeutists
may hold to their classifications and increase the number of subdivisions as may seem to them most useful for their ends, the division of
patients in an asylum or their treatment, these classifications are,
nevertheless, useless for the purposes of forensic medicine, and are to
be wholly rejected, and *in foro* the subordination of each individual
case of proven mental disease under one of the two chief forms of
exaltation and depression which the correct instinct of the legislature
has adopted—mania and idiocy—will be found not only perfectly
sufficient and necessary, because the statutes require it, but also very
practicable, as I have learned from long experience, and shall prove
in what I have yet to say upon the subject. It must not be retorted
that the result of this view is that it may lead to the discarding of
two principal forms of mental disease as superfluous, since the only
material point would be to investigate each case individually, and it
would then be sufficient to declare that N. N. is " mentally diseased,"
or has been so at a former period ; because, besides that the legislators of all times and places have found it necessary to grant a diver-
sity of civil rights to those affected with the different forms of mental
disease—insanity or idiocy (p. 92 Vol. IV.), such a latitude in the
legislatorial definition would give in criminal cases the widest and

most dangerous field for the arbitrary and individual opinion of the forensic physician, to whom it would then be left to call whatever he pleased—mere passion, for instance—" mental disease," whilst the distinct and limited definitions, insanity and idiocy, set needful and healthy limits to all such arbitrariness.*

§ 71. Etiology of Mental Disease.

As in the examination of every sick person, the physician must also in every case of mental disease to be determined legally, endeavour to ascertain the original causes which either probably or nominally might have, or are said to have produced this affection in the individual, and the ascertaining of these may afford information in doubtful cases, although it may seem superfluous in regard to this, to point out that the discovery in any case of any such cause of mental disease by no means justifies the conclusion, that the latter has actually resulted because the former has been in action, yet experience teaches us that this simple dogma is frequently sinned against. I may refer at present to the abuse in this respect that has been made of such etiological causes as cranial injuries, sexual

---

* It is with pleasure that I have observed that recent purely psychiatric authors, such as Flemming, Neumann, &c., completely homologating these views, and correctly recognising the facts, have wholly abandoned the natural system of classification, and have even set themselves decidedly to oppose it. Neumann, in his Lehrbuch der Psychiatrie (Erlangen, 1859) says very correctly: " Let us for once realise the supposition that all the diseases of mankind are so classified as that mental diseases are reckoned as one *genus*, while each individual case represents a *species*. Then we should be at once exempted from the necessity of any further classification, so far as that lays claim to the name of a natural system. And can any really serious objection be made to this supposition ?" When purely psychiatric authors have but little hesitation in casting aside artificial systems of classification, forensic psychonosology must, as already said, go yet further, and pronounce them to be not only of doubtful utility but positively objectionable. " Artificial classifications," says the physician just quoted (*op. cit.* p. 237) "do not really promote true medical diagnosis, and are positively hurtful in forensic psychology. The tendency of physicians to deceive or intimidate the judges by the employment of systematic names (Monomania, Pyromania, &c.) instead of explaining the case to him by a psychological analysis, arises chiefly from the use of artificial systems, and forensic psychology will only then occupy a worthy position at the bar of the Courts of Justice when it has shaken off the fetters of the schools."

development, pregnancy, &c., inasmuch, as in cases in which it is difficult or impossible to prove the existence of mental disturbance, and
especially by physicians who embrace this last practice, and grasp
with pleasure the handles presented to them by the handbooks in
this respect, these etiological causes are fallen back upon, and an
ergo deduced from them.   A great, if not the largest number of
medical reports in regard to juvenile fire-raisers give a warning example of this.   The accused were about the age of puberty, a period
of life which "according to experience" predisposes so strongly to
mental disturbances ; the authors, A., B., C., and D., have also shown,
that at that age a morbid propensity to fire-raising is particularly apt
to be developed, in these accused there was no apparent motive for the
deed, &c.—*ergo*, this must have been a case of " Pyromania "! Such
medical reports are only excellent examples how psychologo-forensic
reports ought *not* to be drawn up.   For every one must acknowledge, that of all possible causes of mental disturbance, there is fortunately not one single one that must necessarily be followed by this
result ; that in thousands of cases cranial injuries, the bodily diseases
in question, pregnancy, puberty, &c., are not followed by this lamentable result, that consequently in any given case the actual or probable
existence of mental disease must first be proved by other facts, after
which the discovery of any one of the causes referred to may, and
must be brought forward in the report as contributing to the development of the case in a psychologo-empirical point of view.   These
causes are partly of a corporeal, and partly of mental origin, and are
as follows :—

A. Somatic etiological causes discoverable during life.

1. *Hereditary predisposition.*—It certainly excites attention, when,
during the investigation of a case, it is ascertained that any of the
relations of the party concerned in the ascending line, or any of his
brothers or sisters have been or are insane, because the hereditary transmission of such diseases is unquestionable.   Nevertheless, the bare
fact, for instance, that the party's father is at present in a lunatic
asylum, is not to be received uncriticised, wholly irrespective of what
has already been urged in regard to non-necessity in general.   We
must rather ascertain, whether this disease has not arisen in the
party's father or mother long after his birth, and from purely
accidental causes; in such a case these facts are deprived of
every diagnostic value.   For instance, the fact that Schraber's
brother (*vide* Case CLX.) was living in a lunatic asylum could

not for one instant mislead us from the general nature of the case
otherwise.

2. *Cranial injuries.*—This is certainly a not unimportant cause,
and one which frequently produces its effect after many years, and
even after apparently trifling injuries. There is, however, scarcely
any other circumstance in practice more abused by culprits, or those
who pretend to be mentally diseased, and often enough some trifling
scar on the head, such as thousands get during their youthful years
is pointed out with ostentation, though not the slightest symptoms
of reaction had ever been produced by the preceding injury.

3. *Sun stroke.*—A cause which but rarely occurs, especially in towns.
Since, however, insolation, when it acts injuriously, and does not
prove suddenly fatal, produces actual inflammation of the brain and
its membranes, consequently, phenomena which are severe and not
likely to deceive: this cause is seldom likely to give rise to
any doubts.

4. *Functional disturbances.*—These, themselves the products of
morbid processes, again react, as few others can do, as causes upon the
cerebral powers. This is peculiarly the case, as is well known, with
disturbances of the abdominal functions, among which we may in
many cases reckon the suppression of the catamenia, though this
latter, for many reasons already alluded to, must only be accepted
with the utmost caution, and with due consideration of the state of
the bodily health otherwise.

5. *Metastatic inflammation of the brain and its membranes*, particularly erysipelatous, rheumatic, and arthritic, however important in
a therapeutic point of view, scarcely ever come to be considered from
a medico-legal point of view, since the severe bodily diseases which
in these cases cause the disturbance of the sensorial functions, cannot
be mistaken.

6. *Cerebral congestions.*—Amongst these are to reckoned, first of
all, and as a most frequent cause of controversy in forensic cases,
drunkenness, and its results, then sleep and somnolence, and the effects
of the vapour of burning coals (Case CLXXVIII.).

7. *The act of parturition.*— This most important circumstance in the consideration of cases of accusation of child-murder in a doubtful mental condition, belongs partly to the rubric
of cerebral congestion, partly to that of sudden disturbance of
the mind by fear, shame, anxiety, despair in those who are
unmarried, and also by excessive joy in those who are mar-

ried (*vide* § 72, under No. 3). Numerous well observed cases*
have unquestionably proved the possibility of the violent corporeal excitement of the act of parturitions producing the most various
disturbances of the sensorium, from simple confusion of the senses,
or complete loss of consciousness, up to an attack of the most violent
mania. This question, as already stated, comes to be considered chiefly
in relation to accusations of child murder. But such accidents
happen scarcely so often as once in ten thousand cases, and form
rare exceptions to the general rule. The rule, however, must be
presupposed, but the exceptions must be proved, and to do this, it is
necessary thoroughly to investigate each individual case. Since it may
be possible, where unlawful deeds have been actively performed, to include the case under the head of " Insanity," or of " Idiocy," where
these have been passively permitted to occur, while under other
circumstances we can only prove the "emotion," and leave it to the
Judge to employ this as a measure in deciding upon the case.
Finally, I must warn against an error into which it is easy to fall in
judging of any particular case, when we disregard the fact, that nothing is more common than the lying pretence of such a state of mind
on the part of such culprits (Case CLXIII.). Truthful information
must in such cases be sought for in the entire circumstances of the
case, particularly the nature of the labour, the general psychological
diagnosis (§§ 61, 62), and the mode in which the child has died.

8. *States of sexual development,* among which are to be reckoned
puberty and *pregnancy.* When pregnancy, partly from purely corporeal influences, partly, especially in unmarried women, from the
co-operation of psychical causes, grief, anxiety, shame, despair, produces an affection of the mind, and where during this general
mental disturbance any illegal action is performed, then the ordinary
means of diagnostic examination, such as are required in any other
case, whatever may have been its etiological cause, will of itself be
sufficient for its explanation. The mere longings of pregnant women
will be psychologo-forensically considered in § 79. The influence
of the period of puberty has been excessively exaggerated, and
Osiander's book, for instance, which has been written with much
more diligence in compilation than in criticism, has done much
harm in this respect. The idle pranks of wanton, ill-bred boys and

* Jörg, Die Zurechnungsfähigkeit der Schwangern und Gebärenden.
Leipzig, 1837, S. 326. Osiander, neue Denkwürdigkeiten, I., S. 134
Kluge, med. Zeitung vom Verein, u. s. w. 1833, No. 22.

girls, as well as the spread by psychical contagion of various forms
of convulsions and the like, which are referrible to a perfectly
different department; but which (the Maid of Orleans among the
rest) Osiander and his disciples ascribe to the insanity of puberty,
are susceptible of, and require a totally different interpretation. That
at the period of sexual development, in which the body becomes
matured, mental maturity is also commenced, that at this epoch the
most remarkable and interesting psychical changes take place, that
the individual then for the first time commences really to place himself in relationship with the world, a perfectly novel idea of the
external world becomes developed, imagination comes powerfully
into action, a novel and hitherto unimagined impulse—the sexual
impulse—is awakened, &c. can be just as little denied as that in individual cases this internal revolution may derange the mental health.
Who, however, can estimate the minute proportion which such anormal cases bear to the normal? This, of itself, is sufficient to
commend caution in all such cases. It is, however, perfectly unscientific and objectionable so to abuse this etiological cause for the
excuse of young and responsible miscreants, as to extend the period of
puberty to a most unseemly degree; for instance, from the tenth
or eleventh to the twentieth year, and giving an emphatic value in
girls to the fact of the commencement of menstruation being somewhat delayed beyond the normal period, or of there being a temporary
suppression of that discharge. I have only to repeat that each individual case must be estimated according to the means already
recommended (§§ 61, 62), and then as a subordinate question it will
be elicited how far the exciting period of puberty, or

9. In other cases to which the same reasoning applies, *any
cerebral-neurosis* that may be present, epilepsy, St. Vitus' dance,
somnambulism (which may be certainly reckoned in this category),
may be shown to be connected with the mental disease, the actual
existence of which must be otherwise proved. In regard to epilepsy,
I have already (p. 100, Vol. IV.) pointed out how perfectly untenable
are those much affected general categories, under which epileptics
are reckoned as responsible agents, or how completely objectionable
it is to declare epilepsy to be a disease which renders those affected
with it irresponsible agents, as is done by the earlier authors and their
modern compilers.* Irrespective of the fact, that according to
Platner's perfectly untenable thesis, *facta epilepticorum quamvis*

* Platner, *Quæstiones*, etc. *Ed.* Neumann, p. 45.

male faciendi et ulciscendi consilio suscepta amentiæ excusatione non carere (!), epilepsy of itself would confer the privilege upon those afflicted with it of committing every possible crime, the fact taught by experience that epilepsy occurs in nature in innumerable degrees of intensity is opposed to the general applicability of any such dogma. For as on the one hand, the limits between purely hysterical and epileptic convulsions cannot be very sharply drawn diagnostically, so on the other, every one knows that A. may have from one to three fits in the year, B. the same number every week, and C. the same every day. All these three are epileptics, but is it likely that the disease will in all three have the same reaction on the mind? Further, it may be granted that deep-rooted epilepsy which has lasted for many years, may at last easily lead to insanity or idiocy; but where the disease is of more recent origin, this is seldom if ever the case. Where is the limit of the influence of duration in the two classes of cases? After this I need scarcely refer finally to the numerous epileptics from Cæsar to Napoleon who have not only remained mentally sound but have made themselves famous by their talents.

10. Excess *in venere* is in itself also to be reckoned among the etiological causes; since, however, it usually—only in one case, that of a young and educated man, have I seen acute mania arise from inordinate masturbation—produces, when it comes into action at all, a condition of the deepest depression both of body and mind, idiocy, it will not generally give an opportunity for any very erroneous conclusions.

§ 72. CONTINUATION.

B. *Psychical etiological causes* occur, in general, much more frequently in forensic practice and are much more difficult of estimation than somatic causes.

1. *The passions and emotions* are a frequent source of crimes, and also of insanity; yet even in this there is a remarkable difference. Passions such as love, homesickness (which, without forcing,,may be reckoned one), avarice, love of play, and, more than all the others put together, unbounded vanity, pride, arrogance, often produce insanity but are seldom the cause of crimes, while, on the other hand, jealousy and irascibility probably lead equally often to both, and covetousness is more prone to lead to crime than to insanity. I shall return to this subject in § 86.

2. It is well known that continued excitement of the brain by *mental exertion too continuously prolonged* at the expense of the bodily functions (digestion, sleep, &c.), may produce a disturbance of the mental functions, and this all the more readily when the object of this employment is some mystical, supernatural subject or some insoluble problem. How many have become insane from striving to square the circle, or to discover perpetual motion (Case CLXXIII.), how many have fallen victims to the mania for table-turning, spirit-rapping, and psychography in the first five years of the latter half of the present century, of which I have already published several instructive examples.* Ordinarily, such cases are not difficult to decide, because the general expression of such insanity is very easily recognised. The case of the woman Glaser, already detailed (Case CLXXI.), proves, however, that religious or mystical insanity, which very frequently occurs in a truly epidemic extent, may also be simulated. In this category, also, must be reckoned the delusions respecting supernatural union with Christ, with angels, &c., which so often turn the heads of an entire population.†

3. Sudden and unexpected *emotional excitement* from the threatened loss of one's dearest possessions, honour, life, freedom, or property may produce such sudden mental confusion and despair as must, according to all psychological experience, be regarded as limiting and destroying the freedom of the will in action, and this has been also recognised by the legislature (§ 86). I have already shown that in not a few cases the emotional excitement during the act of parturition of unmarried women bringing forth in secret belongs to this category.

4. Much has been said in regard to *morbid impulses*, which may occasion, and have frequently occasioned mental disease and, through it, illegal actions. But this opinion, which is not based upon sound criticism, and which is fraught with such important consequences for the doctrine of responsibility, requires a more extended consideration, which will be found in § 88. In the category of psychical causes of anormality in the mental functions we must further reckon

5. *Deafmuteness*, in so far as the original bodily defect in general renders it impossible for the individual so affected to assimilate himself mentally with the external world, as is requisite for the natural mental development and the normal reproduction of the images and ideas

* *Vide* Casper's Vierteljahrschrift, Bd. XI. S. 1, &c.

† *Vide* the remarkable case of Luisa Braune, aged sixteen, related in Casper's Vierteljahrschrift, Bd. IV. S. 26, &c.

received. How much those err, who, sitting at their desk, believe that a compensation for this want in deafmutes may be supplied by instruction, will be exhibited bye and bye (Cases CCXVI. to CCXXI.), every year affording us proof to the contrary. I have already spoken of the subject of deception from mere simulation of deaf-muteness (§ 56, p. 90, vol. IV.).

6. Finally, I have still to mention (once more *vide* p. 95, Vol. IV.), as belonging to this category, those numerous cases in which a whole series of somatic and psychical influences have, during a long course of time, sapped and destroyed the mental health, and may thus produce either insanity or idiocy. I refer to the hundreds of persons, more males than females, who fill the Workhouses, Hospitals, and Refuges, who by their restless vagabondage, aversion to labour, drunkenness, lying out at nights in the open air, excess in *venery*, and defective nourishment, during a long, dissolute, wasted life, sink at first into a condition difficult to estimate, just bordering on insanity, and finally into unmistakable mental disease.

FIRST SECTION.

§ 73. GENERAL.

INSANITY is a disturbance of self-consciousness based upon delusions. Self-consciousness, the recognition of our own personality, never becomes extinct in insanity as it does in idiocy, not even in acute mania, in which there is always a more or less dim perception of self-consciousness. The statements of maniacal patients who have been cured leave no doubt as to this, even if it were not proved by a careful observation of their behaviour during the attack. But self-consciousness is no longer ranged upon its original healthy basis, it is deranged, and our thoughtful mother tongue has done more in this case than make a mere pun where it speaks of derangement. And the cause of this aberration is, delusions of any kind which, from any cause and in any way, have taken root in the mind. The insanity, however, of the erroneous delusion or idea arises not from its falseness but from its taking root. When in the dusk of the evening we at a little distance take trees for men, we at once correct the erroneous idea by more accurate examination or by advancing nearer. We were only possessed for an instant by the false notion, without having lost the power of correcting it by our judgment. Insane delusion only commences when the possibility of this ceases. An Esop may reckon himself an Adonis, Xantippe may imagine herself to be a young, beautiful, and gentle woman. Hundreds of bunglers in the arts have considered themselves masters and geniuses. This is all only opinion or conviction, and does not correspond with the reality as perceived by their fellow men. Such individuals we term fools, without, however, ever thinking of sending them to an asylum: and rightly so; for the delusive idea is only superficial and not deep rooted. The parties themselves do not firmly believe in it; they themselves doubt it. Why else does the old "fool" who thinks himself young and beautiful dye his white locks, why does the old female "fool" rouge her colourless cheeks, if it is not because they doubtingly ask themselves, am I really so beautiful? Now it is not to be denied that, as it is often almost impossible in psychical processes to make a distinct division between healthy and diseased, so also it may be in individual cases very difficult to determine where, as I may say, foolishness ceases to be healthy and becomes morbid. Thus a wise thrift or liberal management of one's property may pass quite imperceptibly into avarice or extravagance, and these may for long, even for a lifetime, keep within the limits of mental health, but may also go beyond these limits and pass into actual insane delusions, when the miser no longer eats or drinks and watches his chest armed for fear of robbers, or the ruined spendthrift, the fancied possessor of principalities, squanders the penny he has begged for. In these, as in every other case, the *actions* of the men are decisive, and each individual case, as such, must be considered and estimated according to general psychological rules.

Accordingly, in my opinion, the nature and character of the delusions have nothing to do with this estimation of the case, least of all in a medico-legal point of view, and the ontological specification of the insanity according to the character (often shifting) of the delusions, as exhibited long ago in the setting up of an insanity of religion, of love, &c., and recently and customarily, especially by the French, and now become very prevalent, I refer to the supposition of a "mania of pride," a "mania of persecution," &c., to which I might easily add a mania of querulousness (Cases CLXXXIV. to CLXXXIX.); suppositions which are indeed of but very dubious psychiatric value, and are all the more objectionable in forensic psychology because all generalisations and setting up of species and varieties lead, as experience has shown, only too easily to errors and to dangerous reasoning in the medico-legal reports, and these mislead and deceive the Judge. The forensic physician ought and cannot but know that suspicion, especially in a person of weak character and under favourable circumstances, may, in the end, lead to insane delusions, and that such a man may imagine himself to be persecuted by his own family or by all the world; that a fancied injury of the conscious rectitude, especially of vain and selfwilled individuals, may be stirred up by passion to insane dog-

matism ; that the vanity which unreasonably overvalues its own
excellence and does not find a corresponding estimate accorded, either
in social position or public recognition, continues to attribute to
itself more and more of those honours and dignities which the world
denies, till at length the delusion of being a Count, Prince, or
Emperor, is fully developed.   The forensic physician can and must
make use of these facts, ascertained by experience, for the psychological explanation of the new and corresponding case which lies
before him.   But here, as always, we have only to deal with the history
of the development of · the individual case from its etiological
relations, and we have nothing to do with the scientifically unjustifiable* setting up of particular species or varieties of insanity which
are not only superfluous but positively injurious for forensic
practice.   Such a subdivision is only justifiable in forensic psychology when the species represents a form which expresses an actually
specific character in the insanity of an individual, because certain
specific spheres, as the faculties of perception or of desire, are specially
affected by the delusion, whereby a most decided influence on the
behaviour and actions of the patient may be and is produced.   Accordingly, *Melancholia* and *Mania* may be properly reckoned as species
of the genus "Insanity."

§ 74. CONTINUATION.—MELANCHOLIA AND AMENTIA OCCULTA.

Melancholia is the result of depressing mental affections, which
are continually reproduced, whether the cause of these affections
have any real existence or be solely based in delusion.   The affectionate father of a family, who since he has been reduced to

---

* Unjustifiable, because the characters of the species in all such cases are
not fixed and exclusive, but the alleged specific character of the delusions is
often changeable.   The insane unmarried woman Stang, who called herself
"von Selvini," asserted that she was married to a man who was "Count
and His Excellency," and babbled continually of her aristocratic connection
and exalted relatives. *But at the same time* she also asserted that the street
in which she lived (one of the most private and aristocratic in Berlin) was
"a perfect *bagno*" of bandits, and that for months she had been persecuted
by robbers and police spies, so that she always went to bed with her clothes
on, and had a double-barrelled pistol beside her.   Here, therefore, was the
"insanity of pride and of persecution" combined in one patient. (*Vide* also
Cases CLXXXII. and CXCVII.)

beggary, has good reason for regarding his family as a prey to penury, and falls into melancholy, is all the less psychologically to be distinguished from the millionaire, who only fancies that his treasures have lost all their value, and that he is ruined, as sooner or later delusions become developed even in cases of the former character.   Thus we see men fall into melancholy from the most various purely bodily causes, which of themselves involuntarily produce a depressing effect on the mind.   The impossibility of delivering one's self by the energetic action of the will from this painful position is characteristic of this general depression, this *quasi*-paralysis of the mental functions.   This condition has been termed Abulia, and it actually exists.   Because the will displays just as many modifications as any other of the mental acts.   No difficulty terrifies the powerful, the iron will, and where the external means are at its command, it moves the world upon its poles.   But the man of feeble character, is feeble because he has no energy of will to fall back upon—he is undecided.   Where action is required, he cannot easily disentangle himself from the confusions of motives *pro* and *contra*, even when the action involved is of the simplest and most inconsequential character, such as the taking of a walk.   Daily experience, and hundreds of historical instances teach us that a man of so feeble a character, may be perfectly normal, or even highly gifted in other psychological respects. There is, however, certainly a still higher degree of defective energy of will, besides that which is so usual, in which the whole power of action of the man is as it were paralysed.   He cannot emancipate himself from this condition, often to his own great damage ; he neglects his official duties, his business, the care of his property, for no other than simply this reason.   Nevertheless, this so-called abulia must not be elevated to the rank of a peculiar species of mental disease.   Here we have the gloomy mental state and delusions of melancholia, then in other cases we have an abulia, a complete temporary paralysis of the will and energy (despair, distraction) produced by the force of circumstances.   In individual cases these will not be difficult to determine, and the man, so to speak, will be easily referred to the place in the psychological rubric to which he belongs, supposing him to be accused of some punishable act of neglect.

Finally, however, the melancholic patient, often after long struggling, adopts a resolution, and, even if it be of the most horrible

character, such as to kill his dearly loved children, and thus deliver
them from penury, he feels satisfied : " since," says Hoffbauer*—
uncommonly true to nature—" it is an observation which everyone
can readily make on himself, and which is just as easily explicable,
that when we find ourselves in a doubtful and alarming position
which we do not see our way out of, we gain mental ease so soon as
we can come to any decisive resolution, even though under other
circumstances this resolution might cost us an incalculable amount of
victory over self. Our only care is then directed to carrying out this
resolution; if this be in our power, we at least experience that
peace which is acquired on the actual arrival of those circumstances
and conditions of life which alarm and terrify us when regarded as
future, but which when present speedily produce mental equanimity,
from the knowledge that all alarm and dread we might experience on
their account are in vain. Just because we find peace in such a resolution, it is not to be expected that any one should swerve from the
resolution suggested by the delusion of a melancholia, provided the circumstances upon which the resolution is based do not alter. From
what I have said, it may be comprehended why a man, during the
period which intervenes between the adopting and the carrying out of
such a resolution, may perhaps act with the most remarkable composure of mind, and yet not swerve from his determination," &c. (I shall
by and by relate the most pertinent illustration of this description, *vide*
Cases CLXXIII. and CLXXIV.). It is to this condition that Hoffman has applied the specific term, so often employed subsequent to
him, of the " *Incitement of an obligatory resolution,*" a term which is
so hazardous, and therefore objectionable, because—besides what has
been already stated against all generalisation—it may without any
straining be applied to the mental condition of any criminal, whose
resolution to carry out any deed of wickedness, performed after a
long struggle, is rendered truly " obligatory" by the impulse of
passion, and the eager desire of the advantages to be obtained
by it.

This " Incitement," &c., of Hoffbauer's does not materially differ
from that state of mind to which E. Platner, ten years before him,
applied the term, so famous or infamous, of *amentia occulta*, thereby
giving occasion to the setting up of a whole series of ontological
hypotheses, which have been the occasion of an untellable number

* Die Psychologie in ihren Hauptanwendungen auf die Rechtspflege.
Halle, 1808, S. 333.

of erroneous and false decisions in medico-legal practice, and have chcifly contributed to bring medical opinions into miscredit. Platner only limited his theory in so far as he restricted it exclusively to violent actions. "*Est igitur amentia occulta nisus et conatus animi oppressi ad actionem violentam, hanc actionem secreto appetentis et molientis, tanquam suæ oppressionis levamen et liberationem.*"* These and subsequent similar appellations (§§ 75, 76) originated in a period when in nosology generally the *nomen morbi* was the most material point, and when it was of more importance to lay down a correct *systema morborum* than to institute more correct and accurate inquiries into the diseases themselves. In those days *Hydrops* was set up as a species, and a. *Hydrops saccatus,* b. *Hydrops abdominis,* &c., were treated of. Then we had as species, *spasmi, profluvia,* even *dolores,* &c.! The child must above all things have a name, and so it was also in psychical nosology. Practical medicine has very properly flung overboard this kind of ontological nosology, and forensic medicine has no less right, indeed it is its duty to do so also. And upon what large amount of actual observations may we ask, upon what actual facts has Platner based his much abused "setting up of species"? Upon two, only two cases, forsooth, the documentary evidence connected with which was submitted to the Faculty of Leipzig and which might, indeed, have been decided without the slightest difficulty, and without the invention of any "new disease." The most important case referred to a good-natured, weakminded, superstitious and hypochondriacal man, who had expressed his suspicions that one of his comrades persecuted him with sympathetic and magical tricks, and in particular, that he caused deadly vapours to blow upon him. To get some rest he had formerly diligently practised putting with a ball of lead, and finally he killed his persecutor by throwing this ball at him ; upon which he went and informed upon himself, declaring, in a perfectly rational speech, how much rather he would be executed at once, than endure any longer the torments to which he had been subjected by his foe. The second case, moreover, was that of a girl, aged seventeen, a fireraiser, one of those common and by no means obscure cases to which I shall by and by (§ 91) recur, and in regard to which Platner himself objected to ascribe complete irresponsibility. And upon *such* a basis has Platner placed his theory, and thus without even the most obvious criticism, which but a trifling practical experience might have sup-

* *Op. cit.* p. 4.

plied, has *amentia occulta* been received into forensic medicine! It cannot, for instance, be too often impressed upon non-professionals (lawyers), that those who are truly and indubitably affected with general insanity, and of course also such patients as labour under melancholia, conceal and keep their delusion out of sight, and this often with the utmost energy and cunning, that they reason logically, and that they can skilfully repel all reproaches in regard to plans and resolutions which have been betrayed, &c., as is proved by numbers of patients in every lunatic asylum, and exhibited by hundreds of suicides insane from melancholia, who have frequently been scarce parted from in a company, where nothing remarkable was observed, when their death has been learned, and an insight obtained into the sufferings they had long struggled against by means of writings left behind them. A man, possessed of the best qualities of mind and heart, respected and beloved by neighbours and friends, and loving his wife and children tenderly, murdered them all in one night without any of the ordinary motives, and without having ever exhibited any of the precursory symptoms of insanity. He was condemned (in England), but while in prison complete insanity became developed and he died mad a year after his condemnation. Taylor, a most praiseworthy man and loving father, who had fallen into poverty, strangled first two, and then in the same night his other two children, "that they might not be cast out to the streets." He shook each of them by the hand before he strangled them. Next day when arrested he made a full confession without attempting any defence. No witness had ever known him to be deranged. But a physician to an asylum came forward with the declaration, that he had under treatment the grandmother and the sister of the accused, the latter of whom had also murdered her children, whereupon the man was acquitted. These were therefore evidently cases of insanity from melancholy, as in our own perfectly analogous cases which shall be detailed presently, accompanied, as is so frequently the case, by the skilful concealment of the feelings and delusions from the outer world, till the moment arrived when the resolution, so long cherished and fostered, was embodied in action. Meanwhile, had we more accurate data in regard to those, as well as many other oft quoted cases; if we knew from documentary evidence the behaviour of these individuals during the time just preceding the deed, as is the case in regard to my own personal observation, it is probable that much of what appears, though only to non-professional people, extraordinary in their conduct would disappear. The little, however, which can be learned in regard to the above case of Platner's, leaves no doubt that this was a perfectly common, every-day case of—to use a modern and equally objectionable ontological term (§ 73)—the "mania of persecution," and after the description given of the motives to the deed, this statement requires no further illustration. And just as little is required to prove the practical danger of a theory, such as that of *amentia occulta*, which provides a convenient cloak for all actual crimes perpetrated under similar circumstances to those detailed, when the agent has been declared "never to have been observed to be deranged" previous to the commission of the deed.

There is, therefore, no true species of Insanity which differs specifically from other lunacies, and to which the term *amentia occulta* could be justifiably applied. This unscientific and dangerous term, therefore, *ought not to be employed in practice, and the examination of the case according to the general diagnostic criteria* (§ 61) *also makes it perfectly superfluous.*

§ 75. CONTINUATION.—MANIA AND MANIA SINE DELIRIO.

However remarkable it may appear not to permit Mania to be reckoned as a distinct class—quite irrespective of the corresponding regulations of our own and other statute books—but rather to account it as belonging to the class Insanity; yet, other most respected authors of former as well as recent times, such as Metzger, Rose, Jacobi, Ideler, &c., have also, with the utmost propriety, pointed out that there are no peculiarly characteristic points of difference, no means of making a differential diagnosis between Insanity and Mania. That the "Raving" does not constitute any true distinction can be seen every instant in the wards of any lunatic asylum, while the fury, the blind destructive impulse, and the brute-like violence of action are not present in most cases of mania. Ideler* says very properly, speaking indeed only of the similarity of mania with what he has termed the mixed affection, but which is all the more truly applicable to insanity; " We cannot regard either the falsification and obscuration of the perceptive powers, nor the confusion and loss of memory, nor the wild imaginations which chase one another through the fancy, nor the contradictions, paralo-

* Lehrbuch der gerichtlichen Psychologie. Berlin, 1857, S. 267.

gisms, tergiversations, and desultory incoherence of thought, nor the obscuration of the reasoning powers as decisive symptoms of mania, and even in the domain of the feelings their inconsistency and perversity are not greater, nor the violence of the foolish impulses of the will more unregulated and unconnected than we find in the highest degree of this affection" (of "insanity"). "Just as little is there any bodily condition capable of conducting us to a decided conclusion," &c. Accordingly, it would not be at all necessary to assume the existence of mania generally as distinct from insanity, did not experience teach us that a large proportion of such insane are specially predisposed to violent actions dangerous to the public, which—independent of the necessary therapeuto-police regulations—supplies a characteristic distinction for forensic practice, precisely as the reverse did in the case of melancholia (§ 74). The opinion, certainly based both in nature and experience, that mania and insanity are not opposed to each other as different genera of disease, would only be overthrown if it were true and proven that there were cases—however rare their occurrence—in which mania was observed without the simultaneous existence of delusions. Such a condition certainly was first mentioned by Ettmüller, who (*Prax.* II., *cap.* 4.) speaks of a *melancholia sine delirio*, in which there was *recta ratio sine delirio*, and after him Pinel, a man of otherwise the greatest merit in the treatment of insanity, who, by his assumption of a species of insanity, *Mania sine delirio*, in his famous work published about fifty years ago, has, in a special manner, been the cause of setting in motion a whole host of learned pens down to the most recent times. Let us in the first place again examine the facts upon which this doctrine is based. Pinel's first case* is stated only to exhibit "*les premières nuances de cette espèce d'alienation.*" The only son of a feeble mother, who had been accustomed to indulge all his whims. Opposition only increased his daring, and he lived amid constant quarrels and disputes. If a horse, a sheep, or a dog annoyed him, he killed it at once. In society, at festivals, he got excited and both gave and received blows. Perfectly rational when quiet, he managed his large property with propriety, fulfilled his social duties, and was beneficent. Wounds, litigations, and fines had been hitherto the sole result of his quarrelsomeness, but one day he got incensed against a woman who had offended him and threw her into

* *Traité Medico-philosophique sur l'alienation mentale*, 2 edit. Paris, 1809, p. 155.

a well. This case may well be termed a horrible newspaper anecdote, but scarcely a medical observation ! What can be proved by these scanty data? was insanity gradually developed in this man as the result of his evil dispositions, perhaps in consequence of other unsuspected causes? what was his conduct during the time just preceding the act, what was it subsequently ? what was his relation to the murdered woman? or was he really only a man of violent passions, who could not, because he would not, command himself? on these and many other points Pinel is silent.—His second case is that of a man, who at the time *des accès de fureur* was seized with a burning sensation in his bowels, violent thirst, and obstinate constipation ; the heat ascended, his face became reddened, his temporal arteries pulsated violently, and he was seized with an irresistible thirst for blood. If cutting implements were at hand he was ready to sacrifice in a kind of fury the first person he came across.   Moreover, even during the attack, he was in full possession of his reason, he answered questions put to him correctly (*directement*) and betrayed no incoherence in his ideas, no evidence of insanity ; he felt, indeed, the horror of his position, he was full of regret, as if he had to reprobate this involuntary inclination.   He once before an attack warned his beloved wife and called to her to flee if she would avoid a violent death. A similar thing happened in the hospital where, at another time, he stabbed himself with a knife in the breast and arm.   Strict watching and a straight jacket have *arrêté le cours de ses projets suicides* (was he kept in a straight jacket for a month, a year, or for the remainder of his life ?). This case is an example of the so-called " Homicidal Monomania," and will be more fittingly considered along with it (§ 93).   The last case of Pinel is, however, the most famous.   The mob, which during the revolution stormed the prisons to free the alleged victims of tyranny, also broke into Pinel's asylum (the Bicêtre) and found a man in fetters who spoke "*plein de sens et de raison*," and bitterly complained that he was confined in fetters along with lunatics, since no fault could be found with him. This he declared to be the most horrible injustice, and he conjured these strangers to free him.   He was set free and carried off triumphantly amid shouts of *Vive la République !* The sight of so many armed men, their furious shouts, and their countenances flushed with wine roused the fury of the lunatic.   He forcibly wrested a sabre from one near him, hewed right and left around him, and had to be taken back to the asylum.   Such is that remarkable chief proof for the existence of *mania sine delirio**! Men such as Reil, Hoffbauer, Mittermaier, Hartmann, &c., have not hesitated to accept this theory, whilst Esquirol, Henke, P. Jessen, and others have opposed it†. The opponents of this theory have endeavoured to interpret this case differently, and to collocate it along with other similar ones under the rubrics of intermittent mania, fixed ideas, morbid irascibility, &c., overlooking, however, the most important point—the clearness and accuracy of the observation—which have been as it were silently presupposed. Who, in all the world, was that fettered patient of Pinel's? What was his *vita anteacta?* How long had he already been in the asylum, and why was he there? How had he behaved himself while there? And, finally, who were the witnesses from whom alone we learn that he spoke *plein de sens et de raison?* A drunken rabble—Pinel calls them *brigands,* and he himself says not one single word as to his personal knowledge and observation of this his patient! Was this man, however, therefore not insane (*sans délire*) because he himself told the mob he was not, and that he might be let loose? Every tyro knows that almost all lunatics speak in this manner. Thus this case also is wholly without any scientific basis. And the same may be said of Reil's case " of mania without insanity," the subject of which, amongst other things, long before he committed the deed of violence in a fit of mania, had a fond-

---

* How carelessly the otherwise so meritorious Pinel deals with his " observations" in favour of this theory of his is also proved by the following. In the note to the above case it is stated, "I have in the first section detailed other instances of *manie sans délire.*" At the part alluded to we find, however, only the following (word for word) :—" A very lively woman, very commendable for her household virtues, had for long given herself up to the most unbridled wrath on the most trifling occasion ; a slight delay in carrying out her commands, the most trifling fault in her children or domestics raised a stormy scene. This unhappy tendency came to an end by the woman becoming insane (! !). Three insane young maidens were admitted into the Hospital ; one of them had become insane at the sight of a pretended white spectre, which some young men had exhibited to her during the night ; another by a violent thunderstorm during the flow of her menses ; and the third by horror at finding herself in a house of ill fame (*mauvais lieu*), whither she had been cunningly enticed." We cannot help asking in astonishment what these perfectly trivial cases are meant to prove, and especially what they are meant to prove in regard to the existence of a species of insanity *mania sine delirio ?*

† *Vide* the critical history of this question as fully given in Henke's Abhandlungen aus dem Gebeite der gerichtl. Medicin., 2 Aufl., Bd. II and V.

ness for throwing stones at people, &c. Every one who knows and has
often experienced how very difficult it is, in unusual cases of insanity
which has been for long coming on and has gradually attained its
full development, and which, as has been already remarked, the
patients so often know how to conceal skilfully even from experts,
how difficult it is in such cases to collect the facts necessary for a
diagnosis, how often even after repeated examinations of the patient
we are still insufficiently informed, and how light is for the first time
thrown upon the case by an examination of the documentary
evidence containing the depositions of many witnesses who have
accurately observed the individual for many years, will also know
how to estimate such stories which are passed off as "cases." There
does not exist one single *well observed and fully related case* which
can be accepted as proof for the actual existence in nature of a
peculiar species of madness, mania without insanity, and such proof
will never be forthcoming, for this supposition is, according to
Ideler's appropriate description, a *contradictio in adjecto.** The
maniacal person is excited to action during his attacks only by insane
delusions, for else they would not possess the characteristics of a maniacal deed, and Pinel's patient would not have attempted to cut down
his liberators had he been really so *plein de sens et de raison!* It is,
however, not enough to strike out of science this untenable hypothesis of a *mania sine delirio ;* practical forensic medicine, adminis-
trative penal justice, have a still more important duty to perform—
to banish it from their domain. Because for the latter it is a still
more dangerous assumption than that of a *mania occulta,* and has
bred more mischief than it, since it has been employed as the most
abominable of all cloaks for crimes committed with the most perfect

* I cannot except from the above judgment the cases of mania without
insanity related by M. Jacobi (Die Hauptformen der Seelenstörungen, I.
Leipzig, 1844). If they be only read correctly and without prejudice, it
will be found that not one single case can be regarded as a possible exception. In the patient No. 1, " transitory, trifling, insane delusions were, it
is confessed, to be observed." In No. 2 there was " no insanity," yet he
persisted "in cutting off his own perfectly healthy finger?"! And was it
not insane delusions that produced "the most violent jeering laughter or
howling weeping" in the midst of the most frightful attack of mania? The
cases 4–34 are classed by Jacobi himself under the head of " mania with
delirium, or confusion, without insanity ;" and accordingly the whole
question with him turns solely upon the disputed meaning of a word. In
all the cases referred to, of mania " without insanity," moreover, according
to the history of the cases, attacks of idiocy or insanity had preexisted.

freedom of the will, inasmuch as acts of violence or murder committed by criminals who were not insane have been declared to be produced by an apparently blind fury or objectless striking at random, the result of a *mania sine delirio*. Nevertheless, I have already related facts to show* how often it happens that murderers at the time of the commission of the deed hack their victims in the most needless manner, and that when brought to confess they have unanimously acknowledged that after giving the first stroke or blow they felt as if mastered by a perfect "fury" in which they have blindly continued to repeat the blows. And the dissection of the bodies have confirmed this statement. The child killed by the house-keeper R. displayed on its body forty-six injuries, the tailor murdered by Haube exhibited forty-two, and the shoemaker killed by Maskendorf displayed twenty-four (cranial) injuries! † There are certainly cases of "fury" without insanity in which the demoniacal nature of the evil principle breaks forth unshackled, but precisely because they prove the existence of fury *without* insanity they necessitate the assumption of criminal responsibility at the time of the commission of the evil deed. *Consequently there is no peculiar species of insanity which may be termed* mania sine delirio. This unscientific and dangerous designation *ought not to be used in practice, and the consideration of each individual case according to the general diagnostic criteria* (§ 61) *renders this perfectly superfluous.*

§ 76. CONTINUATION.—MODE IN WHICH INSANITY ORIGINATES. MANIA TRANSITORIA.

Insanity exhibits several differences in form, according to its mode of origin, course, and psychical limitation, all of which have an important relation to criminal responsibility.   1. As to its *mode of origin;* ordinary cases are not difficult of estimation; in them, in a person who has up to that time been in perfectly sound mental health, on the most various occasions (§ 71) a delusionary disturbance of the mind suddenly breaks out and continues to exist in unmistakable diagnostic distinctness for a longer or shorter time.   In others, perhaps the larger number of cases, the mental disease is gradually developed.   An alteration in the manners and habits of

* Mörderphysiognomien. Berlin, 1854.
† *Vide* Cases LVIII. LXXI. and CVII., Vol. I.

the patient are the usual characteristics of the often still unsuspected disease. The punctual man of business commences to neglect his duties, and has always some remarkable excuses for his peculiar conduct; the steady home-loving man takes to gadding about and roves aimlessly hither and thither; the careful mother neglects her children and busies herself with all sorts of nonsense. Actions come to light more and more remarkable and excitant of misgivings, extraordinary letters to unknown persons, to exalted personages, to Boards, steps taken for the sale of house and property, the speech becomes incoherent, and at length, though it may be only after the lapse of a long period, the existence of complete insanity can be no longer doubted. That form of insanity termed melancholia is specially apt thus imperceptibly to steal on. The interest taken in persons and things till then cherished and loved ceases in a remarkable manner, the cleanly tidy woman neglects her appearance, the ordinary mental employments give place to an aimless idleness, society and amusements, formerly delighted in, are now avoided, and solitude sought after. The patient still displays no incoherence of ideas, and he, whom his friends at the most regard as only suffering in body, sinks gradually, no warning sufficing to rouse him. By and by fears begin to be developed which excite apprehension; the harvest will not be prosperous, the children will die, the property is gone, &c., and finally the hitherto " concealed insanity" (§ 74) becomes evident. Or finally insanity breaks out in a psychically perfectly healthy man as the *result of one of those causes which experience has correctly recognised as efficient,* suddenly indeed, but without following its usual course it exhausts itself in one single attack, and when that has ceased the mental disturbance has also entirely ceased, and often never reappears during the whole course of life. (I may also point out, only by the way, that a similar threefold mode of origin also prevails in the domain of bodily disease.) Such was the case of the privy councillor Lemke, whose case was published by Heim seven-and-forty years ago,* a case which has gained undeserved notoriety because it was the occasion of setting up another—and one of the most dangerous—species of insanity, transitory mania, *mania transitoria.* That generally respected man had gone to bed at night in perfect bodily and mental health, during the night he was shaken up by his wife, with whom he had lived most happily for many years, she having heard him

* Horn's Archiv, &c., 1817. Bd. I. S. 73.

breathing stertorously in his sleep, he fell upon her, hauled her out of bed, dragged her to the window and made gestures as if to throw her out, till his family breaking in delivered her. The family physician, who was instantly summoned, ordered what was necessary. L. became quiet, fell asleep again, and next morning had no recollection of the horrible occurrence, and, as I know from personal knowledge, to the end of his long life was never again attacked by any similar fit of " transitory mania." Now, as this case occurred in a sleeping person, so also the most of the cases coming under this category* have been observed in those somnolent (§ 84), who, when half awake, have broken into the most violent actions and committed illegal† deeds. In other cases it was the act of parturition which has produced a fit of mania as rapid in its origin as in its cessation ; while in not a few other cases irritation of the bowels has produced consensual disturbance of the mind, which disappeared on the removal of the former, &c. Now it is known to be indubitable that bodily conditions, such as sleep, intestinal irritation, parturition, sunstroke, &c., may originate a sudden and transitory cerebral reaction with maniacal symptoms, which again disappears on the removal of the cause. But it would be an offence against the rules of general pathology to elevate an attack of insanity of this character, which is only the symptom of a certain condition, to the rank of a peculiar species of insanity, and this all the more that it is impossible to regard the mere duration of a disease, in which alone the transitory mania differs from others, as a character sufficiently characteristic to separate one species from others similar. I need say nothing more as to the danger of this supposition, since nothing is easier, and it has often happened, than to ascribe an outbreak of furious passion in a man who has been perfectly sane both before and after the commission of the criminal act in that state, to an attack of *mania transitoria*, which excludes responsibility. And when Heim (*loc. cit.*), in publishing the case of Lemke, adds with concern, " it is

* A comparison of the cases collected in periodicals, collections, works of compilation, &c., exhibits the fact that many cases are reckoned under the head of *mania transitoria*, which belong partly to ordinary insanity and partly to mere furious passion. *Vide*, among others, the cases in Pyl's Aufsätzen, &c., Bd. VIII. S. 236 ; in Henke's Zeitschrift, Bd. XIV. S. 134 and 165 ; and in Marc's work, die Geisteskrankheiten, Bd. II. S. 374.

† A collection of cases may be found in P. Jessen's work, Versuch einer wissenschaftl. Begründung der Psychologie. Berlin, 1855, S. 670–691.

beyond doubt that many have been cruelly martyred by the executioner's hands, or have lost their lives in dungeons or penal institutes, who were wholly innocent and had only the misfortune to be seized by such an attack of mania to which the best of men is liable" —later experience has shown that the reverse has been the result of this newly discovered disease, namely, that by the abuse of this doctrine in penal law a far greater number of criminals have had the good fortune not to lose their lives! *Consequently, there is no such peculiar species of mania, no so-called mania transitoria. This unscientific and dangerous term ought not to be employed in practice, and the investigation of each individual case, according to the general diagnostic criteria, renders it wholly superfluous.* Who, for instance, would have objected to declare privy councillor Lemke irresponsible, had his case come to trial and been investigated according to the general principles of psychological diagnosis?

§ 77. CASES ILLUSTRATIVE OF THE SO-CALLED AMENTIA OCCULTA AND MANIA TRANSITORIA.

CASE CLXXIII.—MELANCHOLIA.—BLAICH, THE MURDERER OF HIS CHILDREN.

On the 17th of January, 18—, the master joiner, B., with a razor cut the throats of his two legitimate children, Paul, aged four years, and Charles, aged one year and a half, in consequence of which they died almost immediately. This dreadful deed excited all the more astonishment in his wife and friends, as from his previous character, habits of life, and relations to his children, such an act would not have been expected of him. All the witnesses were perfectly unanimous in regard to the data mentioned, their statements gave a distinct portrait of the accused, and in combination with the results of my own examination, which perfectly agreed with it, supplied a sure basis for a psychological opinion. B. had been married for five years to the mother of his children, and in that period had begotten four children, of which the second died soon after its birth, and the last was not born till after the commission of the deed. His marriage was a very happy one as deposed by his wife, and confirmed by all his friends. He loved both his children while living in an unusual degree, and as his wife deposed, he had always done all in his power to support his family. In regard to this, the witness R.

stated, that B. had the children always extremely neatly clad, that he exhibited them with pride to other people, and often bought sweet-meat for them, though "money was scarce with him." Such a tenderness of feeling was indeed to be expected of a man such as the accused had always shown himself to be. The evidence of his former master, E., dating from the year 1845, described him as "always an honest, quiet, diligent, industrious, and in every respect a moral man," so that he gave him his fullest confidence. The same character, quiet, orderly, diligent, sober, living peaceably for himself, loving his wife and his children, was given to him unanimously by all the witnesses, particularly M., who had known him from his youth up, and who declared that "he loved his children almost too well." According to the evidence of this witness, a remarkable alteration in the bodily and mental condition of the accused, dated from the year 1845, and this he himself confirmed. In the summer of that year he received during a tumult, in which he had no active part, an accidental blow from a cudgel on the head. Since that time he had complained much of pain, giddiness, and weakness in his head, and it struck M. that B. from that time commenced to indulge in curious inquiries, and to have "fixed ideas." In particular *he declared he had discovered the "perpetuum mobile,"* or continually racked his brains about it; he was continually making drawings upon his joiner's bench, and also other sketches in regard to it, *which he carefully concealed;* he told M. he had now found it out, and would get the three tons of gold which had been offered in England for this invention, and with this he would build a new church in his native city, &c. The representations of his friends had so little effect in dispossessing B. of this idea, that up to the last he was constantly busy with it, and had actually constructed a machine, which, however, did not come up to the ideal; this incited him to puzzling his brains anew, but moved his wife to burn the machine to divert his attention from the matter. M. deposed, that he had constructed such a machine in his workshop, and that he had worked at it for a whole night with the door locked and the window hung with a curtain. Such conduct could not but astonish his friends. The merchant R. declared, that he had always given him the impression of an eccentric man, who was very conceited of his knowledge and abilities, and on that account the witness had been long anxious, and had repeatedly said to others that B. would yet lose his senses. Widow S. had for many years back seen him always moving about the house in deep thought, and had often observed his speech to be un-connected, and it also had not escaped his wife since her marriage with him, that he was constantly puzzling his brains about something or other, and was lost in thought. This mental condition could not but be increased by the depressed position in which he then was. His earnings being small he fell into debt, and had to pawn various articles. His mental state continued to become more remarkable. During the last eight days preceding the commission of the deed he moved restlessly about the workshop, did not get on with his work, and was constantly staring before him ; he also seemed to M. to be pale and miserable-looking, and he therefore recommended him to consult a physician. The witnesses G. and S. had heard him dur-ing the last few days " talking incoherently and knocking things about." He stared long at one spot so that his eyes " started." His wife confirmed this report of his conduct at this time, and added, that he had quick breathing, heat, restlessness during sleep, strong fever, expectoration of blood, and complained of his head and chest. His face was also flushed; he walked restlessly up and down the room, scarcely answered her inquiries, and repeatedly declared that he was too deeply injured by his comrades, they had martyred his soul, and then, pointing to his head, he hinted that there was something there which he could not get over. He himself confirmed all this, and deposed that, on account of his heat, he could scarcely get enough of water to drink, and in spite of the cold (in January) he had always slept with his window open all night, because he felt as if he would choke. His brain felt as if it oscillated like the pendulum of a clock. In this condition the journeyman joiner F. saw him only a few hours before the commission of the deed. This was carried out thus : taking advantage of the momentary absence of his wife, he set the children in front of him and cut their throats with a razor, laying the bodies beside each other on the floor. He himself says, he does not know why he did this; he " could not get over the ill-feeling against himself and his father." Immediately after the commission of the deed he attempted to cut his own throat, but had not strength enough to do it. He then went into the garret, and tried to kill himself with an axe, but this also did not succeed. He next tried to hang himself but was speedily discovered, and although already asphyxiated, he was recalled to life by a physician, and sent off to the Charité Hospital. After his unimportant external wounds had been healed, he complained, four weeks subsequently, again of giddiness, singing in the ears, winking of the eyes, heat in the head, and an uneasy sensation of confusion, nevertheless, he was able to be dismissed " cured,"on the 14th of March. At the recognition of the dead bodies we saw him throw himself upon them exclaiming, " Ah ! my poor children !" then, however, his voice failed him, he was moved convulsively and could only answer, after the lapse of some time, when he called out " Ah ! what has happened to me, what has become of the good man, Ah, Ah ! I that have been so good and true," but he appeared so much affected, that the business had to be put a stop to.

The result of my own examination is as follows :—" B. is a man aged thirty-four years, of middle size, pale countenance, normally formed skull, and with features expressive of openness and good-nature. As to his bodily condition, he complains constantly, as formerly, of pressure and a not very clearly described feeling of uneasiness in the breast, and a similar feeling in his head, sometimes as if both of them would burst, and of restlessness, sleep and heavy dreams. His digestive functions are regular, but his pulse remarkably hurried. The physical examination of his chest revealed that B. suffers from hypertrophy of the heart, which is sufficiently explanatory of his former and present symptons, both physical and psychical, cough, expectoration of blood, rapidity of pulse, restless sleep, weight in the chest and head, and sensation of anxiety and restlessness.   Further, the scars of the injuries formerly received were distinctly observed by me on his head.   In regard to his crime, he has expressed himself to me precisely as he had done at his previous precognition. He described his tender love for his children ' greater than was probably ever felt before.' He wonders ' what has come over him that was always so upright and honest, and has done all in his power for his family ;' and he says this and other similar statements in a tone of the most perfect conviction, which, as well as his whole appearance, is completely opposed to there being the slightest suspicion of his feigning.   Questions such as, has he never thought how punishable his deed was? he answers by stating that he has never been able to think about it, that his state of mind was too horrible, that his comrades in the workshop at R. had been too hard upon him, &c. I have also conversed with him about his *perpetuum mobile*, and it is extremely characteristic to see how B.'s whole appearance immediately changes whenever this subject is approached, and particularly when the possibility of its success is denied. He instantly brightens up and declares that nothing is impossible to men who possess the proper qualification for their task. America was 'discovered' contrary to the judgment and opinion of the contemporaries of Columbus, and he related quite correctly the story of Columbus and the egg; he was already on the way to make the discovery, but had convinced himself that as yet he had not been successful, &c. At the same time he also declared, thereby confirming what is stated in the documentary evidence as to his estimation of himself, that he has often seen how his fellow-workmen in making a table could not get it to do, while he has taken just one board and then another and a third one, and then showed how they all fitted. Finally, I must remark, that B. shows scarcely a trace of repentance, properly so-called, unless we reckon as such his complaints as to his fate.

" From time immemorial, in judging of cases of doubtful criminal responsibility, it has properly been regarded the most important criterion, whether the deed committed could have been preconceived of the perpetrator? that is, whether it might be regarded as the result of his general character? No detailed deduction is needful to prove that this has decidedly not been the case in regard to B. So affectionate a father does not kill his children, particularly without any rational cause. Such a contradiction of the holiest and most powerful laws of nature permits the most inexperienced to conclude, and rightly so, that at the time of the commission of the deed the perpetrator was in a state of mind in which the recognition and the influence of these natural laws was abolished, in a condition of deranged psychical activity in which the powers of perception and will were both altered. And deeds, such as the present, are examples which justify the generally extremely dangerous dogma that it is possible from the mere nature of the action to deduce the irresponsibility of the agent. Extremely dangerous, I say, since in most cases it is the duty of the psycho-legal judges of the criminal first to prove the existence of mental derangement and its causes, and then to deduce from that that the deed has been occasioned by and in that condition, whilst in this case it is and may be assumed that the deed itself proves the existence of mental derangement. But also the origin of such a condition in B. is, à priori, not difficult to prove. One mental cause and two bodily causes in this case have acted together in producing a confusion of the rational powers. I refer to the cranial injury received some years before and the cardiac disease, and also the vanity of the accused, which many witnesses have testified to.

The influence of any one of these causes alone has frequently driven men out of their mind and in B. all these have acted together. In regard to the effect of the cranial injuries, I can refer to the ordinary experience of their results, which is well enough known even to nonprofessional people, while they also are probably not less aware that cardiac hypertrophy is very apt to produce a deeply hypochondriacal, melancholic state of mind, which gradually increases *pari passu* with the increase of the disease. Such a patient, who, blinded by vanity, feels himself called to higher things, falls upon the idea of discovering perpetual motion; the less success attends his endeavours, so much the more he busies himself in subtleties; the more—as was latterly the case—diseased and reduced in body he becomes, so much the less successful become his endeavours. Thus arises a continual struggle between the will to do and the inability to execute, in which hundreds of minds more powerful and less disposed to derangement than that of B. have succumbed and lost their balance. And when that state is arrived at in which they have become incapable of considering the (statutory) consequences of their actions which are then executed from mere instinctive impulse, which excludes the applicability of the statutes, this condition is termed, not indeed by the Penal Code, which contains no proper definition of it, but by the General Common Law, 'idiocy' (a term not properly in accordance, however, with ordinary medical phraseology). In this sense of the term I must give it as my opinion that at the time of the commission of the deed accused was deprived by 'Idiocy*' of the freedom of voluntary action in the sense of § 40 of the Penal Code. B. was transferred to an asylum.

CASE CLXXIV.—MELANCHOLIA.—DIETRICH, THE MURDERER OF
HIS SON.

The following case, in which the insanity existing *previous* to the deed was only much less evident even to his nearest connections, otherwise very much resembled the foregoing. The weaver D., a small feeble man aged fifty-three, got up one morning while his family were already at work in an adjoining apartment, brought an axe out of the kitchen which was close to it, went to the bed where

* The defective nature of the statutory definitions of insanity and idiocy (*vide* p. 92, Vol. IV.) daily compels the Prussian forensic physician to designate forensically cases of insanity as cases of idiocy.

his youngest son was still asleep, and smashed the boy's skull by repeated blows with the axe. He calmly told the deed that had been done almost before eye-witnesses, and quietly allowed himself to be arrested. I had occasion to examine the accused at the very commencement of the preliminary investigation and without obtaining any previous information from the depositions of the witnesses. At this examination I discovered, in the first place, a distinct swelling of the left lobe of the liver, with all its usual concomitant symptoms, such as the peculiar complexion, &c., which I need not enter upon here. During the very first night following his arrest he had violent attacks of hematemesis in the prison; his digestion was completely gone, and his bowels were so sluggish, that he had to be almost immediately sent to the Lazarette. Far from complaining of his ailments, everything had to be interrogated out of him, and his replies were given with the same equanimity, the same perfectly tranquil apathy with which he constantly answered every question put to him regarding his crime. The alteration in his inner man, which was well enough known to himself, he was less inclined to ascribe to his abdominal disease than to a bite which a fly had given him upon the back of his left hand about a year ago, the result of this being the development of a very malignant ulcer (cattle typhus—*milzbrand*—inoculation?) which suppurated for a long time, and the scar of which is still visible. From this time he dated the commencement of an internal agony which, in his usual sententious manner, he described as horribly tormenting. This attacked him, he said, five or six weeks previous to the commission of the deed, and persecuted him continuously up to that time. He could get no rest for the thought that he and his family, from their alleged poverty and want of food, were threatened with death from hunger at no distant date. My representation that, according to what I had learned, his poverty was by no means so great, since both his wife and his two eldest children contributed to the support of the family, and they had, moreover, butcher's meat as a daily part of their food, could not convince him. On the day before the commission of the deed, he said, he had a piece of cloth to deliver, but he observed that it was quite dirty, and that several yards were wanting; this convinced him still more that he was perfectly unfit for work, and his anxiety was all the more increased that rent time was drawing near. It now became always the clearer to him that it would be better for himself and his family to be out of the world. This thought continually tormented him all through his sleepless

nights.   And thus, according to his own confession, he was led to
perpetrate quietly and coolly the crime on the twenty-third of July.
He says he never thought of its results, and on my representation
that he had a severe punishment to expect, he constantly replied
with the utmost apathy, that he deserved "three or four times as
much," and that his crime was "incomprehensible and horrible."
But this was, I stated, "not the barbarous equanimity of the cold
and hard-hearted criminal, as every one must confess who is acquainted with persons of this class, and has compared their appearance
with that of D.   It was rather the equanimity resulting from the
utmost apathy, the state of being completely done with one's self and
the world, the morbid equanimity of despair.   Thus he was at the
recognition of the body for one instant affected, but nothing less
than repentant or deeply moved, &c."   In spite of all these data,
however, I felt compelled, in the absence of any information as to
D.'s previous life, to withhold any decisive opinion as to the case.
And, in fact, a subsequent examination of the documentary evidence
supplied me in this case also with the decisive facts.   It was testified
that D. was an extremely reserved and lonely living man, who
was at peace with everybody in the house where he had lived for ten
years "peacefully, orderly, and diligent."   He had always punctually
paid his rent ; and in regard to his fear already referred to, it was
important that his landlord deposed that he had never been hurried
in regard to it.   No one had ever observed any dispute, quarrel, or
violence, within this quiet family, in which the youngest (the murdered)
son was reckoned his father's pet.   Further, one witness made the
important statement that he had seen D. sitting at his work work-day
and Sunday from early morn till late.   That of the manufacturer for
whom D. worked was still more important, that the piece of cloth
delivered by him the day before the commission of the crime was
neither dirty nor deficient by several yards, and that D. would not
be convinced of this even when it was measured before him!   I
then, in my report, went on to mention the trade of the accused, his
sedentary mode of life, the severe abdominal disease with which he
was affected, its character, the insoluble contradiction between his
love for the child and its murder, his behaviour at and subsequent
to the commission of the deed, his notorious insanity *previous* to
that, at the same time taking notice of the doctrine of *amentia
occulta*, a doctrine which might have been so easily and so erroneously applied here, and everything in regard to it which has been

already stated in § 74., and then replied to the question before me
as follows :—"That the weaver D. was mentally deranged at the
time when the crime was perpetrated, that he still labours under this
mental disturbance, and that he committed the crime when in that
state of mind termed 'Idiocy' in the statutory sense of the word,
and that he is not to be regarded as a responsible agent." D. was
sent to a lunatic asylum, where he died paralytic about a year
subsequently.*

CASE CLXXV. — MELANCHOLIA.—INFLICTION OF A CRANIAL
INJURY.

This case also belongs to the category of the so-called *amentia
occulta;* it was, however, peculiar from the whole conduct of the perpetrator, especially subsequent to the deed. In the primary court he
was condemned to two months' penal servitude; in the court of appeal,
however, the advocate for the defence moved, that his mental condition
should be inquired into, and the following unusual question was laid
before us, "Has Schmatte now, and had he at the time of the commission of the crime the undisturbed use of his *moral faculties?*"
The accused was one evening in the W. workshop in company
with the journeyman St. and several other persons. Suddenly (!)
Schmatte seized a lath and struck the former on the head, wounding
him severely. St. has deposed, that he could not otherwise explain
the extraordinary conduct of Schmatte, "who struck him without

* A subsequent more horrible case was precisely similar to the above, and
apparently a still more authentic proof in favour of the existence of a so-
called *amentia occulta,* a supposition which, however, was shown, by a
sufficient investigation of the individual case, not to be in the least necessary
even in this one. It likewise concerned a moral, peaceable, and irreproachable man, who loved all his four children with a most affecting tenderness,
and cut the throats of all four one morning, killing two of them on the spot.
No witness had ever suspected any mental derangement in him previous to
the commission of the deed. He had premeditated the deed long beforehand,
carried it out with the utmost calmness, and appeared after the deed and in
the course of the investigation so calm, cool, and apparently rational, that
the most earnest appeal was required to convince the dubious judicial
officials of the actual existence of deep melancholy and insane delusions
based upon it. Among other probative facts, I will only now mention that
on searching his home there was found a perfectly insane will written in
his own handwriting eight days before the commission of the crime. He
was sent to an asylum.

the slightest provocation, and without the slightest previous altercation, except by supposing that he was not always altogether himself." Latterly St. has further declared, that the accused was in the habit of behaving in such a manner, that he was held to be affected with " silent madness." Similar statements were deposed to by the witness E., that there had been no altercation, and he only remarked that Schmatte was apparently much excited and offended, and that he blustered out the words " command" and " hoax," while he was in general "remarkable for his melancholy and sententious brevity." The accused himself, who never denied the deed, expressed himself in regard to it in the following words:. " What was the occasion on my part, I no longer know ; I was very often vexed and imposed upon by St. when we were together. He also very often had disputes with my fellow journeymen, and was generally very quarrelsome. He also desired to stir me up to give the master a sound beating; to this, however, I would not consent. Probably (*sic !*) he had also again imposed upon me on the day on which I struck him, so that in a fit of passion I struck him—I know not how it happened. Moreover, I have been truly sorry for the occurrence after I became aware of what had happened, and I have offered St. full reparation. Immediately after the occurrence I went off secretly from Berlin without taking any money with me, although I had eighty-five thalers (£12 15s.) in my cupboard. By the way I slept in thickets in the open air, and fed upon sloes, &c. Thus I rushed into the wide world, I knew not whither, I had taken the mischance so much to heart." Thus he got to Hamburgh, whither he alleged he did not wish to go, but to Osnaburg; he also passed through Hamburgh to Hunteburg to his family, whence he returned to Berlin, after receiving a letter, stating that St. had been perfectly satisfied with the reparation offered. Schmatte states, that he has for long suffered from various diseases, particularly giddiness, and that he has frequently been "quite senseless." He has been under treatment in the hospitals here for gouty pains and small-pox. His statement that nine years ago he was insane is confirmed by an official communication from Hunteburg, dated the 13th of April. According to the official communication from his native place, the accused on the occasion just described arrived at Hunteburg from Berlin "perfectly naked and in a deranged state of mind," and had to be put under medical treatment. This document further certifies, that Schmatte bears a high character in his native place. Further remarkable is a somewhat confused letter from the accused to the Royal Court of Justiciary, dated the 8th of June, in which he announces a change in his residence, and states, that he looked forward with composure to his punishment, even if it were the gallows. In this letter he says, amongst other things, "I keep myself as much as possible by myself, when anger becomes violent a man never does any good, and often he knows not what he does. Anger has never done any good, because I have for some time frequently laboured under spasms of the heart and chest which rise to my head, so that I become quite stupid, and would fall if I did not seize hold of something; I willingly avoid all quarrelsome men, when I can, but sometimes I cannot," &c. Finally, according to the records of the inquiry, "the mental powers of S. appeared much blunted and obscured, and his power of judgment certainly very limited," while no symptom of actual derangement was observed during the whole examination.

"The accused is thirty-five years of age, of moderately powerful build and middling constitution. The formation of his skull presents nothing remarkable, while his, on the whole soft and good-natured, expression is certainly very peculiar, particularly from the small deep-set eyes and dull, timid look. And not less so is Schmatte's voice, which has a hollow tone, easily observed in his slow, low, and drawling speech. There are no objective morbid symptoms observable on the body of the accused, while he still asserts, that he sometimes suffers from giddiness, tearing pain, restless sleep and spasms, morbid symptoms which he describes correctly enough. When questioned as to the occasion of his crime, he expresses himself precisely as he has done in the documentary evidence, and it is not difficult to recognise in his open undisguised statements evidence of his educated and moral character. It is, he said, one of the most horrible crimes to set a man in opposition to those set over him, and this was precisely what St. attempted to do with him. Nevertheless, he could not give even me any special reason how it happened that he assaulted St. on the evening in question. He referred with a perfect recollection to the history of his former illness, and his opinions in regard to ordinary circumstances were in unison with his amount of education. There was not the slightest symptom of any fixed idea, or any evident derangement observed during his examination. It would, however, be erroneous *for that reason* to deny the possibility of the freedom of the will having been disturbed, not only at a bygone period, but also at present in Schmatte, as well as in similar cases. Those cases in which a disturbance of the mental functions prevails, so to speak, in the interior of a man without evincing itself exteriorly are so frequent, that every lunatic asylum contains more than one example, so that it does not seem in the least necessary for practical forensic medicine to embrace a doctrine so dangerous as the rightly notorious one of the supposed existence of a species of mental disease, termed "*amentia occulta.*" We feel all the more disposed, indeed justified, in assuming the existence of such a deep internal derangement when doubts are already thrown upon his mental condition by his antecedents, and the circumstances connected with the crime of which he is accused. This, however, is precisely the state of matters in regard to the present case. It is proved in the documentary evidence, that some years ago he laboured under a melancholic affection of the mind, and was dismissed from hospital uncured. That his subsequent conduct attracted the attention of his comrades, and that he was, therefore, supposed to be affected with 'silent madness' has been already pointed out. And in regard to these depositions, it is specially to be remembered, that they have reference to the period immediately preceding the commission of the crime, inasmuch as they have been emitted by his fellow-workmen in the W. workshop. The deed itself bears the impress of being the last link in an isolated chain of thought within the mind of the culprit. He himself, indeed, asserts, that he is irascible and easily excited, but not one of the witnesses has made any similar statement regarding him; neither is there in the evidence any actual proof of violence of temper or imbecility, whilst on the contrary Schmatte is rather described as peaceable, and it is officially testified, that in his native place he was in the best repute. When such a man, without the slightest provocation, flies at his neighbour, attacks him after a fashion which to ordinary minds seems dangerous to life, and, moreover, does this before witnesses, the strictest psychological judge must confess that in such a case the limits of healthy action are overstepped. In considering the case thus, we cannot omit stating that the apparent *causa facinoris* is not so totally wanting as the witnesses have alleged, for the accused himself confesses that he indulged a certain amount of animosity against St., and the words 'command and hoax' were also heard. Presupposing, nevertheless, that there really had been such an ill-will on the part of the accused against the accuser, his crime must then only be held to approach somewhat more closely the normal laws of voluntary action, for it requires no detail to prove, that even in such a case a man of perfectly undisturbed intellect would not have taken his illegal revenge before several witnesses, and would have been still less likely to do so, without some recent and immediately preceding irritation on the part of his opponent. Could there still be any doubt as to this, namely, as to whether the accused has acted thus during a fit of derangement, in a moment of morbid excitement acting on his latent and still uncured mental derangement, this doubt must be removed by the consideration of the behaviour of Schmatte after the deed; of course, I do not refer to his flight in itself, as an endeavour to escape the punishment he knew he had incurred, any more than it is to be regarded as a contradiction, that I speak of the recognition of a punishment incurred, involving as this does the consciousness of the illegality of the crime committed, since the psychological experience of innumerable cases of unquestionable and decidedly insane persons has proved a similar fact, and shown that an obscure intimation of right and wrong, of good and evil, occurs even amid the disturbance of the freedom of the will. The flight itself, however, from its peculiar concomitant circumstances, is well worthy of consideration. In possession of—for him—a considerable sum of money, which was more than sufficient to take him comfortably to his native place, he leaves Berlin without taking any money with him, and wanders about upon the country roads, sleeping out all night in October, and living on sloes, &c., so that his statement that he arrived at Hamburgh, which was not the desired end of his wanderings, he knows not how, is deserving of belief. If it were possible still to assume this to have been a lie or a simulation, this supposition must be at once set aside by the official attestation that he had actually arrived at his native place in a state of mental derangement for which he had been medically treated, but ' without much result.' His conduct and statements as above detailed, justify the assumption that even yet he has not been quite cured of his melancholia, mental affection." In my report, accordingly, I answered the judicial query by stating, " that the accused is not now, and was not at the time the assault was committed upon St., in the full possession of his moral faculties."

CASE CLXXVI.—MURDER OF A BOY DURING A FIT OF
MELANCHOLIA.

A short notice of this case, extracted from a very copious report,
may obtain a place here as a proof, not only that the most intimate
acquaintances do not recognise and dispute the existence of manifest
insanity, but chiefly as an example of the skilful premeditation,
even to the most minute details, with which the insane know how to
prepare for the execution of any illegal deed.   Johann Gnieser, aged
fifty-two, a small, feeble man, quite paralytic on the right side,
formerly a dealer in furniture, now living on his money, was " weary
of his life, and wished to put an end to it." He had an affection
for the son of one of his friends, and the boy was in the habit of
assisting him in his small housekeeping once or twice a week when
requested.   According to his statement, he made an attempt to cut
his throat—the scar of which I saw—but this failed; an attempt to
drown himself also failed, " because people were in the neighbourhood" (!).   He then took it into his head to kill the boy.   He strewed
some dominoes about the chopping block in the wood cellar, because
he thought that the boy, when he went into it with him to cut
wood, would stoop to pick them up, and as he could then " hit him
better," he would strike him dead from behind with the axe.   As he
thought, so it happened.   From his numerous, always identical
declarations, I quote the following, which gives an epitome of the
whole case :—" I was weary of my life, and wished to escape from
this world.   I could not take my own life, and in my sleepless
nights, in which I was constantly tormented with the desire of
escaping from this world, the thought struck me that I would
murder the boy H.   Yesterday at noon he came to me at my
request.   I had, previously, placed some fire-wood in a basket, and
laid my kitchen axe on the top of it.   H. went down first into the
cellar; I followed him.   I took up the axe, and as H. stooped to
pick up some dominoes I struck him with the left hand a blow
with the axe on the hinder part of his head, intending to kill him.
H. sank down, head foremost, breathing stertorously and groaning.
I saw that he was not quite dead; I gave him two or three more
blows with the axe.   Then I flung this away, went out of the cellar,
shut the door behind me, and went straight to the police, where I
declared what I had done,"—and, in truth, he entered the police

chambers with the word "I have killed a boy, and wish to be executed as speedily as possible!"—"I am sensible," he further deposed,
"that I have done wrong, but I could not do otherwise. The thought,
ought I, or ought I not? had made me so uneasy, that I thought if I did
it I would most speedily get out of the world. The boy had never done
me any ill, nor his parents either " (this was confirmed), "but I had
to take him, because I had no other. For three weeks I have intended
this. I also considered that Wednesday or Saturday would be the most
suitable days for the deed, as there was no afternoon school upon
those days." (The deed was actually committed on Saturday afternoon, about three o'clock.) In regard, now, to the statements of the
witnesses, his brother-in-law deposed that Gnieser had been formerly
much addicted to drinking. He had always appeared to him to be
"a very simple man," but never "insane or an idiot." Eleven days
before the commission of the crime, Gnieser showed his nephew a
note addressed to Gnieser, Rentiér, with the remark that he was
sometimes written to as furniture dealer and sometimes as Rentiér,
and if the "police councillor" D. knew that he bore two titles he would
have him taken up. According to the statement of the nephew he
had also other "foolish notions." The father of the boy never
observed any mental derangement about him, *not even on the day
the crime was committed*, when he came to fetch the boy. This was
also confirmed by the boy's sister. "He was quite calm, and just
as usual." On the other hand, his niece and the witnesses N. and
R., with whom he had much intercourse, and who never saw him
drunk, not even on the day the crime was committed, believed that
he was "not right in his head," because his "talk was often so
intricate and confused" that the witnesses could not understand him.
Moreover, they knew him only as a soft-hearted man, *who could not
bear to see any one suffering*, and, up to the last four weeks, cheerful
and good-humoured. In the course of his examinations, Gnieser had
behaved himself very remarkably; of this, I only give the following
as instances: in the middle of the examination he demanded what
o'clock it was, giving as his reason "because of half-past five, when
I am disposed to fall asleep." Another time he congratulated himself that the examination was put a stop to, and said, "at another
time he would be longer at their service." In regard to his share of
an obligation for one thousand three hundred and ninety thalers
(£208 10s.) it was quite impossible to get any distinct answer from
him, and it was equally impossible, from his constant digressions, to

carry on anything like regular conversation with him. And I found him precisely the same in the course of my repeated examinations;—" he carries himself bent forward, and is, as he states, subsequent to small-pox, and from his childhood quite paralytic on the whole of his right side, his right hand is atrophied and contracted, his right leg shortened, so that he limps and gives one the impression generally of a very infirm man. He confesses that his bodily health is good, and this is confirmed by examination. There is nothing anormal in the conformation of his skull, except that the posterior part of the head is somewhat flat. The glance of his blue eye has, as well as his whole physiognomy, something winning and good-humoured about it. He usually carries his head bent forwards, and only at times looks up. It is very difficult to keep up a conversation with him. His usual answers are, Hm!—yes, and no. At times, when he thinks he has said something incontestable, for instance, that it is easy to die on the scaffold, he raises his head, and speaks in a tone of the deepest conviction," &c. It is not unimportant to state that G., suddenly in the midst of a precognition, brought forward the assertion that his relations with the boy were unchaste and that he killed him lest he should let out the secret. But soon thereafter, and consistently in all subsequent precognitions, he declared that the priest had exhorted him to confess whether sensuality had any connexion with the crime. He then thought, "I must give some reason, and then things will go on faster, and death will be the sooner attained!" The father of the boy, when questioned in regard to this, had decidedly denied the existence of any such connexion, declaring that from his openness his son would, certainly, not have concealed it from him.—In my report, I laid down the views explained in this chapter, and from them deduced the opinion "that Gnieser could not be regarded as a responsible agent either generally or at the time of the commission of the crime."—He was sent to an asylum, in which he died.

CASE CLXXVII.—SEVERE INJURY OF A CHILD.—SEEMING "TRANSITORY MANIA."

The perpetrator of the peculiar "severe injury," described in Case CXVIII. (p. 31, Vol. IV., Deprivation of Speech), was a school-master, who had abused a little girl in school with the greatest rudeness and passionate violence, and was, therefore, brought to

trial. The trifling fault of the child seemed to be so disproportionate to the behaviour of the accused, that the defence of sudden
mental disturbance was all the more readily set up that it was
certain that the accused had been seven and three years previously
treated in the lunatic ward of the Charité Hospital for actual mania,
regarding which, it was recorded in the sick reports of that institution, "that this morbid mental excitement was chiefly the result of his
extremely violent and passionate disposition." The case, accordingly,
was certainly not a usual one. Was the ill-treatment perpetrated
during a renewed attack of mania, or was it only the result of the
well-known violent and excitable temperament of the accused?
Moreover, there was no other testimony obtainable in regard to the
circumstances immediately preceding and accompanying the deed but
that of a few young school children. "The circumstance," we state in
in our report, "that M. has twice already laboured under mania,
must certainly excite suspicion, since it is well known, on the one
hand, how easily mania is re-excited by trifling excitement, even
when recovery is apparently perfect, because it only slumbers and
has not been wholly removed, and, on the other hand, it is also well
known that this complaint, even when perfectly cured, readily relapses.
In the present case, however, I cannot assume that either the one or
the other has taken place. That M. was actually perfectly cured of his
mental derangement is proved by the circumstance that he had been
restored to his position as civic teacher, and there is no proof in the
documentary evidence that during the long period of three years,
passed in this position, he had ever displayed any trace of mental
derangement. On the other hand, the deed of which he is accused
does not display the character of a maniacal relapse; because such
an attack, once occurred, would not have instantaneously disappeared,
but would have lasted for a longer or a shorter time, which was not
the case. Further, the manner in which M., immediately and with
a clear judgment perceiving the wrongfulness of his action, urgently
and repeatedly begged the injured girl to say nothing of her ill-treatment to her mother, proves that he did not ill-treat the child during
an attack of mania, but solely by being carried away by his passionate
temperament. Finally, as to his present mental condition, in it, also,
there is not a trace of mental alienation to be discovered. He is
perfectly calm, perfectly collected, he endeavours, as he did on his
trial, to palliate his deed, and denies all gross ill-usage of the child,
exhibiting himself as in every respect a perfectly conscious and

mentally sound individual." Accordingly, the question put before us had to be answered thus,—"that M. is to be regarded as
responsible for the ill-treatment of the girl Elisa of which he is
accused." M. was condemned, was not replaced as a teacher by the
Board, and for the last six years no similar complaint has been
lodged against him.

CASE CLXXVIII.—SUDDEN TRANSITORY MANIA, PRODUCED BY
SOMATIC CAUSES.

An extremely interesting case. The accused was the wholly irreproachable shipowner D., aged twenty-nine, a man of whom the witnesses at the trial, to which I was cited to give my opinion, unanimously
testified that he was perfectly staid and peaceable. He was accused
of destroying the property of others and of assaulting the officers of
justice. Very early on the morning of New-year's day 18— he had
gone into a beer-house and drunk a cup of coffee; no one present at
the time remarked that he was at all tipsy. After remaining quite
quiet for a time, he sprang suddenly up, ran into the kitchen to the
girl there, and declared to her that he was the devil, Satan, that she
must do his bidding and come at once into the public room. Then he
returned to the room, commenced quarrelling with the guests, broke
the seats, and attacked the host with the leg of one of them. He not
only insulted the constables, who were called in, with his tongue,
declaring that they had nothing to do with him, that he was emperor,
the sole emperor, &c., but he also actually assaulted them, and in
particular struck so forcibly against one of the helmets that he bent
its spike. He was bound, during which he still behaved himself
furiously, and he was then brought to prison. After he had slept
he was perfectly quiet, and declared that he had no recollection of
the previous night. In the preliminary precognition and at the trial
he asserted that he was troubled with fits of blood to the head,
particularly when he *heated his cabin with coal or peat*, and it has
thus happened that when he went out he had to hold on by something to keep himself from falling. On the last night of the year
he had again heated his cabin with coal, and sat in it reading a
romance of chivalry, and thenceforward he had no remembrance of
this night. I declared at the public trial that the case was susceptible of a three-fold explanation: a passionate or evil disposition, &c.,
but this, however, according to the evidence, could not be assumed

to exist in the accused; or design and simulation, for which, however, no possible motive existed, and against which, also, there was the fact of the remarkable development of muscular power which he had exhibited at the commencement of the attack; or, finally, the sudden occurrence of some mental derangement. In support of the latter supposition I do not require to go back upon the fancied existence of a peculiar species, adopted by many, *mania transitoria*, because the case itself presents sufficient points of support in favour of the supposition of a sudden outbreak of mental derangement. In favour of this we have the existing predisposition of D. to cranial congestion, and the fact of his passing a night in a small close cabin, filled with the vapour of burning coal with its well known narcotic action. Considering this circumstance, as well as the isolated nature of the deed of which the prisoner is accused, his character, the absence of any motive, &c., I must assume that D. at the time of the commission of the deed was in an irresponsible condition. The public prosecutor, thereafter, let the case drop.

CASE CLXXIX.—ATTEMPT AT MURDER IN A DOUBTFUL MENTAL CONDITION.

Another very remarkable and unusual case, since it affords a new and striking example how necessary for the foundation of a medico-legal psychological opinion is an acquaintance with the documentary evidence as well as one's own occasional observations (p. 187, Vol. II., and 124, Vol. IV.), and also another remarkable proof how cleverly lunatics can, for years, conceal and suppress their insanity, and finally, how, just for that reason, a perfectly erroneous opinion may be arrived at in the absence of any information in regard to the former life and conduct of the party in question. I shall relate the case chronologically, as it was gradually developed, especially as concerns myself, as will be seen from the two following reports.

I. *Report of sixth August*, 18—.—"On the twenty-first of July of this year, the accused, the cook H., had a violent quarrel with his mother, who was alleged to have been drunk; she abused him, and struck him on the hand with a stewpan; upon that, he seized a loaded pistol, nominally only to defend himself. His brother, who now arrived, considered it necessary to summon two policemen, and as one of them proceeded to arrest H., he called out 'I will shoot dead whoever touches me.' In fact, he immediately plucked the pistol from his own breast, aimed it at that of the policeman, and fired; fortunately, the cap alone exploded and Schmidt remained uninjured. At his precognition, as well as to myself, H. has disputed the whole of this statement, and constantly asserts that he only kept the pistol in his hand in self defence. His responsibility is disputed, specially because he was treated for mental derangement in the Charité Hospital seven years ago. According to the sick reports he had become insane in consequence of his bankruptcy, and was received into the lunatic wards as 'maniacal.' Previous to his admission he fancied himself to be prosecuted as a political criminal, he uttered the most violent threats against his prosecutor, and armed himself, showing at the same time a disposition to religious fanaticism, prayed much, and read the Bible the whole night long, &c. It must be remarked, however, that this conduct was observed previous to his admission into the asylum, while in it from the very day of his reception, it is recorded 'there were, indeed, no morbid phenomena observed in him,' and during the whole of his five months' residence there reference is only made to H.'s peculiar perversity of disposition. In the end of August he left the asylum, never returning after obtaining leave of absence for one day. Since this time neither his mother nor his brother have ever observed any insanity or outbreak of fury in him; the former only says that he seemed to be sunk in 'religious melancholy,' since he constantly read much in the Bible and went to church three times every Sunday. The culprit is thirty-six years of age, of good bodily health, respectable demeanour, calm, proper behaviour, free, open look, and displays nothing remarkable in his exterior. And the examination has also revealed not the slightest trace of anything anomalous in regard to his mental condition. No delusions could be discovered either in regard to religion or anything else; his thoughts are logical, his perceptions distinct, his answers consistent, his memory unenfeebled, and he knows how to avert cleverly the accusation from himself. It is evidently no proof of religious melancholy that he, as he confesses, still occupies himself with much pleasure in reading the Bible. Even though he has been, seven years ago, certainly for no long time, actually mentally deranged, it does not follow that he still labours under mental disturbance, and this all the more that in the long interval that has elapsed from that to the present time no new outbreak has been observed. The deed also, of which he is accused, and his behaviour at its commission, however remarkable and unusual both certainly were, do not exhibit the characteristics of insanity, and no insane expressions were heard either by the two police officials or the examining judge, since to the former he 'appeared to have the full use of his reason,' and the latter found him to be 'perfectly responsible.' He himself said to me that he was 'violently excited' on the occasion referred to, and he thus gives the key to his behaviour at the time of the commission of the deed." Accordingly, in the absence of any opposing facts and in consonance with my then knowledge of the case I could not give any other opinion than "that H., neither at the time of the commission of the crime, nor at present, has suffered or suffers from insanity or idiocy (§ 40, Penal Code), but is to be regarded as a responsible agent."

II. Eleven weeks after the giving in of the foregoing report, H. was brought before the jury court. The advocate for the defence brought forward the plea of irresponsibility, and supported it by handing in documents in the hand-writing of the accused. These documents were laid before me in December. They consisted of a tolerably large fasciculus of letters, some without any address, and others addressed to ladies of exalted position, and of sheets of rhymes full of the greatest nonsense, and the filthiest and most repulsive obscenity! Fortunately most of them were dated, and it was thus ascertained that these documents had been written during the last five or six years! Upon the 10th of December I gave in the following second report :—" In my report, dated the 6th of August of this year, in accordance with the then state of the documentary evidence and the results of my own personal examination of the accused, such as I then found him, I was obliged to come to the conclusion, that H., neither at the time of the commission of the crime, nor at present, has suffered, or suffers from insanity or idiocy (§ 40,' Penal Code), but is to be regarded as a responsible agent. The deed of which he was accused certainly bore the impress of a furious burst of wrath ; since, however, the limits between mere passion, always a responsible affection, and actual mental derangement are so extremely difficult to define, the latter cannot be of itself assumed as an explanation or efficient exculpatory plea for any illegal action, except when there is other proof of the existence of actual mental derangement previous to, at the time of, or subsequent to the commission of the deed. And the forensic physician dare not deviate from this fundamental dogma of a sound forensic psychology. At the time my report was drawn up, however, no proofs such as those referred to were known to exist. H., indeed, had, seven years previous to the commission of the deed, been received as a maniac into a lunatic asylum, but even this fact lost its significance by the consideration, that in the asylum itself, as I have related, the physicians had not observed any actual disease, and what must appear extremely important, that also during the last seven years, and up to the very time of the commission of the deed, neither *his own family* nor *his master's* had ever again observed any actual mental derangement in him. Moreover, in the course of my examination of H., he did not exhibit the slightest trace of insanity, I therefore thought that I dared not come to any other conclusion than that already detailed. Subsequently, as is well known, new facts have been added to the documentary evidence, and have come to my knowledge, which give *quite a different appearance to the whole matter*, and by a new and remarkable instance have confirmed the old experience, that those indubitably insane know how to keep secret and conceal their complaint, often with the most wonderful consistency, from others, even their nearest neighbours, and from physicians, &c. The documents in the hand-writing of the accused now made known, which are dated during the last few years up to within a few days of the commission of the deed, afford the most indubitable proof that he labours under periodic insanity. These epistolary effusions of a deranged brain, which were so secretly preserved by him, contain, however, also proof that *H. was certainly, only a few days before the commission of the crime, involved in one of his periodic attacks of insanity*, and this places in a different light the formerly certainly remarkable character of his crime, as a motiveless, furious outbreak of passion, and for the reasons already given, this cannot now be regarded as merely a burst of anger in a man of sound mind, but violent passions, whilst his family never observed anything remarkable about him, except that he read much in the Bible, and spent whole nights in doing so; whilst it is also to be assumed, that in the course of the discharge of his household duties he never behaved himself as one actually insane. H., as we now know, for years, at certain times wrote continuously, secretly, and for himself alone, those insane letters and poems already referred to, whose revolting and low sensual style particularly, would, had it come to the knowledge of his respectable family, have certainly drawn upon him the most deserved reproofs. *Consequently he was insane for years without appearing so.* That this is the case still might be assumed, *à priori*, since for the last seven years no attempt has been to cure him, and such complaints are not usually cured by the powers of nature alone, without the somatic and moral curative discipline of regular medical treatment. But an apparently trifling, but very instructive trait, dating from the most recent days of H.'s prison life, proves distinctly, *à posteriori*, that this mental derangement *still subsists*, though now, as formerly, he is able, in some measure, to suppress it. I refer to a letter written from prison to his wife, and dated the 29th of November of this year, which has just been added to the documentary evidence. In this long communication H. describes his position, inquires after his child, speaks of the state of the inquiry concerning him, trusts in God, &c., in clear and connected phraseology, and then suddenly in the middle of his letter he writes, 'Ah! think of me, think of me, for I am a poor rhymer; men often called me a poet, in prison I am kept on a diet'—'farewell, sweet life, God and heaven will forgive, thou shalt not steal nor commit adultery (!?), the Lord of heaven will avenge me '*—and immediately thereafter he concludes, subscribing himself 'Thy husband, languishing in prison.' But few words are required to characterize the events in the mental life of such a man. Such statements in the letter of a prisoner, who must be aware that a serious charge is impending over him, would appear extremely suspicious, even if he had never previously been suspected of mental derangement; but, how much more must this be the case in regard to the accused, in whom they actually supply proof of the present existence of insanity within him. In accordance with what has just been stated, I now, after obtaining more correct information, give it as my opinion, that H. has laboured under insanity (§ 40, Penal Code) for many years, that the time when the deed was committed fell within a period of insanity, that he is not yet free from his insanity, and that consequently he neither was a responsible agent at the time he committed the deed, nor can be considered responsible even yet." In consequence of this opinion, the charge was departed from.

§ 78. INSANITY.—CONTINUATION.—LUCID INTERVALS.

STATUTORY REGULATIONS.

GENERAL COMMON LAW, § 20, TIT. 12, PART I.—*Persons who*

* In the original German these unconnected sentences form a sort of doggrel rhyme.—TRANSL.

are only occasionally deprived of the exercise of their reason, may make valid preparations for death during a lucid interval.

IBIDEM, § 147.—*If the Judge be aware that the testator occasionally laboured under a want of reason, he must be fully convinced that the testator was in full possession of his reason when he wrote or dictated his testament.*

IBID. § 148.—*If this should be doubtful, the Judge must consult an expert; if the matter permits of no delay, the Judge must undertake the matter, &c.* (respects the construction of the protocol).

GENERAL JUDICIAL REGULATION, § 9, TIT. 3, PART II.—*Persons who only occasionally, and with certain intermissions, labour under a deprivation of reason, must not in general be permitted to enter into onerous engagements, but must be placed under trustees. When, however, a peculiar case arises, in which such a person has, during a lucid interval, entered into such an engagement, and the matter is so urgent, that the formality of placing him under trustees cannot be waited for without subjecting him to personal loss, the Judge must fully convince himself, in every case, with the aid of a physician, that the contracting party is actually at present in a lucid interval, and that his reasoning powers are sufficiently clear to form a proper conception of his action.*

(FRENCH.) RHENISH CIVIL CODE, § 489.—*He who is of full age, and who is usually* (en état habituel) *in a state of idiocy, insanity, or mania, must be interdicted, even though subject to periodic lucid intervals* (*according to § 496, the party requiring it will be examined by the Judge, with the assistance of a judicial clerk*).

The second difference in the form in which insanity occurs, besides that in regard to its mode of origin (§ 76), respects the course of the disease. In regard to this, insanity is, at one time, and in most cases, continuous, and at others intermitting, that is, its attacks alternate with periods during which the person formerly insane actually, or at least apparently, recovers the free use of his reason, only to relapse presently into insanity. Of course, the fact of insanity running such a course is of decided importance in regard to the practice of (law and) forensic medicine, in so far as the question is forced upon us whether, and how far, any one affected with insanity can be made legally responsible for any civil or criminal action committed during the period of such a *lucidum intervallum ?* The solution of this question, which is one of the most difficult when abstractly considered, has been rendered easy by generalization. Experience shows, it is said, that an insane person, even when apparently clear and collected, always has delusions concealed in the background, which break forth upon a suitable occasion and put an end to the lucid interval. *Ergo* an insane person, even during the lucid interval of his disease, is still to be looked upon as an insane person, and regarded as such "forenso-psychologically." Inversely, it has just as often been declared that when a man, though at other times subject to attacks of insanity, shows himself to be at any particular time, in a certain degree, free from mental derangement, as is, indeed, conveyed in the expression "clear or lucid interval," he must then be regarded as responsible for his actions during that time. A single glance at the ambiguous regulations of the Prussian Statute Books given above, shows that the legislature does not always regard this question from one and the same point of view. Whilst, according to the Rhenish Code, any one only periodically insane must be placed under trust like any other imbecile, lunatic, or maniac; in the rest of the monarchy a valid will may be made during a lucid interval, and even onerous engagements may, according to the judicial regulations, be entered into under certain circumstances, whilst "in general" this is not permitted; but he must be placed under trust! The Prussian Penal Code, and, so far as I know, all the other recent Penal Codes, take not the slightest cognizance of lucid intervals or of periodic insanity as such, but only require proof of insanity (or idiocy, &c.) at the time of the commission of the criminal action. A middle way is taken by the Hanoverian Penal Code in the regulation. Art. 83, *sub* 2, "All those are exempted from criminal punishment who labour under any derangement or disease of the mind, by which they are deprived of the exercise of their reason. If the crime be perpetrated intentionally during a lucid interval, the condition referred to can only be regarded as a mitigating circumstance, the punishment, however, cannot be inflicted upon those who have relapsed into the condition referred to." The English law declares, according to Knaggs (*op. cit.* p. 53), that when a lunatic has lucid intervals he is responsible for what he does during such interval just as if he were always sane.—But these difficulties refer more to the legal than to the medical view of the case. The physician, on the other hand, has other and very important ones to treat. No one will dispute the fact that when a (former) lunatic has been thoroughly and permanently cured, when he has actually been placed in the *status quo ante,* he is then to be treated precisely like any other person of sound mental health, just as a man may have at one time a bodily disease which may entirely disappear, leaving no trace behind. But when is a lunatic thoroughly and permanently cured? When does the time arrive when we have no longer a doubt lest the disease be only remittent, and the patient only enjoying a lucid interval? Asylums, which have dismissed their patients too soon as "cured," know what to say as to relapses! It is not for nothing that in large well-managed institutions convalescent wards have been established, like quarantine ports, in which those apparently cured may be for a long time subjected to a strict discipline and keen observation before they are restored to freedom. For experience teaches us that there is scarcely any diagnostic criterion whereby the actual cure may be distinguished from the mere slumbering of insanity during a lucid interval. It is important and *almost* sufficient when the party examined can now recognize his delusions as such, speaks of them with calmness, and acts accordingly, whilst in the opposite case we recognize the pseudo-lucid interval. In investigating the mental condition of lunatics, and those who have been such in those cases of daily occurrence where the institution or removal of a trust is treated of, ample opportunity is afforded of observing individuals exemplifying both conditions, the true and the pseudo-lucid interval. But I say, *almost* sufficient, because even the most experienced may be deceived by the cunning of such men, who, I again repeat, know well how to conceal the delusions still luxuriating in their minds in their endeavour to attain their ends, *e.g.,* to get out of an asylum or to escape from under trust, the more they have recovered from the general excitement of their attack of insanity and the more they have in general attained a certain amount of collectedness. Burrows, whose power of observation cannot be denied, dismissed a young lord from his private asylum, who, for months, had appeared to be cured of his attack of mania, and who wrote the most rational letters to his mother, &c. Even at his own country seat he behaved himself rationally for a long time, till one morning he got up early, ran into the neighbouring village and returned with torn and dirty clothes. His mother gave him a slight scolding, when he seized the tongs from the fire-place and struck her dead! A Prussian nobleman, who had become insane from unbounded pride, and had been long under treatment in a famous private asylum, was dismissed, apparently cured. Immediately after his arrival in Berlin he paid me a visit, and in the course of a long conversation I found him to be quite another man from what he had formerly been, composed, rational, and perfectly clear. Accidentally, he unbuttoned his overcoat, and on the frock coat beneath I espied a star of pasteboard and gilt paper, the "order," which he said, laughing self-complacently, and instantly quite altered, "he had received on account of his relationship with the Hohenzollerns!" This was no case of cure, not even of a true lucid interval. Experience has also, indisputably, proved that the diagnosis cannot be based upon the duration of the apparent lucidity. Mere lucid intervals, without actual cure, may last for a longer or shorter time, and relapses frequently occur after the lapse of a very long time.

These difficulties often present themselves, in a most important manner, to the physician engaged in examining these judicial cases, in which the civil responsibility of the party is concerned (§ 58), and the careful consideration of all the circumstances of the case is all that can be recommended for their guidance. In these cases, however, in which the criminal responsibility of a lunatic is disputed, in which he seeks to avert the consequences of an illegal deed, which the witnesses for the prosecution testify to have been committed in a state of perfect mental integrity, by asserting, or permitting the advocate for the defence and the physician to assert, that he was only in a lucid interval—in these cases, I say, the difficulty is of far less importance; because in such a case the forensic physician, in accordance with the almost universal regulations of the Penal Code, and with the nature of the matter itself, according to which he has to determine "whether the agent *was insane* (*or imbecile*) *at the time when the deed was committed?*" has only to keep the individual case, and this alone, steadily before him, and when he has examined the deed and the agent by the aid of the general diagnostic rules already given (§ 61, &c.), he will generally be able to discover, if not always with certainty, at least with more or less probability, whether the deed has been committed in a state of mental derangement, or with the perfect freedom of the will. When the latter has been proved with more or less certainty to have been the case " at the time of the commission of the deed," it may then be left to the Judge to say whether the former existence of mental derangement *previous* to "the time of the commission of the deed" is to be regarded as a palliative circumstance or not. Finally, experience teaches us in regard to the whole question of a lucid interval, what

I have never seen stated, that, practically speaking, it is, in so far of little importance that it is but seldom mentioned *in foro*. At least, of hundreds of criminal cases which I have reported on, I have never had to do with one in which the question of a lucid interval came in question at all. In criminal cases the accused, or their advocates, in cases at all suitable, go much farther, inasmuch as they assert that they were mentally deranged at the time of the commission of the deed, or, when they can make the former existence of mental derangement appear at all credible, they proceed from this premise to allege further, not that upon this day in question they enjoyed a lucid interval, but, rather, that since that time "they have never been quite right in the head," &c., and thus the case becomes an ordinary one of disputed criminal responsibility.

§ 79. CONTINUATION.—FIXED DELUSION.—MONOMANIA.

There is, finally, a third variety in the form of insanity besides those presented by its mode of origin and course (§§ 76, 78), viz. that caused by its psychical limitation. We accordingly distinguish, perfectly agreeably to nature, between general insanity (madness, craziness, *amentia, dementia*) and the merely onesided, psychically limited, so-called fixed delusion, or fixed idea, partial insanity. In the one form the logical chain which embraces the faculty of thinking is ruptured, and the unshackled thoughts and ideas press and throng one another in unceasing violence, self-consciousness is deranged (§ 73), perpetual phantasms influence the conduct and actions of the patient more or less tyrannically according to the degree of the development in each individual case, so that, when in but a low degree, he is still able to conceal his delusions. Where the delusion is limited or fixed, on the other hand, the mind is only trammelled by one single delusory idea, or by one small class of connected delusions, whilst in every other respect there is no want of normal activity, so that beyond the limited circle of his delusion, the man does not merely seem, but actually is, rational. The French physicians consider the term monomania etymologically quite distinctive of a fixed delusion. But this appellation has been already much employed, both by them and also in other countries, in a much wider signification, inasmuch as it has been employed to denote certain varieties in the character of insanity as well as, and in particular, the so-called "impulses" (§ 88), so that an erotic, a religious, a homicidal monomania, or a monomania of persecution have been spoken of. The literature of the science swarms with examples of monomania, in its restricted sense, the fixed delusion, very naturally, since it is of extremely frequent occurrence in actual life, indeed, much more frequent than is usually assumed, if all slight, easily controlled, anormal ideas—fancies to which the mind has, by degrees, become accustomed, and which by the law of the association of ideas continually reassert themselves, if all so-called " whims, caprices," and the like, are to be reckoned as fixed ideas, as they must. What else was it that made Kant only able to speak fluently from his professorial chair when he could fix his eyes upon the head of one of his students sitting in a certain spot, and made him unable to go on if that place was by chance empty ? Similar facts are known of various men most renowned in arts and science. But the fixed idea may have the actual stamp of a true insane delusion, not merely of a caprice, and yet the integrity of the mind may still continue. The fixed idea of that unfortunate young man, whose history I have elsewhere related,* was, that he continually blushed, and was thus rendered remarkable and exposed to ridicule; this idea had affected him from childhood to his twentieth year, nevertheless, he had passed through all his examinations, &c., satisfactorily, yet at length it mastered him and drove him to suicide. Of two other men, whose physician I also was up to their old age, the one had the fixed idea that he was dangerous to others and that he must, therefore, avoid all contact with them. I, myself, have often seen this good-humoured, moral, amiable man, who was unmarried, and lived with another family, carefully empty out all his chamber utensils before leaving in the morning, so that not a single drop might remain by which the family of his host might be poisoned. I have seen him go far out of his way if he saw any one coming with a child, preferring to avoid them than to cause any mischance to them. At the same time he was a respectable merchant, an able trustee, &c. The other, a subaltern official, had from his youthful years laboured under the extraordinary " monomania," that whenever he saw whips hanging out at a belt-maker's he was immediately seized with sexual desire which had at once to be gratified !† Another example not yet too well known, is

* Denkwürdigkeiten zur medic. Statistik u. Staatsarzneikunde. Berlin, 1846 ; "Biographie eines fixen Wahn," S. 165.

† Hoffbauer (op. cit. S. 351, 353, and 362) makes out of such cases not only a new species, which he calls " blind psychological mastering impulse ;"

too remarkable not to be related instead of a hundred others. An Englishman left his property to his housekeeper, with the stipulation that she should cause some parts of his bowels to be made into fiddle-strings, that others should be sublimed into smelling salts, and that the rest of his body should be " vitrified" and made into optical lenses ! He added, " I know that the world will consider this to be done in a spirit of singularity, but I have a great aversion to funeral pomp, and I wish my body to be converted to useful purposes." The will was disputed, but was judicially held to be valid, since it was proved that the testator had been always accounted a man of sound reason, and an admirable merchant, &c.* The English Judge, accordingly, recognised the complete civil responsibility of one who was partially insane. This question, as well as that of the criminal responsibility of such men, is of frequent occurrence in forensic practice, and has very often engaged my attention. The supporters of the untenable and inflexible theory that all mental derangements take their *exclusive* origin in the sins and passions of mankind, desire to exclude from criminal responsibility even those criminals who are only partially insane, because it cannot be ascertained how far the emotion passing beyond the limits of the fixed idea may have sympathetically involved the healthy faculties of the mind. The same doctrine is taught by the partisans of the lax practice, whose unscientific and dangerous ultra-philanthropy I have already had repeated occasion to controvert; finally, similar views are also maintained by the most recent psychiatric authors, who are more and more inclined to regard every mental disturbance as necessarily produced by somatic causes. Psychological experience, however, the only certain guide, gives us here the right answer in our hand. She shews us thousands of cases, like the above, in which not the slightest general mental reaction was ever observed during the whole life of the individuals affected, she shews us others in which the reverse is the case, in which the individual is hurried on by his partial insanity to actions which bear the decided stamp of insanity. And the result .

not only does he reckon other fixed ideas, such as when a man takes into his head that he must cut his throat with a razor, under the head of another species, " blind impulse to an action," but he also employs another case, in which a man, living in happy circumstances, formed the intention of drowning himself, and actually did drown himself, to construct a third species, saying, " I will designate *this* case with the term, ' sudden enforced intention !'" How many imitations Hoffbauer has had in this mania for generalisation !   * Knaggs, *op. cit.* p. 48.

of the examination of these cases is this, that so long as the individual is able to recognise the fixed idea which fetters him to be such,
so long as he is able to master it, though he may be unable to root
it out, just so long as the restraining reason overrides the fixed idea,
the individual is in general perfectly responsible both civilly and
also criminally, and of this there are thousands of examples. The
diagnostic criterion of such cases is also very simple, such individuals
can endure the mention of their fixed idea. They confess it, laugh
at it, and joke about it themselves, as is very often seen, but they
cannot part with it. When, however, as only too often happens,
the fixed idea has rooted itself more deeply in the mind, as is
specially the case when it is not a mere fancy, such as the thought
of mutilating one's self, but springs from the basis of some emotion,
vanity, dogmatism, jealousy, &c., when it grows, and is always more
and more nourished by this emotion, when in other cases the limited
mental disturbance arising from somatic anomalies increases *pari
passu* with the increase of the bodily disease, when, finally, it hurries on
the patient to illegal actions, *undertaken from its own point of view,*
this is sufficient proof that the patient has ceased to have the mastery
of his fixed idea, and that, on the other hand, it has got the mastery
of him, and he who was formerly only partially insane, must now be
declared to be, as he actually is, wholly insane. Such patients,
however, cannot endure mention to be made of their delusion without at once reacting morbidly. A servant of the court of Justice of
this city, who was officially testified to be "a quiet, diligent, and
accurate man, fully capable of the discharge of his duty," and was
daily occupied with the delivery of a large number of letters and
legal documents, had for seven years been possessed with the fixed idea
that he was successor to the throne of the country. At the last
change in the monarchy, being convinced that he was "supplanted,"
he continued to wait for years in calm resignation till the throne
should again become vacant! At length, he commenced to hand in
the most extraordinary letters, and his mental condition .had to be
enquired into. At an examination, during which he appeared perfectly rational, explaining to me the extent of his business, &c., I
intentionally made use of the expression "most high command,"
which ought to exist in relation to the messenger of the royal court
of justice. Immediately his behaviour changed, he was put out and
restless, "that is all stupid stuff, he only was entitled to give his
most high commands," &c. (compare this with the cases in the

following paragraphs).   However much I am convinced that this explanation of the two forms in which partial insanity occurs is psychologically correct and practically advantageous, yet I must again repeat, even in regard to the fixed idea, *that the investigation of each individual case according to the general diagnostic rules* (§ 61) *is a most important matter*, and its results will be found to be in unison with the views here laid down.

The maintenance of this dogma brings out as vain and practically unprofitable all such questions as, is any partially insane individual to be held responsible for any illegal action unconnected with his fixed idea? *e. g.*, the merchant above mentioned for a forgery—as well as all similar abstract questions such as, are deaf mutes civilly responsible or not? because such abstract questions may be just as properly answered affirmatively as negatively.   Each can only be decided according to its own peculiar circumstances.

The psychical *longings of pregnant women* also belong to the category of fixed ideas, in their inner nature they are nothing else than fixed delusious which the pregnant women, as experience teaches, can master very successfully, but which may also hurry them into illegal actions, undertaken from their own peculiar point of view.   Here also a due consideration of each individual instance will throw light upon the case (*vide* the remarkable Case CCIV.).   I do not require to point out that the forensic physician must in all such cases be on his guard against the mere simulation of the longings of pregnancy, and also, that pregnancy itself is not to be looked upon as a license to commit offences and crimes.

§ 80. Illustrative Cases.

Case CLXXX. — Mastering of the fixed Idea.— Civil Responsibility.

The shoemaker, N., was declared to be an "idiot" in the common law sense of the word, that is, "unable to consider the results of his deeds," and placed under trust on account of religious insanity, subsequently he moved for its removal, and was brought to me for the examination of the then state of his mind.   He was fifty-six years of age and in good bodily health.   His religious delusions had by no means ceased to exist, but they had not the same power over him as formerly.   Being repeatedly interrogated as to them, he declared that he would keep all "these things" to himself, that he was convinced that he ought not to speak of them to anyone, and he would take care for the future not to do so, since the reverse had brought upon him his present misfortune,—thus exhibiting a relative amount of cure. Nevertheless, I, of course, continually endeavoured to recur to these delusions, and succeeded too, though not without occasional opposition from the party examined. Thus, in answer to my questions, he detailed to me, as he had formerly done, the history of his death, which was alleged to have occurred some time previously, and declared in an earnest and solemn tone of voice how that God was in the habit of appearing to him from time to time in the form of a reverend old man. How completely penetrated he was with the idea of his personal relation to God is shown by the following :—I found upon his table, partly in his own handwriting, and partly in that of his wife, a number of sheets full of extracts from the public newspapers, and solely relating to accidents, fires, floods, death from lightning, and the like. When asked concerning them, he replied in an elevated, solemn, singing tone of voice, that when he was declared to be imbecile, he had declared that God would punish the world for it, and the newspapers had given him numerous proofs of the correctness of his prophecy, which he had collected for his own private satisfaction, without in any way bringing them forward. When the erroneousness of his train of thought was pointed out, he confessed that it was not only on his account, but also for their own sin, that God punished mankind so often. "It seems superfluous to quote any more facts from the inner life of N., since those already given indubitably prove that he is still affected with religious insanity. However, he never persists long in conversing on this subject, nor does he mix it up with other matters, but dismisses the subject at once with unconcealed design, and then talks rationally and connectedly of other matters, such as of his former trade as a shoemaker, or his present employment as house-agent. In the latter respect I have found in him an amount of practical knowledge, quite unusual in his station, and everything relating to which has been placed before me in writing has been such as you would only expect from a man in his position and a responsible agent. He has obtained a knowledge of the statutes bearing upon his business, and of judicial forms, &c., and memorials and complaints, &c., which he has shown me in his own handwriting, are, though bad both in orthography and style, yet composed in such a manner that no one would ever suppose the author to be mentally deranged. According to all that I have stated, it follows that the religious delusions of N. are at present confined to a limited sphere ; that he has the mastery of them, and that at present they only exist as fixed ideas. Therefore, it is nothing remarkable, and only in accordance with ordinary experience, that outside of his fixed ideas he exhibits himself as a man in perfect mental health. In order to obtain proof how N. would behave should his fixed delusion come into collision with his life and actions as a citizen, I put the question to him, How would he act if, for instance, in drawing up a purchase contract God should appear to him, and command him to act otherwise than the interests of the parties concerned, and his own, seemed to require ? He earnestly replied to me that this could never happen, as God only admonished him to gain his living honestly and uprightly, but otherwise, while acting thus, lets him do as he likes, and for this he gave an example relating to the purchase of a house, which, if it be true, does his character honour. It deserves, moreover, to be considered that it can scarcely be seen how his delusions, which only affect the deeper thoughts of the mind, can have any disturbing influence upon the business which he carries on, and that there is no recent evidence in opposition to the statement of N., that he now confined his religious delusions in his own breast. According to all that has just been stated, N. has not presented himself to me as one 'wholly deprived of the use of his reason,' nor is he ' incapable of considering the results of his actions.'* He is, therefore, so far restored that his trustees may be removed." This took place seventeen years ago. N. has not since come before me, and from that I conclude that my opinion was correct. A complete series of similar cases of disputed civil responsibility, where there were fixed ideas, were decided on the same grounds, and with the same results.

CASE CLXXXI.—MURDER OF A SUPPOSED RIVAL.

On the 15th of September, Hoffman stabbed with a bread-knife his neighbour and acquaintance, the labourer Hundt. At his first examination he declared in exculpation that he had only had a scuffle with Hundt, who had " seduced his wife, and committed fornication with her," a disgrace which he could not endure after

---

* That is, neither a lunatic nor an idiot, according to the definition of the General Common Law.

having begotten ten children with his wife during twenty-eight years of a married life. He had, indeed, never seen his wife together with Hundt, yet "he would swear ten oaths, that they had both committed whoredom, and there was every proof of it." In consequence of this suspicion of an adulterous connexion between his wife and Hundt, Hoffman had often, especially lately, had quarrels and disputes with the former, and he confessed at his first examination that on the day the crime was committed he had a fight with Hundt and had given him a blow with his fist, but this was all he had done; in particular, he had not stabbed him. "No man can come forward and say that I had anything in my hand; if he swear so, he swears falsely." When it was represented to him that Hundt had died of his injury, he replied, "So, he is dead, then he must have stabbed himself dangerously; if he be dead I shall be sorry; I did not intend to kill him. I could not endure him, because he committed whoredom with my wife, but I did not intend to kill him, I only intended to give him a good beating," &c. When the knife was shown him, and he was asked if that was his knife, he replied, "I don't know, I have many knives like that." At the recognition of the body he behaved in a bold and indifferent manner, and displayed not the slightest trace of repentance. Also, fourteen days subsequently, at the examination on the 29th September, he denied having inflicted the injury, and asseverated, "What he had said was as true as Amen in the churches; he retracted nothing: he could not now recollect what had happened between him and Hundt, it was all written down already." "What shall I say then," he concluded with, "you know it all already; I cannot relate it again to-day, my head is too heavy; I am as cold as ice; the hemorrhoids have risen to my head and breast, so that I can hardly hold out for pain; I must urgently request that I be unshackled during the day, that I may move about in prison." The Judge conducting the investigation has recorded at this stage of the proceedings, that Hoffman's illness did not seem to be feigned; that directly when he came in he appeared to be in much suffering, at length he shivered over his whole body, and his condition was such that he seemed every moment likely to faint. Also at the examination on the 19th of October the accused said, "I have nothing to confess, what I have said, that I hold fast to," adding thereto the following remarkable declaration, "if my wife comes hither, make her show the great rent in her petticoat, and you will see how far the vileness of a whore can go." He asserted further that Hundt and his wife had appointments with each other, and that they were together " day and night *toujours*." Hundt had " brought his death on himself solely by this whoredom," and he did not die from the stab.

Ludwig Hoffmann was married to his wife twenty-eight years ago, and had begotten out of her ten children, who are yet alive. Up to Whitsunday of the present year he was employed latterly as a commission agent; about Whitsunday (six months before the commission of the crime) he commenced to reproach his wife, already fifty-one years of age, with having constant adulterous intercourse with the workman Hundt, their fellow-lodger in the house; this she declares to be wholly unfounded "and will support it with a thousand oaths." During the last six weeks (before the deed) he made his bed upon the floor at the room door, and slept every night with his head upon a pillow lying on a board supported by the door, so that if the door was opened during the night he must fall down. Besides, he had a padlock placed upon the door, and before that he fastened a string upon it, on which he tied knots, which he counted over in the morning. A few weeks before the deed he had promised his son-in-law B. a thaler (three shillings) if he would act as a spy by night. Nevertheless, he continually asserted that his wife went by night to Hundt, and when his daughter jestingly said that her mother must have gone out by the window—which was three stories from the ground—" yes," he said, " you are right, the mother can climb." The night previous to the commission of the deed the accused did not go to bed at all, but spent the night sitting with his head upon the table. Next forenoon he went into the married woman B.'s room and said to her, " to-day I shall challenge the scoundrel to a duel, here I have a dagger," striking the breast-pocket of his overcoat. After eleven o'clock the house inspector saw him in the court, fighting in the air with his stick. He appeared to the said F. to be somewhat tipsy, and this made him warn Hundt, since their mutual relations were well known. After dinner, Hoffman set his seat back to the bed and laid down his head upon the latter. In five minutes, however, he sprang up in such a hurry that he stumbled over his daughter, he called Hundt out of his room and demanded why he would put him in the papers, he had committed whoredom with his wife for three quarters of a year, and now he would have it out with him. After an exchange of abuse between them, Hoffmann stabbed Hundt, receiving, however, a blow on the head with a besom shank; and it is worthy of consideration that the deed was done almost under the eyes of several of the dwellers in the same house; he then went back into his own room, whence he was speedily taken off to prison. "The unchastity of my wife with Hundt," he deposed, at his first precognition, "led me to the deed, Hundt was my worst foe, and I determined to send him out of the world;" but almost immediately he withdrew his confession, alleging that he only desired to give Hundt "a good beating" as a remembrance, and hitherto from pure *dread of punishment* has perseveringly denied using a knife against the deceased. This long-cherished desire to avenge himself of Hundt is also proved by the deposition of Hoffmann's wife, according to which the accused, during the last six weeks, had often threatened "to murder one of the scoundrels," by which he meant his own wife, Hundt, and B., and according to it also he had carried off the knife from among the household utensils and had concealed it (in his coat) for, at least, six weeks. "Hoffman is at present sixty-six years of age, and still robust, and apparently powerful for his years. His small eyes have a somewhat pinched together, piercing look, which, along with a constant smirk about the mouth, give him an ironic expression. His carriage is somewhat decided, his speech is brief and concise, when not, as at times it is, reserved and monosyllabic. He is irritable, and specially sensitive to contradiction, when he readily becomes either violent or silent, so that further conversation is rendered impossible. His bodily functions are normally performed, and he is at present to be regarded as in good bodily health, as he himself confesses in his quiet moments. His complexion is normal, but he has the drunkard's copper-nose. All the witnesses declare him to be a contentious and irascible man, who was in continual discord with his wife and family. It appears, however, that these statements only have reference to recent times; at least, his wife and the inspector F. state this, and that previously they had nothing to complain of him in general. Both, however, confess to an (important) alteration in his conduct for half a year before the commission of the crime (Whitsunday of this year). Hoffmann commenced to be much addicted to drinking, he drank repeatedly every day, and especially 'during the last six weeks he came home drunk almost daily; he also ate almost nothing, and seemed to live upon brandy.' Now it came to pass that every one dreaded him and got out of his way. It was, continued his wife, as if Hoffman had lost his reason. His delusion in regard to my connexion with Hundt became a fixed idea which could not be removed. In former years he used to be jealous, but soon quieted down when he ceased to see the man; with Hundt, however, it was different, he lived beside us, he saw him daily, and his suspicion was thus continually excited afresh, &c. In prison, the behaviour of the accused changed very much in a few weeks. Already on the fourth of November the minister Bl. declared that Hoffmann had prayed him to free him from the wrath of his wife, who came to him in prison every evening surrounded by firebrands and tormented him horribly, till he called out her name, whereupon she vanished, leaving no trace behind. Similar occurrences and treatment are alleged by Hoffmann to have constantly happened in prison up to the present time, and this 'shameful usage' first moved him to avow it on the tenth of November. May he 'be cut to pieces upon the scaffold; two, ten, fifteen women are always watching him, like an American ape, through the observation pane, which is most horrible for him; they divine from his breath what he thinks, and what his eye winks, they blow up his straw sack till he is quite choked, and with his breath everything that lies upon his heart escapes too,' &c. At my first visit I commenced my conversation by alluding to an unimportant wound on his nose, and Hoffmann instantly replied in a hasty and bombastical manner, 'that is just it—they have knocked in my nose —these spies positively force their way into my interior; it is horrible. The spying goes on through a pane in the roof, but also through every hole in the wall,' &c. He denies having any visions of small animals (as in the *delirium potatorum*). When asked about the concealing of the knife, he confessed it with the utmost candour, but always recurred to the above stories, and it is to be remarked that his whole air, as well as his mode of expressing himself, conveyed the *impression of the deepest conviction* on his part. Precisely the same may be said of all the many subsequent conversations which I have had with Hoffmann. He never enters into any discussion as to his alleged experiences, which is extremely characteristic; when this is attempted he nods with his head as if giving an affirmative reply ironically. He has repeatedly flung water by night into the depression for the window above the door; because, said he to me, last night there were fifteen there. When I visited him on the thirty-first of December, I found the window of his cell covered by a besom; he would not explain the reason of this but only nodded, adding monosyllabically I would soon know. 'Don't you hear them,' said he, another time, *apropos* of some loud talking among the prisoners, 'there they go again,' &c. Is the insanity visible in the demeanour and statements of Hoffmann mere simulation or not? Every one who has had occasion to observe him for any length of time will, like the judicial deputies, have no doubt about it, *no doubt, namely, that Hoffmann is by no means simulating, but is actually insane.* I have already spoken of the deep conviction visible in all his statements. He is firmly convinced of the truth of all his allegations of his visions. In making this statement I place little or no value upon his occasional noisiness during the night, or upon the facts mentioned above of his having placed a besom over his window, &c., because in these matters design and studied cunning might be very well presupposed, although in connexion with all the rest, this behaviour is very remarkable; but the conduct of the accused, as above described, when his insane delusions are attempted to be contradicted, is of much more value. A counterfeit would, as experience teaches, have acted quite differently. While it is in his interest to use every possible persuasion to make the Judge and the forensic physician believe in the correctness of his insane delusions, Hoffmann disdains any such procedure, he constantly asserts that he 'is in full possession of his reason,' and on every attempt at contradiction he ceases to dispute and nods assent to every objection; in doing this his every feature seems to say ' why should I dispute with you? it is of no consequence to me whether you believe my statements or not.' To interpret this procedure of Hoffmann's in a different manner, and to assume it to be the result of a deeper cunning which avoided the appearance of feigning, would be all the more forced, as there are existing reasons which make the occurrence of insanity in him explicable enough. These are his jealousy and drunkenness," &c. (Here follows the development of the influence of these causes.) " It is a fact that particularly during the last six weeks he was drunk almost every day, ate nothing, and seemed to live on brandy. It seems all the more superfluous here further to detail the injurious influence of such a manner of life upon the mind of a man already suffering from a tormenting passion, as I may rather venture to assert that the preservation of a normal, healthy frame of mind under such circumstances would be much less in accordance with experience. On the fifteenth of September Hoffmann was in a state of mind generally similar to that described. The first step standing in any close relation to the deed now about to be considered, was the concealing of the instrument of murder for weeks previously. This action evidently manifests not only an intention to perform the deed, but seems also to prove an appropriate premeditation. But this is only apparent. Innumerable instances of persons indubitably deranged have shewn how well able they are, when they are meditating any illegal action, to make the necessary provisions, often most judiciously, often indeed with the greatest cunning. In other cases, again, they exhibit in those very preparations the stamp of their insanity, and this was precisely the case with Hoffmann, who, for weeks, concealed a knife taken from among his household utensils, which must speedily be missed, and which he scarcely concealed having carried off. If a responsible criminal would have sought to make himself more secure in this matter, he would still less have done as Hoffmann did, threatened and published abroad for weeks after he had determined to do this deed of violence 'that he would murder one of the scoundrels.' As in this, so also in other matters he proved that even before the commission of the deed it did not seem of much importance to him to make such provisions as might possibly protect him from discovery and punishment—a most important circumstance in judging of cases of disputed responsibility—in so far, namely, as he committed the deed almost under the eyes of several witnesses, whilst under the existing domiciliary relations it would not have been difficult for him to have got at Hundt in secret. Taken singly, none of these speeches or actions are absolutely probative, but they are of the utmost importance for the decision of the case when taken together and coupled with the behaviour of the accused at and after the commission of the crime. After having never gone to bed on the previous night, he next morning gave formal notice of the deed to B., inasmuch as he spoke of the 'dagger' which he had in his coat-pocket, and with which he would 'to-day challenge the scoundrel.' He drinks as usual, fights with a stick in the court in an apparently tipsy condition, so that F. formally warned Hundt against him, and after dinner, when the long cherished resolution had matured, he springs hastily up, stumbling over his daughter by the way, begins to quarrel with Hundt and inflicts upon him a mortal injury. Such conduct at the commission of the deed is just as readily explicable under the condition of a mental derangement, such as described as already existing long previous to the deed, as it is difficult to reconcile it with the assumption of a normal mental condition in the agent. Suspicions are, however, excited by the unquestionable fact that Hoffmann's crime did not want a *causa facinoris*, which he had so openly and consistently alleged, and that this passion of jealousy was one which so naturally gives rise to hate and revenge. Under the existing circumstances, however, there is no doubt but that in regard to the supposed guilty parties this *causa facinoris* rests entirely on a delusion, which his wife has very correctly termed a fixed idea. The fixed idea of itself does not absolutely exclude responsibility. Whoever, however, acts immorally and illegally from the point of view of a fixed delusion, thereby proves that his reason has lost the mastery of the fixed idea, and then the limited delusion (fixed idea, monomania) has ceased to be merely a limited one. Suspicion may finally be excited by the fact that the culprit at first denied the deed, and this cannot be otherwise regarded than as an attempt to escape the punishment due to his crime, and this in its turn permits the conclusion being drawn that he himself was conscious of its penal character. But in this we have only a repetition of what has frequently been observed in men who, after committing crimes when in a state of most evident mental derangement, have taken similar precautions, have hid, fled, or concealed themselves, &c. For the consciousness of the wickedness of a deed of violence is by no means always extinguished in those mentally deranged, and they then very naturally endeavour to escape judicial punishment, whereby it remains to be considered that the responsibility of an individual is not to be decided according to the question, did he know that he meditated an evil deed? but that there is another question still more decisive, namely, were there any internal influences effective enough to prevent him committing a deed recognised by him to be punishable? This was just such a case. In the foregoing I think I have succeeded in proving—1. That Hoffmann at present does not merely simulate insanity but is actually insane: 2. That an ever increasing jealousy and drunkenness have occasioned this insanity even previous to the commission of the crime: that he committed the deed in this state of mind and from the point of view of his insane delusions; and I have in conclusion to answer the questions put before me as follows:—*ad* 1. That Hoffmann is now in an irresponsible condition: *ad* 2. That he was in an irresponsible condition at the time of the commission of the crime." Hoffmann, long after, died paralytic in an asylum.

Case CLXXXII.—Attempt at Murder in a state of fixed
Delusion.

The subject of this case was a Doctor of Politics, named S. A
short time previous to the commencement of my examination of his
mental condition, he had in broad day, on the open street, shot
at a young man who was unknown to him; he missed him, however, and without attempting anything further, he was alleged to
have requested him to go with him immediately before the court
of justice, while he himself was instantly arrested. "The accused
is a man of forty-three years of age, tolerably thin, of medium
stature, dark complexion, well-marked Jewish physiognomy, in
which the deep lying, strongly-shadowed eyes are conspicuous,
with long black hair hanging rather wildly about him. His
bodily condition is quite satisfactory. 'Corporeally,' he said to
me, spontaneously adding—very significantly—voluntarily, 'and also
mentally, I am quite sound.' According to his statement, he has led
an unusual life. Formerly a Rabbi, he suddenly became a Christian,
studied law and politics, and for long employed himself in preparing
young legal officials for their final state examination, in which he says
he had on an average seventy pupils, and an annual income of five
thousand thalers (£750). At the same time, he says, he also became
an author, and with the same air of ill-concealed modesty and smiling
self-complaisance with which he speaks of his knowledge and of his
mental accomplishments, he expresses himself as to his writings, of
the general recognition which they received, and the expectations which
he thus excited in the scientific world. Meanwhile, his sphere of
action did not content him, and he went to Zurich, where he became
a university teacher; 'of course,' he says, he could not be contented there, when he could only count five pupils for the seventy he
had at Berlin. Nevertheless, according to his statement to me, he
enjoyed comparative peace at Zurich. His former co-religionists, he
says, since his conversion to Christianity, had continually persecuted him, but rejoiced at his want of success in Zurich they left him
at peace. He now went to America to become an advocate, but found
difficulties in the way of acquiring the necessary civil rights, and
after one year he returned to Europe; in the first place back to Switzer-
land and afterwards to Berlin. At first, he says, he was received
with general respect; 'I went about,' he said to me to-day, in so
many words 'only with ministers and presidents;' his knowledge and

his writings he says, procured him general applause and recognition; he had, 'indeed, reason to believe that the court itself was not devoid of a similar feeling in regard to him.' All this, however, he alleges became changed, he knows not why or wherefore! And now commenced a system of persecution in regard to which he expresses himself in the most absurd words. In a conversation with me, he told me that his persecutors constantly watched him through holes in the roof of his chamber, prevented him from working, and at last stole his thoughts, and when he had prepared manuscripts which would have brought him ten gold Fredericks (£8), they have been rendered worthless by the enemy abstracting the best of them before they were printed. In a subsequent conversation he denied this system of espionage, but alleged that a number of strangers, some of them of the highest rank, had taken lodgings opposite him, that he had often seen carriages drive up to the door, out of which perfect strangers descended, and that all these were watchers, persecutors, spies. In order to get rest, he had applied to the 'ministers and presidents' but was finally obliged to rely upon himself. Therefore he took pistols with him when he went out, and as he saw a young man approaching him with a long beard, and generally, as the accused unwillingly asserts, a very suspicious-like appearance, who was one of his thousand spies, he fired upon him. He is in the habit of adding laughingly, that he had not hit him, and to-day when again spoken to about his crime, he said he was now in prison as an accused person, instead of being an accuser; he is, however, not the least disturbed about his lot, but rather too chatty, laughing, and cheerful.

"It is not difficult to prove that Dr. S. is mentally deranged, the root of his delusion being vanity, and that he committed the deed under the influence of his delusion. He is unquestionably a man of rare attainments, even though there were nothing in the documentary evidence to prove this. Whoever, however, has taught and worked for years as a Rabbi; whoever has been able to assemble and keep together a large audience as a lecturer on law and politics: who speaks many languages, and has finally appeared as a scientific author—is certainly entitled to the title I have given him. But S. overestimated his knowledge and attainments, he had too high an opinion of himself. The greatest in the land are said to have patronised him, he consorted only with ministers and presidents, and it is psychologically a very natural and usual thing that very soon no place was good enough for such a man. In Berlin, in confessedly

very good pecuniary circumstances, his position as a private teacher
was, as he himself told me, a too subordinate one, and he gave up
this brilliant position to occupy a more elevated academic sphere of
labour, in which the result must appear to be more uncertain. And so
it actually was; again he became dissatisfied, and emigrated to America, &c. Unquestionably, all these changes did not improve his material position, indeed, he has thereby, to judge from his dirty and needy
appearance, become more and more reduced in circumstances, and that
happened to him which is of daily occurrence in similar cases, when
men with immoderate claims upon the world find them disallowed—
namely, that he commenced to seek for the cause of his fancied neglect in external circumstances, in enemies, enviers, and persecutors,
because vanity prevented him from seeing clearly that the cause of
the disproportion between his claims and their acknowledgement lay
in himself. This idea continued more and more to gain the mastery
over him, till it became in him a true fixed idea. When there is
thus an internal truthfulness in the statements of the accused which
renders them credible, so also his very nature and the manner in
which he performed the deed forbid the assumption of any mere
feigning, for which, moreover, there is not the slightest imaginable
motive. That his own assertion, that he is of sound mind, does
not prove the reverse, requires no illustration, though it is a
statement which we constantly find made by those actually insane,
while feigners have almost never the courage to come forward with
such an assertion. Accordingly, it appears to be indubitable that
Dr. S. has been actually deranged for a long time, and that he has
laboured under, and still labours under the delusion, that he is continually watched and persecuted by enemies and spies. And the
crime itself of which he stands accused bears completely the impress
of this delusion, and of it alone. He, a man versed in the law, who
must have known what he had to expect if he shot at any one, exposes himself to all these consequences with such composure of mind,
that he did not even lurk unobserved in the dark for one of ' the
thousand spies,' and afterwards flee away, or, if discovered, deny
everything, &c.; but in clear daylight, upon a crowded street, he fires
at a person wholly unknown to him, and then in the most ingenuous
manner invites him to go with him at once before the Judge. How
much insanity is betokened in these proceedings ! Could he, a man
learned in the law, believe that the Judge would acquit him, that
his righting of himself, and righting himself in such a manner

too, which might so readily have proved fatal, could be recognised as lawful? Could he hope to reacquire his lost peace of mind, even if he had actually put an end to one of his 'thousand persecutors?' Unquestionably, we have here a deed which wants every rational motive, and which has been solely carried out from the point of view of a morbid delusion, in which the perpetrator has not considered the results of his deed because he was not in a position to consider them. The undeniable intention to commit the deed, proved by the fact that he took a loaded pistol out with him, confessedly with the intention of righting himself, cannot be accepted as contradictory proof, since it is well known that lunatics of this character constantly commit insane and illegal actions, of which they have often long previously formed the intention, and for the execution of which they often systematically prepare with great skill and cunning," &c.

Dr. S. was also sent to an asylum.

CASE CLXXXIII.—BLASPHEMY, PROCEEDING FROM HALLU-
CINATIONS.

The following case presents an extremely peculiar form of fixed delusion of a religious character, or of a very peculiar form of constraint produced by a hallucination. The tailor S., in broad daylight, on the open street, and so loud as to cause a scandal to all the passers-by, while he himself was unquestionably not drunk, had given vent to the lowest expressions regarding the person of Jesus Christ, wherefore he was brought up for examination on account of his blasphemy. " At his examination he made precisely the same exculpatory statements as he did to myself subsequently, and which I shall presently quote. S. is apparently, and as he alleges, perfectly sound in body, he is fifty-five years of age, evangelical, and born at Waltershausen, in Gotha. According to his statement he has during the last few years been able to make but a scanty living as a tailor, and this circumstance, as well as the constant sedentary life of his trade, may have assisted in producing that disturbance of his mental faculties which is now apparent as unquestionably existing in him. For though S. expresses himself about ordinary matters, such as his business as a tailor, his former circumstances, &c., perfectly clearly and rationally, yet whenever the conversation takes a religious turn his mental derangement becomes at once apparent. He is in prison because he has lodged 'an evangelical-Lutheran-Christian protest,' and has protested, and must protest, so long as these 'rampart waggons' or 'fiery waggons' are not destroyed, by which he means the railroads, as since their formation all justice has vanished from the world, and in particular his 'evangelical-Lutheran rights' have been continually abridged. His otherwise pale countenance flushes at these moments, his usually lustreless eyes sparkle, his goodhumoured expression becomes wrathful, his brow wrinkles, his movements become lively, and he speaks and acts as one deeply irritated. From the above description it is evident that it is not possible to make him sensible of the absurdity of these and similar statements, such as, for instance, that his birth-place Waltershausen is the place where 'Moses' finally dwelt after leaving Egypt. The offence with which he is charged is most intimately connected with these delusions. According to his statement, a voice continually follows him, which *calls to him that he must kneel down and pray*. When this cannot be instantly done, as when he is on the public street, when he is engaged with his customers or in making purchases, &c., then *this voice constrains him to utter in a loud voice the blasphemous expressions referred to*, which he assures me he is quite unable to restrain. All idea of feigning must be excluded, for, besides that it is impossible to perceive what motive S. could have for so rare and abominable an offence, it has been unquestionably ascertained from observation that no one is more pervaded with horror at his crime than he is himself. With the demeanour of one deeply convinced of the truth of his assertion, he declares that he cannot find words to express 'the infamous vileness' which that voice compels him to utter. He is wrathful with himself because though he must sit still in prison, instead of working, yet he can get no peace for his vileness, &c. That illusions of the senses, and particularly of that of hearing, are frequent concomitants of insanity is a fact generally acknowledged. In the case of S. also the existence of the hallucination only confirms the actual presence of mental derangement. And though it is certainly rare that a man should be involuntarily compelled by such an hallucination to make use of certain expressions, yet the case is not unexampled, and I could give from my own medical experience examples of analogous cases, and it is evident that no criminal intention, but only the morbid impulse of a deranged mind of a man, such as the accused, who has from predilection busied himself with holy and religious matters, compels him to make use of such expressions."

I declared S. to be "insane" according to § 40 of the Penal Code.

There is another class of insane people, of scarcely less frequent occurrence than those who imagine themselves persecuted and tormented by the rest of the world, those *insane from pure dogmatism*, insane querulants. I have already in § 73 (p. 193, Vol. IV.) shown how little I am inclined to construct from this insane querulousness a peculiar species of insanity. It would, indeed, be quite impracticable to do so in these cases, because in each individual patient the characteristics of this peculiar delusion are mingled and blended with those of others, such as the so-called insanity of "Pride" or of "Persecution." But I, nevertheless, hold myself in duty bound to point out the frequent occurrence of such cases, so that the support of analogy with many similar cases may be afforded to those who may fall in with others. The origin of such a delusion is easily explained. The consciousness of right is one of the deepest feelings of mankind. The consciousness of the individual that all his rights are guaranteed to him, and must continue so, is based by him upon the state, which is the protector of the rights of all, as where this conscious possession of rights is in any measure shaken, the state is dissolved, and it is for this reason that any actual or fancied injury to one's rights is so deeply felt. This is quite peculiarly the case in men of a limited understanding, and in those of precisely the opposite character who possess, or in their vanity fancy they possess, much higher mental endowments; in the former, because they cannot perceive the reasons which produce this shock to their consciousness of right, and in the latter because in their selfishness they have accumulated *à priori* rights which neither society nor the law can recognize as such, and the possession of which, therefore, the mouth-piece of the latter, the Judge, must deny them. Therefore we often meet with individuals who, when what they consider as rights belonging to them have been denied them by consistent and repeated judicial decisions, have their minds thus deeply and permanently shaken and more and more depressed. In their desire, becoming ever more impetuous, to attain and achieve their fancied rights, they dissipate their means, they importune the courts of justice, from the lowest to the highest, with continually fresh memorials and complaints, they study the common law day and night, and ruin their bodily and mental health ever more and more. It is very natural, therefore, and experience proves it, that such men should finally, after years spent in vain lawsuits and complaints, actually suffer in their reasoning powers, that the thought that they are right and all the rest of the world wrong, should finally develop itself first as a fixed delusion, which then frequently, often after many years' duration, increases to complete insanity. Then they hurl written documents, full of the most insane and vilest insults, at the judicial boards, which they describe as "corrupted and having underhand dealings with their opponents," they openly oppose the execution of any judicial measures, an execution, imprisonment, &c., and now, if not sooner, their mental condition has to be inquired into. I have only to add that this peculiar insane dogmatism is not to be found only in educated men possessed of legal knowledge, but occurs even in the lowest classes, and also among women. From among a large number of similar cases, I have selected the following as proofs of what I have stated :—

CASE CLXXXIV.—AN INSANE QUERULANT.

L., Doctor of Laws, aged forty years, had, more than twenty years ago, by his conduct drawn the attention of the authorities upon himself. He was arrested in 1819 in Leipzig, for a street brawl with a journeyman tailor, who was singing a song which L. took as a personal insult. During the years 1825, 1827, 1828, and 1830, he had insulted many persons by words or deeds, because, though they decidedly denied it, he fancied himself insulted and his honour wounded by them. Indeed, in February 1828 he caused a public scandal in the theatre at Leipzig, which ended by his stabbing with a dagger a strange man whom he fancied had insulted him, and for this he had to suffer eight weeks' imprisonment. Besides taking the law into his own hands on these occasions, he also denounced many persons during the years mentioned on account of offences alleged to have been committed upon him; indeed, on one day (the twenty-sixth of January, 1828) he handed in no fewer than three different accusations, and it is remarkable that in all his numerous complaints he quotes the Royal Saxon mandate of the year 1712 against revenging one's self. In the year 1834 he was banished from Dresden on account of defective legitimation. In 1837, in a complaint on account of denial of state rights, he made use of such insulting language, that he had to be punished by a fine. In the same year L. had displayed such offensive conduct at

Teplitz in the theatre, and on the streets, that he had to be banished from this town also. And now the excesses committed by him increased in frequency, particularly in the theatres, because he constantly believed himself to be insulted by the glances, looks, laughter, &c., of strange men, which he took to refer to himself. In the year 1838 he was again banished from Dresden, and we find from the documentary evidence that, as formerly, he gave vent to numerous complaints and remonstrances, which are partly characterized by a pettifogging sharpness, but also already bear the stamp of actual mental derangement. This is exhibited still more clearly in a letter to the city Police Commission at Dresden, in the year 1840, in which he assures them "that in 1837 he was the handsomest man in Dresden, and certainly the man most favoured by the ladies, and that some ladies from love for him, and others from scorned affection, had fallen at his side into an ecstasy;" in order to prove this he demanded in the letter mentioned that "the handsomest man in the Police Commission," the Director of Police, should go with him to the theatre, and he opined, he was convinced that every eye would be turned on him (L.), while not one would glance at the Director. In the year 1842, though the documentary evidence gives us no particular information in regard to this, his condition must have got much worse, as he was sent to the public asylum at Jena, out of which, however, he was dismissed as "cured" on the twenty-third of December of the same year. He went once more to Dresden, where he lived with his brother, an advocate, till July 1844, when he was again banished on account of his continued molestation of the Grand Ducal authorities of Weimar, he being a subject of the king of Prussia. His brother, who had repeatedly paid his debts, relates of this period "the unbridled outbreaks of his presumedly morbid pride, and of his arrogance, which could be restrained in their expression by no motive whatever." Upon the nineteenth of November, 1844, the accused wrote to the magistrates at Erfurt, and begs them, to set aside a report of this character, to certify officially that during his residence in the town hospital there, in the year 1841, no one, either from the magistrates or otherwise, had meddled with his genitals or measured his penis while he slept. He also sent several other letters to the authorities regarding this circumstance; but in April of the same year he again commenced to insult people on the public streets, wherefore sentence of banishment was passed upon him, against which he anew presented innumerable remonstrances. It is also deserving of notice that he stated to a police official that he had heard upon the promenade that the minister of justice had declared that he had too small a penis, and that he would take the minister to law about it.   About the same time, and always occasioned by mortifications said to have been inflicted upon him, numerous written accusations were lodged by him againt the calculator W., Cand. Baron L., President v. Z., Dr. W., &c., in which there was no want of quotations from the Statute Books and Handbooks of Penal Law.   On the twenty-eighth of August, 18—, L. finally left Dresden, and repaired to Berlin, whither the requisition for an inquiry into his mental condition by the Royal Board of Justice of the former city was transmitted.   "L. is a tolerably small and thin man, with plentiful dark hair over-shadowing his small and flat forehead, a pale grey complexion and somewhat piercing look.   The distorted smile, with which he almost always speaks, is an unmistakeable indication of wickedness and cunning.   Corporeally, he may be said to be relatively healthy.   He lives in a small apartment in which are two mirrors, one of which he keeps always covered with a cloth because, as he stated in answer to me, it dazzles him, which cannot be reasonably true.   Leading into an adjoining apartment is a door hung with double curtains; and over these curtains L. has hung sheets of paper, sewn together, because, as he alleges, the talking in the adjoining room would otherwise disturb him.   His wood he keeps locked in a secretaire, and his table is covered with writings in his own hand, which I could not more closely examine from his distrust.   I represented myself to him as a physician who had been requested in writing by a gentleman in Dresden, one of his acquaintances, I presumed, to enquire after his health.   He immediately stepped backwards and gave me to understand his surprise at this Dresden gentleman's 'very remarkable commission, which offended him grievously.'   Upon my representation that such a request was quite a usual affair for a well known physician, he repeated his expression of surprise with ill-concealed resentment, remarking, that he must regard it as a serious affront to enquire after any one's health.   My rejoinder that the question 'how are you?' was a most usual greeting and certainly anything but an insult, he would not admit, and he stated that steps must have been taken at the Ministry of Justice in Berlin, by which he gave me distinctly to understand that he regarded me as an enemy.   He opined that I was only employed 'to take the chestnuts out of the fire,' and he warned me to have nothing to do with 'that,' as he could assure me that a physician by such conduct to him had nearly brought himself to penal imprisonment. His desire to obtain the name of the writer of the letter became more and more urgent, and when, at length, pretending that I could not exactly remember, I suggested the purely fictitious name of 'Brückner,' he became more violent and declared that everything was now clear to him, that he saw through the whole intrigue, and he urged me to confess whether I was not employed to ascertain whether he was deranged or sane, and in saying so he did not fail to express his surprise and indignation. In the course of a long conversation I let out a few of the names mentioned in the documentary evidence, such as those of the Messrs. v. M. and P. ; upon which he detailed the respective circumstances as erroneous and quite unfounded accusations against him, and then immediately recurred to the offence caused by my visit to him. I could not withdraw L. from this subject, and I was finally forced to take my leave of him.

"Dr. L., according to his antecedents, as exhibited in the documentary evidence, and as his present conduct shows, is one of those men in whom a disproportion between an imaginary high personal value and the estimation accorded by the world, has deranged their normal thoughts and feelings. This source of mental derangement, morbid vanity, is well known to be of frequent occurrence. In regard to it it is unimportant whether the individual sets a more extravagant value upon his external appearance or his internal worth. With Dr. L. the former seems to be the case (however little his actual external appearance corresponds with his relative idea), as is shown by the above-mentioned letter to the Directors of the Dresden police. Closely connected with this are his whims in regard to his sexual organs. He regards himself as a handsome man, favoured by the fair sex, and thinks himself capable of justifying this favour in every respect. Any doubt in regard to this offends him. In such a state of morbid irritability no illegal step, taken from the point of view of his morbid logic, could surprise us. In his quieter moments he disdained to 'avenge himself' and lodged accusations; when more excited, however, he did not hesitate to do so, and threatened, and struck, indeed, even stabbed with a dagger. Psychological experience has been long acquainted with querulants and grumblers such as he, and that in no small number. There is nothing at all wonderful in the innumerable written complaints lodged by him, a lawyer by education. When several of these exhibit the stamp of mental

s

derangement, the fact that others are apparently quite rational in their construction, is not opposed to the supposition that he himself is deranged, since his insanity is not always, and especially has not originally been, general, but only a fixed delusion having reference to his imaginary mortifications, the actions in connexion with which need not necessarily exhibit the phenomena of general mental derangement. The documentary evidence affords no information as to the time when this partial delusion passed into general insanity, inasmuch as it speaks only generally of his residence in asylums but not of his behaviour in them : it is certain, however, that at present with the persistence of his fixed idea, as shown above, by his malevolence towards his fellow-men and continual imagination of attacks upon his honour and his rights, his mental derangement is no longer partial, since the description of his household arrangements, as already given, points to the existence of other delusions than these. In any case it is indubitable that L. has no clear and correct idea of his relative position to the external world, that he is continually involved in delusion regarding it, and that accordingly he is, as I finally declare to be my opinion in accordance with the terminology of the Prussian General Common Law, in such a mental condition as renders him incapable of considering the consequences of his actions.

CASE CLXXXV.—AN INSANE QUERULANT.

This case was that of the wife of a master joiner, whose mental condition had to be enquired into, because she had at last heaped the lowest invectives upon the Royal Chamber of Justice (Supreme Court), and in this the following unusual query was put before us : " Whether it is to be assumed that the accused is in a state of monomania, and therefore irresponsible?" This woman was fifty-eight years of age ; in her external appearance there was nothing remarkable, except a complexion indicative of abdominal congestion. At my first conversation with her, she entered at once upon those complaints and accusations against her (divorced) husband, which had been the cause of innumerable complaints before many courts, and had, indeed, given rise to a penal sentence. With characteristic loquacity she constantly repeated that her husband had defrauded her of twenty thousand thalers (£3,000), which she alleged she had won in the lottery, that her money had all been taken from her and deposited in court, and was now illegally withheld from her. Every contradiction, every representation as to the improbability of her statements, only made her more violent. It was interesting to see how she brought forward judicial instructions wholly irrelative to the matter, indeed, mere summonses to trial, &c., as proofs of her rights, "and that she was not deranged," and that she indeed deduced her rights from isolated judicial documents which actually proved the reverse. This conduct had now lasted for fifteen years, and we could not, therefore, hesitate to assume, especially in connexion with many similar experiences, that the "monomania" and "irresponsibility" in question were proven.

CASE CLXXXVI.—SHOEMAKER K., AN INSANE QUERULANT.

The subject of this case was one of the most obstinate of dogmatists. Ten years' previous to our examination into his mental condition, he had to pay judicial costs for the regulation of his father's inheritance. The execution served on him was, however, fruitless: it was subsequently discovered that his oath of manifestation (that he possessed nothing but what he had on his body) had been falsely taken, since amongst other things he had concealed a silver watch, &c., and he was punished for perjury. And now commenced a continuous series of petitions up to personal demand, and of complaints whose tendency was to prove his innocence (amongst other things he was in the habit of asserting that he had not sworn falsely, since he "had the watch upon his body," that, moreover, the oath ought not to have been exacted, since he was not due any costs, &c.). Innumerable fresh complaints, after he had been warned against further querulousness, indeed, after he had been fined for gross abuse of court, complaints lodged with the minister of justice and the Ober-Tribunal against the court engaged on the matter, and renewed abuse, because of renewed punishment, &c., led finally, after the lapse of ten years, to our examination of K.— " Shoemaker K—," I said in my report, " whom I found working quietly and apparently quite rationally, is thirty-seven years of age, and with nothing remarkable about him, except a pale complexion, which is, nevertheless, frequent in men of his trade, and a dull, languid look. He is sound in body, except an alleged pain in the back, the result as he alleges of his long penal imprisonment. It was not difficult to bring this ' unwearied querulant,' as he is termed in the documentary evidence, to the subject of the investigation. With the same persistency which he

s 2

displayed in his numerous complaints and remonstrances, he also declared to me that he had been already trampled on by the sale of his paternal property, and for this he alleged a reason which sounded apparently correct. With the same consistency he demanded that the documentary evidence should be laid before the Supreme Court, because he expected there at last justice would be done him. He cannot be convinced that the judicial officials in G., and the president of the High Court of Justice there, can not be such objectionable characters as he described them in his latest written complaint. He makes all his statements with composure and apparent clearness and consistency. It would be, however, a gross error to deduce his mental soundness from that alone. Men such as he is, are of no infrequent occurrence. To their original irritability an important lawsuit is superadded to their great hurt. They regard themselves as injured, appeal, lose again; they are, moreover, burdened with costs; every possible legal remedy is now tried, and all in vain. The more they ought to be convinced of the erroneousness of their legal views, so much the more is their conviction confirmed, that only the ill-will, enmity, and corruption of the Judge are opposed to them, and their dogmatism becomes, the more their anger is excited against Judges, Courts, King, &c., so much the more a fixed idea." It was now detailed, that the insults heaped upon the judicial boards and officials in former, as well as in more recent times, bore the character of deeds done from the point of view of a fixed delusion, and this detail was reasoned out by means of the views developed in the text, after which the irresponsibility of K. was assumed.

CASES CLXXXVII. to CLXXXIX.—THREE SIMILAR CASES.

CLXXXVII.—The first referred to a chancery servitor, whose repeated petitions for a settled appointment, &c., had to be disregarded. He then pursued the same course as those before him; wrote continuous complaints, and at last gross abuse to the Supreme Court, and this occasioned the necessity for inquiring into his mental condition. He was perfectly clear and rational upon every other point. But when brought upon the subject of his so-called claim, he at once became violent, brought forward the most absurd reasons for his alleged claim to an appointment, and could not be convinced that if, as he alleged, twenty were appointed before him, there were not other sufficient reasons for that, than "a twenty-fold, intentional chicanery against him," &c.

CLXXXVIII.—In the course of many years the surveyor R. was repeatedly brought before us for examination; he had written whole volumes of complaints and petitions, and was an apparently quite rational, though very irritable man, and one who made an honest livelihood by the preparation of plans in his own department. His fixed delusion was that he imagined himself to be injured in his rights and defrauded of his payment for work done for the Royal General Commission in St., and for the Royal High Court of Justice in X., wherefore, and for the gross insults heaped upon these courts and their chiefs, he was repeatedly punished by imprisonment. At our last examination of him, in answer to a question in reference to these points, that he had complained of officials of high standing and acknowledged rectitude as if they were scoundrels, &c., he replied, that he had said nothing which he could not clearly prove. Heaven itself had justified his statements. For according to a Royal Cabinet order all dishonest officials were to be dismissed; now it happened, that the former chief president in X., the Baron v. Z., who was in collusion against him with the High Court of Justice there, and with the General Commission at St., died on the night of the 31st of March, and was therefore truly "dismissed." When it was pointed out to him that this was all very well, as a play upon the word, but could never be regarded as any real proof, he said : "I only say this—I have given proof sufficient," &c. When given to understand that he could not possibly be in earnest in supposing that in any well-ordered state, perfectly independent judicial boards in different courts and places could be composed, as he alleges, of dishonest men, of "scoundrels," who would intentionally deprive him of his rights, he confessed, indeed, that the whole of these boards were not dishonest, but that their higher officials were so, and on being further questioned he, as usual, would not yield upon this point, namely, that relationship and other ties bound the chiefs of the various courts together against him. The money due to him for his work had been certainly misreckoned, of that he was convinced, and he gave us plainly to understand, that these courts and their chiefs had applied it to their own use. Long conversations with R. always ended in the same result. This case, however, presents still one more point of interest. At the last examination just referred to, after I had not seen R. for a long time, I found that his former partial delusion had spread wider and developed itself into general insanity. R. had now hallucinations of the most various characters. He saw corpses with white and with black noses; he saw the Saviour; he came by night in his shirt and drawers into the bedroom of his hosts, and, according to their statement, he entertained them with an account of his visions and other "stupid stuff," &c. He must now be declared to be "incapable of considering the results of his actions," that is, an "idiot," according to the terminology of the Common Law.

CLXXXIX.—Of another case, that of a coarse man from the lower classes, I will only shortly state, that after many years' complaints, turning upon the sale of a mill, cunning legal deductions, &c., which had been in vain, after he also had been punished for abusing the courts of justice, after personal application had been rejected, it reached the interesting climax that he wrote fresh letters of abuse to the highest earthly Judge, and when he was punished anew for high treason, he directed his insane complaints against the highest Judge in heaven and gave vent to the vilest blasphemies! This gave occasion to a fresh examination, the result of which could not be doubtful.

§ 81. INSANITY. — CONTINUATION. — THE INSANITY OF DRUNKENNESS.

STATUTORY REGULATIONS.

GENERAL COMMON LAW, § 28, TIT. 4, PART I. *Persons who are deprived of the use of their reason by drunkenness are, so long as this drunkenness lasts, to be regarded as insane.*

PENAL CODE, § 119. (only regards the punishment to be inflicted upon those found drunk.)

The Prussian Civil Code and other statute books with it class the mental condition of drunkenness in the same category with insanity, to which it naturally belongs. Meanwhile, however, the Prussian Penal Code takes no cognizance whatever of drunkenness in regard to the question of irresponsibility, indeed, the very word is never mentioned in this respect; it therefore requires, at least *implicite*, that intoxication be included under the head of "Insanity" (or "Idiocy") as in § 40. It follows, therefore, that all that has been said as to insanity in general, is also applicable to that special form of it produced by alcoholic intoxication, and it would, therefore, appear to be scarcely necessary to enter upon the further consideration of it.

Experience, in fact, teaches us that this question, from a medico-legal point of view, has not the importance generally theoretically attri-buted to it by authors; because, as we daily see in Berlin, the Judges decide in most cases in regard to the punishableness (responsibility) or unpunishableness of illegal actions committed during a fit of drunkenness, by themselves and without consulting any forensic. physician, who, moreover, could seldom give his opinion from per-sonal observation, but only with reference to the evidence in regard to a long past condition; this condition is, however, very well known to the Judges, and they only require to be sufficiently informed by the witnesses, as to whether the perpetrator at the time of the com-mission of the deed was "senselessly" drunk or not. According to this, it seemed that the existence of two kinds of drunkenness is very properly generally adopted, a drunkenness which "leaves a man the use of his senses," and one which "deprives him of them." Ac-cordingly two, three, and four stages of drunkenness have been laid down. And I cannot omit quoting a witty Neapolitan proverb, which very concisely and correctly points out the characteristics of these three stages :—" The first glasses that you drink are lamb's blood, they gently soothe; the next are tiger's blood, they drive to fury ; the last are swine's blood, after them man rolls in the dirt! " Every one knows that intoxication from its first commencement to its closing phenomena passes through various stages. But these might, if it were profitable, be quite conveniently comprehended in two stages or grades. In the first stage, that of tipsiness, the ex-citement of the circulatory and nervous systems by the narcotic fluid renders both the mental and bodily actions more lively ; the silent become talkative, the quiet gesticulate ; the flow of thought is in-creased; the ideas crowd upon one another, and, as the excited mind now casts aside the bonds of morality, personal interest and custom, the drunkard displays both indecencies and vulgarities, and lets loose his tongue in regard to matters of fact or defects of character, which it has hitherto been his interest to conceal, in confirmation of the ancient proverb *in vino veritas*. The characteristics of this stage are the volatility in its designs, and indiscretion in its actions, which are all the less likely to be violent in their character, that the drunkard as yet, often contrary to his own nature, is cheerful and good-humoured, and inclined rather to embrace all the world as dear brothers than to strike at them. He is still master of his senses, and he is still able to correct those deceptions which are commencing;

he turns out of one street into which he has wandered, in order to get into his own, which he can still recognise; he can also especially see very well when the bottles or glasses are empty, &c. It is otherwise in the more advanced second stage. The more the intoxication increases, so much the more increases with it the actual purely corporeal pressure on the brain, which explains the difficulty of muscular exertion, so also in like manner increases the constraint and oppression of the mental faculties. The drunkard loses his senses, with them disappears the consciousness of his relation to the external world, the passions break forth unchained, violence is the psychological characteristic of this stage; the drunkard falls into a veritable state of mania with all its characteristics. It appears unnecessary to assume more than these two stages of intoxication. But, what is of more importance than the setting up of certain stages of intoxication, be they two, three, or four, is the fact that there is no need for any thing of the kind for psychologo-forensic purposes, because it is impossible to trace with any certainty the limits of these different stages generally, as well as in regard to individual men. In the latter respect, as is well known and requires no detail, neither the kind of drink (amount of alcohol in it), nor its quantity is decisive, because the manner of life, custom, bodily constitution, &c., occasion the utmost varieties of effects. Consequently, there is nothing else for it but *to consider the practical circumstances of each case,* and this principle of the individualisation of each case, which in my opinion ought to be firmly maintained throughout the whole of forensic psychology, is, for the reasons given, indubitably never more correctly applicable than in regard to intoxication. In this respect, and to guard against the mere simulation of a senseless condition of intoxication which is daily attempted by accused parties, the physician, when his opinion is requested, must pay attention to those circumstances which more immediately fall within his sphere of observation, in order to discover whether the amount of that kind of drink alleged to have produced this intoxicative insanity, is such as would be, consistently with experience, likely or not likely to do so in this individual, who perhaps daily drinks the double of it, or who never drinks at all, or who labours under determinations of blood to various organs, &c. If in accordance with this the drunken condition at the time of the commission of the deed be thus actually determined, then the whole case simply falls to be considered under the head of that general category of mental derangement to which it belongs, (temporary) insanity ; and it is now all the more to be considered according to the general diagnostic rules (§ 61). Because, since every form of insanity does not of itself exclude responsibility (§§ 78, 79), so least of all does intoxication in general, in which, I may repeat, it is very difficult to define the precise stage in which the drunken man was at the moment of the commission of the deed, and in regard to this the witnesses themselves are not always able to give correct information. Therefore it is necessary in this, as in every other case, to pay attention to the possible existence of any *causa facinoris*, the psychological individuality of the culprit, and his behaviour previous to and after the commission of the deed. A man well known for his loyal and patriotic sentiments, and who had proved their truth by action in serious times, gave vent when drunk to the vilest abuse of the King. A moral, and peaceable artisan, when drunk with beer, killed his beloved brother-in-law by stabbing him in the lungs with a dagger.* In such cases, the actual existence of intoxicative insanity must be assumed, whilst on the other hand, we continually see accusations, which occur so often in a large town, of injury and ill-treatment of police officials alleged to have been committtted in a state of senseless drunkenness, end very properly in condemnation by the Judge, if it has only been proved that the accused has recognised the officials as such, and has first exchanged words with him relative to this official position, &c.†

§ 82. CONTINUATION.—DIPSOMANIA.

Men who have become habitual drunkards (" *Trunkfällige*," Clarus), act in a threefold manner. Either—and this is the most common—they have such a mastery over their vice as never to be altogether overcome by it. They drink daily, without ever being actually drunken ; weakening thereby their digestion and their mental energy more and more, and, finally, fall victims to bodily diseases arising from chronic alcoholic poisoning. Or, for the same reason, they fall into that form of periodic insanity which is only too well known as the delirium of drunkards (*delirium potatorum, del. tremens*),

* *Vide* Case XLIX. p. 122, Vol. I.

† It is very peculiar that, according to the statement of the very experienced reviewer of the Thanatological Part of this Handbook in the British and Foreign Review, October, 1857, p. 389, the English law protects a drunken man from the results of his civil, but not of his criminal actions !

and which possesses no more specific interest for forensic medicine than any other form of insanity that may arise from drunkenness, which is so frequent a cause of mental diseases in general. Because the estimation of illegal deeds done during a fit of delirium tremens differs in no respect from that of any other deed committed through insanity. Or however finally, the habitual drunkard, in certainly very rare instances in general falls into that form of periodic intoxicative insanity which Bruhl-Cramer* has termed *Trunksucht*, Erdmann † *Saufsucht* (Dipsomania), and of the actual occurrence of which Clarus,‡ Fuchs,§ Rademacher,‖ and others, have given proofs, to which I add the following from my own observation. A young and well-educated man enjoyed the perfect confidence of a grand ducal house, and amongst other departments was intrusted with the chief superintendence of the wine-cellar. In this position he had accustomed himself to the use of wine, and as usual had gradually progressed to the use of stronger alcoholic drinks, and at last fallen into dipsomania. About every three months this large, stout, uncommonly powerful, and unmarried man of thirty, was seized with a frightful desire for drink. He caused baskets full of wine, pale ale, and rum, to be brought into his room, into which henceforth only his maid-servant and I, his physician, were permitted to enter, and he then drank day and night continuously, without ever coming out of his state of deep intoxication, till sickness and vomiting came on, and then he drank not one single glass of all the drink that stood around him. Then, as if returned from a short business journey, or as if convalescent from an illness, he again appeared in the house of the prince, in which for years his drunkenness remained a secret, for in the intervening periods he never drank more than a few glasses of wine at his master's table. He died young, but I shall never forget the supplicating entreaties and tears with which this unfortunate man besought me to free him from his misery; I can testify that this long since forgotten man was neither devoid of the most earnest desire to reform, nor of moral disgust at himself. A similar case from a much

---

* Ueber die Trunksucht und eine rationelle Heilung derselben. Berlin, 1819.

† Beiträge zur Kentniss des Innern von Russland. Dorpat, 1823.

‡ Beiträge zur Erkenntniss und Beurtheilung zweifelhafter Seelenzustande. Leipzig, 1828, S. 130.

§ Henke's Zeitschrift, &c., 1837, 3, S. 55.

‖ Erfahrun sheilkunde. Berlin, 1843, S. 753.

lower sphere was that of a woman, the history of whose dissection has been already given.* She was the wife of a spirit rectifier, she had given herself up to drunkenness and had become dipsomaniacal. When the fit came upon her, she left the house with as much money as she could get hold of, and extra clad. Then she drank in the public-house first her money, and then one article of clothing after the other, till she was at length picked up and sent home half-naked and deeply intoxicated. How she at length drank ethereal oils and prussic acid, and thus poisoned herself, I have already related. Her husband and all the witnesses proved that in the intervening periods she did not "drink," and regulated her household properly, &c. After facts such as these, which cannot be done away with, I cannot agree with its opponents (Heinroth and Ideler for instance), that there is no such "disease" as "dipsomania." Ideler's comprehensive criticism,† full of moral wrath, raises its voice after the pattern of Heinroth's well known theory of sin in the doctrine of mental diseases,‡ specially against the notion of necessity, the compulsory use of stimulants from bodily causes, and in that only sees the result of vicious habit, &c. With apparent justice, Ideler has appealed to "the vast results obtained by temperance societies," as opposed to the assumption that there exists any such disease as dipsomania. "How dare any one," says he, "henceforth speak of a physical necessity, by which dipsomania, like any other severe bodily disease, has borne down the opposition of the will to complete impotency, after we have seen the controlling power with which the moral example of large societies rouses even the feeblest characters to emulation, and has brought back millions of perishing drunkards from their beastly drunkenness to absolute temperance"? But the permanence of these "vast results of temperance societies" has, as is only too well known, been by no means undisputed. How great has been the shipwreck of even the most famous temperance societies, we may learn from L. Pappenheim's§ Handbook of Sanitary Police, he having had occasion to observe for several years the perfectly resultless endeavours of the Upper Silesian Temperance Society during the forties of this century. The case has been precisely the same in Ireland, &c. After such examples as these, the appeal to the results

* Case CCIII. Vol. II.   † Op. cit. S. 321.
‡ Vide the paragraph on Dipsomania, in Heinroth's System der psychisch-gerichtl. Medicin, S. 263.
§ Handbuch der Sanitäts-polizei. I. Berlin, 1858.

of temperance societies in regard to the question of dipsomania is perfectly untenable. Besides the increased consumption of tea by the abstainers from alcohol, the *Teatotallers* (*Nichtsalstheetrinker*, sic in orig.) and temperance societies of England—the consumption of opium in the three kingdoms has also very largely increased, and it is only the name of the intoxicating agent that has been changed by many! The chief point, however, in any criticism of the results of temperance societies, which on the whole have been most blessed and pleasing, the chief point for our object is the indisputable fact, that these results to the "millions of perishing drunkards" have never been *individually* proved by any one, and that it yet remains to be proved, that cases of dipsomania (always isolated and rare) have been actually cured by purely voluntary abstinence. It is certainly sufficiently apparent from my whole statement of the doctrine of responsibility, that I do not in the least belong to that class of authors whose practice, lax, untenable, and dangerous to the public weal, it is always to elevate what is corporeal and to deny the power of the mental energy to control the immoral passions and desires. On the other hand, for reasons which will presently be given, I have not here to enter upon the pathology of dipsomania, otherwise it might not be difficult to explain the patho-genesis of this disease by a reference to the general laws of nervine physics, and the doctrines of stimulation and overstimulation.* I must not and shall not

---

* I will not even point out that Dipsomania, in accordance with these laws, may be more or less identified with opium-eating, because its opponents may also refer this to a "vice," although this assertion certainly does not explain those numerous cases of bodily disease in which the sufferers have gradually become accustomed to the use of this pain-soothing remedy, and have then found it impossible to do without it. But a recent extremely remarkable and unique case, related by Büchner (Archiv f. Pathol. Anatom. u. Physiol. 1859, xvi. S. 556), of the urgent necessity for chloroform inhalation to subdue constantly recurring fits of gallstone colic—a treatment which the patient had adopted instead of the use of opium—affords an unequivocal proof of the truth of the existence of a morbid impulse to intoxication from purely physical causes without the slightest trace of "vice." The reporter states: "On the average, every four or six weeks, I found the patient lying senseless from chloroform. The stupefaction induced by each inhalation lasted only a few minutes, so that the patient was necessitated to pour fresh chloroform on the handkerchief every ten or fifteen minutes, and again to hold it before his mouth and nose. After having done this repeatedly for a few hours, the patient enjoyed a few hours' quiet sleep, after which the same performance was repeated, and thus several days and nights were continuously passed without the ingestion of the slightest particle of nourishment. When

dispute the occasional occurrence of cases such as those referred to. But the pith of the question, which has been quite overlooked, is that *the dispute as to the occurrence of this so-called dipsomania is of extremely subordinate, indeed, relatively, of no importance for medicolegal* (science and) *practice.* For this question is only of consequence in those countries in which a distinction is made between blamable and unblamable drunkenness, and where the physician may be asked, whether the intoxication of a dipsomaniac is blamable (intentional) or unblamable (unintentional)? understanding by unintentional drunkenness, that produced by the accidental swallowing of strong alcoholic drinks, by the administration of narcotics by a third party, by sojourning in an apartment the air of which is impregnated with alcoholic vapours, &c. Thus, according to the former Prussian Penal Code, he "who has placed himself in such a condition by drunkenness, &c., as that the freedom of his action is thereby removed or limited, shall be held responsible for any offence committed in this condition, *according to the measure of the blame attachable to him.*" The present Penal Code, however, includes neither this nor any other regulation in regard to the responsibility of drunkards. And thus in our country, as well as in all those with a similar Penal Code, dipsomania is not a forensic subject at all, and in any cases which may occur, the intoxication of a dipsomaniac is to be treated like any other case of ordinary drunkenness (§ 81). But even when the Penal Codes make particular mention of an "intentional" intoxication, as is done in Austria, Würtemburg, Saxony, Baden, &c., by this it is, of course, not meant that the drinker took his glass in hand for the mere purpose of getting intoxicated, but that he made himself drunk for the purpose of executing some crime previously resolved upon during the existence of a protective state intoxication. This is not, and can never be the case with any dipsomaniac, for he drinks and becomes intoxicated only from the compulsion of an internal necessity, and never for any other reason. Dipsomania, therefore, possesses far more pathological than forensic interest, and its further consideration may, therefore, be left to writers upon nosology.

more chloroform was refused to him, he got into a state of unbounded wrath and fury, in which he broke all the household furniture, and at last had recourse to his stock of collodium, which stood always ready," &c. During the intervening periods, this man (a photographer) was diligent, quiet, and rational, " till, after the lapse of a few weeks, the same scene was renewed."

§ 83. ILLUSTRATIVE CASES.

CASE CXC.—INJURY INFLICTED WHILE IN A STATE OF INTOXI-
CATION, COMPLICATED WITH INTERNAL CONGESTION.

The shoemaker Ernest stabbed his neighbour, the woman Straube, with a shoemaker's knife, in the left side, on the afternoon of the 2nd of September, but without inflicting any important injury. A short time before the commission of the deed he had returned home in a state of intoxication, as testified by several eyewitnesses, and speaking of the woman Straube, had said, "there stands the gay coloured spadille," he then ran into the court crying out, "this day more must be similarly served," and amongst other threats, referring to the woman St.'s son, he had said, " I will also do for the one-armed man." Immediately previous to the deed, the woman B. had seen him sitting at the window of his room striking at the air with his hand, and to her warning to take care and not fall out, he replied, " he had done nothing to her." He did not resist his arrest, which took place immediately. At his admission into prison the chief surgeon of the prison found that he laboured under "swollen and inflamed hemorrhoidal tumours." At his precognition Ernest stated that on the 2nd of September he had worked till noon and had then gone out to buy leather. On the way to do this he had drunk two silver groschen's (2½d.) worth of brandy, after that he bought no leather, why, he knew not, but went into another shop where he drank another silver groschen's (1¼d.) worth of brandy, and from that time he knew nothing of what had happened till he regained his consciousness in prison next morning. To his fellow-prisoner, and also to myself subsequently, he made the same statement. His wife, and several acquaintances who were examined, unanimously deposed that this hitherto wholly irreproachable man was quiet and peaceable, and only became violent and furious when he had been drinking. His wife declared that he drank daily one gill of brandy, but at twice, "because he was too weakly to take the whole at once." The court, therefore, considered it necessary to lay before me the question, "whether there was any reason to doubt his responsibility at the time of the commission of the deed?" " The accused," I stated, "is thirty-nine years of age, very pale complexion, a thin weakly man, with but little hair, and with an open but somewhat timid look. His physiognomy is expressive of feebleness and good humour. His speech is somewhat sluggish as well as his whole behaviour and appearance. All his functions are normal, and he no longer has any bodily ailment. He states, however, that he has laboured under severe hemorrhoids, which have often produced headache, congestion, pain in the anus, and tenesmus, particularly when the hemorrhoidal tumours were forced out. These statements possess the coherence of truth, and are all the more credible that the accused from his sedentary mode of life may be regarded as disposed to hemorrhoidal affections. At my first intercourse with him he held firm to the statement that he did not know the cause of his arrest and that he could not remember what had happened on the previous day. Subsequently he always returned to this subject, stating that he knew from the precognitions what he had done and expressing his sorrow for the deed. His statements bear the impress of openness. I have no doubt, from all that I have learned from the documentary evidence and from his own statements and behaviour, that .he was not in the undisturbed possession of his mental powers on the 2nd of September previous to and at the commission of the deed. He constantly repeats that he cannot stand any brandy, at least more than his usual quantity, and all the witnesses have confirmed this. There is, however, also, one particular circumstance which is of some weight in deciding as to his mental condition on the 2nd of September. It is consistent with experience, and it is easily explicable physiologically, that the coincidence of drinking with congestive conditions produces a relatively rapid and violent intoxication. Now we learn from the documentary evidence that E. shortly before committing the deed had drank three silver groschen's ($3\frac{3}{4}d$.) worth of brandy, and also that on the 2nd of September he was found by the surgeon to the prison to be labouring under 'swollen and inflamed hemorrhoidal tumours' at the anus, and this hemorrhoidal affection was very probably the cause of the bodily and mental depression which, according to the testimony of his wife, had lasted for several days before the commission of the deed. The coincidence, however, of these two circumstances might, in a feeble and irritable man such as E. is, very readily produce a condition of mental excitement and incapacity for considerate action, and the words and actions already described prove that he was actually thrown into such a condition. This explains how this ordinarily peaceable and industrious man, who has never been punished, came to execute such a dreadful deed, which might have resulted in the most severe consequences to him, since he might possibly have dangerously wounded both St. and her

son, a deed which he would certainly never—at least not to the same
extent—have executed when in the full possession of his faculties."
Accordingly I answered affirmatively the question put before me.

Case CXCI.—A Careless Bankrupt.—Disputed Civil Responsibility.

A merchant, formerly very rich, had become addicted to drinking
and was now a bankrupt. It appeared that he had played the most
inconsiderate pranks in his business, and proceedings were taken
against him for careless bankruptcy; these, however, were opposed
by the plea of non-responsibility, and this plea was now to be
examined into. I saw him in prison. "Z. is a powerful, burly
man of thirty-six years of age, who openly confesses to have fallen
into habits of drunkenness. On conversing with him as to his present position, and the alteration produced in his circumstances by his
descent from wealth to bankruptcy, he confessed, with some openness
and indifference, how inconsiderately he had acted in business matters,
how, when he was tipsy, it was all the same to him whether he
bought, for instance, fifty dozen shawls all of the same colour, paying
for them in ready money, or proper merchandise, &c. He does not
despair yet, however, and comforts himself with the conviction that
'God will help him.' During a subsequent conversation, when brought
upon the subject of his relations as a husband and the father of a family,
it was evident that he had an affection for his wife and children, but
when his extraordinary conduct in lavishing his means and completely
neglecting his much-loved family was pointed out to him, he had no
other excuse to offer but drunkenness and careless indifference. The
accused has exhibited nothing in his behaviour which could justify
me in assuming the existence of any actual mental derangement.
Unquestionably, however, he is an example of a man completely
degraded morally by drunkenness and dissolute living, and of such a
degree of frivolity of character as is seldom to be met with. In
regard to the declaration that has been made that the manœuvres
that the accused has carried out in his business would scarcely have
arisen in his own head, I will not deny that Z. is a man who, from
his unbounded frivolity and drunkenness, might very readily be led
to commit illegal actions, and might then be easily made the victim
of worse men who have thus abused him. But this does not justify
me in assuming that he is so far deranged as to render the statutes

inapplicable. Not even a certain amount of weakness of memory, which I have observed in him, and which is the indubitable result of drunkenness, could justify me in drawing this conclusion. Because it is possible that he may yet pluck up courage,—and experience has often enough shown that similar subjects, when they have entered upon new paths with a firm and good will, have again regained their former mental and moral elevation. In the same degree as Z. is now aware that he has acted foolishly and inconsiderately in his trade dealings, he must have been formerly aware of it when he was sober, and if he was conscious of this, he knew then, as he knows now, that he could have "acted" otherwise, if he had earnestly desired to do so. Considering now that there is no evidence either of insanity or idiocy, in the Common-Law sense of the terms, nor of any other form of mental derangement, I have to state that the merchant Z. is not to be regarded as devoid of responsibility.

§ 84. INSANITY.—CONTINUATION.—INSANITY OF SOMNOLENCE.—SOMNAMBULISM.

The mental conditions produced by sleep are as well known generally as they are inexplicable physiologically. From the very nature of the matter, however, it is only in the very rarest cases that they come to be discussed *in foro*. *Dreams* are purely phantasmagoric conceptions arising spontaneously in the brain, which continues to act during sleep and during the so-called dreamy waking, without any stimulation produced by the external world through the senses. The basis of these conceptions is partly the remembrance of former impressions, which are reproduced in thousands of different modifications and fantastic combinations, and partly subjective bodily sensations (nightmare, &c.), which give occasion to the most wonderful vagaries of the brain. How strange it is that only certain senses, particularly sight, hearing scarcely ever, and smell and taste still less often, are active in dream life, but neither this nor many other points connected with the physiology of dreaming can be further entered upon here, but must be left for psychology to explain. The dreaming state passes quite insensibly into that of *Somnolence*, that middle state betwixt sleeping and waking, in which the connection with the outer world is neither that of sleep nor waking. The dreaming state is wholly sleep, somnolence (*Schlaftrunkenheit*) is half sleep, half waking. In it the senses are neither awake nor quite

roused, but are surrounded by a cloud of dream phantasms; the somnolent man sees and hears self-made phantoms instead of real objects;
he hears a shot fired, and dreams of it, while it was only a stool that
fell. He reasons logically, as is well known to be the case also in
dreams, in regard to the impressions supposed to be felt, and may,
since muscular action is not prevented by sleep, act in the most
illegal manner. The famous case of Bernard Schidmaidzig,* who saw
in a dream a fearful looking white spectre coming towards him, struck
at it while half awake and killed his wife; that of the young man
who suffered from horrible dreams, particularly on bright moonlight
nights, and who during one of these, when his father got up by
night and he heard the door jar, sprang up, seized his double-
barrelled gun, and shot his father through the breast;† the precisely
similar case of the young man of property;‡ that of the man who,
oppressed by a dream in which he seemed to be struggling with a
wolf, killed his friend sleeping beside him by stabbing him with a
knife;§ the case related by Taylor, in which a pedlar, who carried a
swordstick with him, and had fallen asleep on the high road, was
awakened by a casual passer-by shaking him by the shoulders, and
who drew his sword and fatally wounded the stranger;‖ these and
other older but similar cases are mournful proofs that the most
horrible deeds may be perpetrated in the dreamy state of somnolence.
But it is undisputed, and so evident as to require no illustration,
that such deeds committed in this condition may be regarded as
proceeding from that derangement of self-consciousness arising from
delusion, which I (§ 73, p. 192, Vol. IV.) have shown to be the
essence of "insanity," and consequently dreaming and somnolence
come to be considered solely under the head of insanity in regard to
the question of responsibility, because the actions of the somnolent
are not influenced by the laws of morality, nor his relations to the
external world, nor the consciousness of these influences, but only
by his own dark and obscure forebodings and feelings. For that
reason it will not be difficult to make a correct diagnosis, in any
case that may occur, when the plea of somnolence is only brought
forward to obtain immunity for some crime committed with full

* Klein's Annalen der Gesetzgebung, Bd. VIII.
† Henke's Zeitschrift, 1851, S. 346.
‡ Casper's Vierteljahrschrift, XII. 2, S. 327.
§ Œsterr. Zeitschrift f. pract. Heilkunde, Bd. I. S. 42.
‖ Knaggs, *op. cit.* p. 52.

responsibility. The general diagnostic rules already (§ 61) laid down will at once supply the means of rightly deciding as to the existence of this peculiar and remarkable condition.

Somnambulism is a condition closely allied to the foregoing. It is well known to medical men of experience that children are frequently observed to get up at night, particularly in bright moonlight nights, and go to their mother's bed or into another room, &c., before they can be brought back to rest. In one family of five children I have observed this to happen with every one, and in them it disappeared with their growth, as it usually does in children. When we know, however, and who does not know it, how very uncommon somnambulism is in adults, then the dozens of narratives of the most remarkable and incredible pieces of work performed by swimming, climbing, hewing, engraving, playing, and writing somnambulists must seem all the more remarkable, and excite all the more critical doubts, that the larger proportion of them, and that no small number, date from previous centuries, and the present and recent times are very poor in such cases. All this points with tolerable certainty to the defective observation, and the credulity or deceit of former times. For the avoidance of the latter the general psychologo-physiological examination of the case (§ 61) will be of more use than the bandaging of the eyes of the doubtful somnambulist, which has been recommended, but could scarcely have ever been put to the proof, or than calling to him by name, &c. How extremely doubtful the following case—one of the early ones—seems. A servant in Halle, who was a somnambulist, fell in love with a girl, who promised to marry him. But another sweetheart of the girl's excited his jealousy, and the idea that the latter passed the night with the girl grew always stronger in him. One night he got up, went out at his garret window, passed over the roof of the houses to the window of a neighbouring one, through which he descended into the room and murdered the sleeping girl with a knife which he had taken with him. He went back in the same way. At the precognition he represented the whole affair as a dream which he had dreamed.* Thus this was murder from jealousy! and committed in a fit of somnambulism? Did the girl sleep with an open window, or did he break the glass to get in without awaking the girl? Was the statement of the culprit sufficient proof of the fact of the somnambulism? I do not doubt but that a thorough sifting of the case

* Stelzer, über den Willen, S. 273.

would have ended in a very different result. Moreover, the somnambulist is of course also a dreamer, a man in a somnolent condition,
consequently, any cases which may occur are to be decided accordingly
(as cases of insanity).

§ 85.   ILLUSTRATIVE CASES.

CASE CXCII.—COHABITATION PERMITTED IN A STATE OF ALLEGED SOMNOLENCE.

The following case is rather a rare medico-legal curiosity. The
brewer's servant H. was accused by the Restaurateur F. of having, in
the night between the twenty-eighth and twenty-ninth of May, gone
to bed with his (the accuser's) wife, and had connexion with her.
Mrs. F., as she gets up very early, actively superintends her household affairs, and never gets to bed till late, is said to be a very sound
sleeper, and this was the case on the night referred to, and thus it
happened that H. had his way with her.   "By an order, dated the
twenty-first of this month, I have been requested to state whether
the § 144, No. 2, of the Penal Code is applicable to the deed
committed by the accused?   This paragraph threatens with penal
servitude any unchastity committed on any one in an unconscious or
involuntary condition.   But Mrs. F. cannot be regarded as having
been in either of these conditions at the time of the commission of
the deed referred to.   At her precognition, for example, she stated
that she 'at once' felt that some one lay upon her, and had placed
his genitals in connection with hers, that upon this she roused herself and asked, 'husband, is it you?'   By this deposition Mrs. F.
has distinctly proved that she was perfectly conscious, since she felt
that some man lay on the top of her and asked him, whether he was
her husband?   Indeed, the doubt expressed in this question, proves
that she, which is perfectly credible, perceived a difference in the
personality of her bed-fellow, consequently, she must have been conscious and could not have been in a state of deep sleep or somnolence in which consciousness is in abeyance.   But if the presence of
consciousness cannot be denied, then an involuntary condition is not
to be supposed in an adult, young (twenty-nine years of age) and
healthy woman, and this requires no further explanation.   Accordingly I gave it as my opinion, that the married woman F. was not
in an unconscious or involuntary condition at the time of the
cohabitation, which is the subject of the accusation."

CASE CXCIII.—A STATE SIMILAR TO SOMNAMBULISM.

A peculiar condition analogous to somnambulism was observed in the case of a boy, aged fourteen years, and gave occasion to the judicial query, "is this boy in a deranged state of mind?" He appeared behind his age both in growth and appearance. His head in particular was remarkable by being flattened behind; his dark bristly hair covered his forehead, and his look was timid, blank, and ordinarily directed to one point. A constant smile completed this picture of stupidity. According to the statement of the father, the boy was for years in the habit of leaving his home in the evening of every month during the increase of the moon, and running about for two days and two nights, roaming about shelterless. From the night-watchers, who had several times arrested him, it was learned that he spent these nights in churchyards, new buildings, &c. The father could give no reason for these rovings, since the boy had at home "a good bed and every attention." Every means, even binding him firmly, had been tried in vain, for the boy always burst his bonds, and breaking the windows made his escape into the open air. From himself, however, I could learn nothing either in regard to this roaming propensity nor anything else, however trifling, for to every question he only replied by a silly smile, and could scarcely give his own name correctly. I had to answer the judicial query affirmatively.

§ 86. INSANITY.—CONTINUATION.—PASSIONS AND EMOTIONS.
STATUTORY REGULATIONS.

GENERAL COMMON LAW, § 29, Tit. 4, Part I. (Like insane people are to be considered.) *Those, who by terror, fear, anger, or any other violent emotion, are placed in such a condition as to be deprived of the use of their reason.*

PENAL CODE, § 41. *No crime or offence exists if the deed was committed in self-defence. Self-defence is that form of defence necessary to avert any actual illegal attack from one's self or others. The plea of self-defence is also to be taken into consideration even when the agent has from surprise, fear, or terror, gone beyond the bounds of mere defence.*

§ 177. *If a homicide has been, not by any fault of his own or by any ill-treatment or serious affront inflicted by himself or any of his belongings, irritated to anger by the deceased and thus hurried into the deed upon the spot, the punishment of penal servitude for life is not to be applied, and the sentence shall be imprisonment for not less than two years.*

§ 196. *If in the case of any ill-treatment or infliction of any bodily injury the agent has been, not by any fault of his own or by any ill-treatment or serious affront inflicted by himself or any of his belongings, irritated to anger by the party injured, and thus hurried into the deed upon the spot, or if it be ascertained that there are any other extenuating circumstances, then in case of the death, &c.* (the regulations as to the punishment follow).

I have already (§ 72, p.189, Vol. IV.) enumerated the passions as causes of insanity, and I have now to detail their influence upon those illegal actions of persons of sound mind, which are exclusively occasioned by passions and emotions. It is certain and indisputable, because every one has had sufficient proof of its truth in his own experience, that every one can regulate the congenital and partial inclinations of his feelings and desires (passions), or indeed even their extreme excitement, which may suddenly arise and as suddenly cease (emotions). And it is also as generally indisputable that he must govern them. The opposite idea would very speedily lead to the breaking up of society. Therefore, it is indubitable that mere passionate or emotional excitement cannot exclude responsibility. But it is another question, whether there may not be circumstances which compel us to suppose that the general possibility of governing the passions may, in individual cases, be cancelled? By which, in order to prevent all mistake, it may be remarked, that, as is self-evident, it is not meant to include such passions as do not act with the suddenness of emotions, but more chronically, and which are to be regarded more as vices than passions. No one has ever reckoned it to the advantage of the gambler that his passion has made him squander his means and brought him, at last, to a careless bankruptcy, or of the miser, that from his passion for gain, he has given his child a prey to slow starvation! But as regards those passions which act suddenly, any further examination of this question could be of but subordinate value for the practice of forensic medicine, since every Statute Book, from the Roman one downwards, has long since decided it positively in the affirmative, so that the Judge finds a sufficient basis for his sentence in the statutes themselves, and does not require the co-operation and assistance of the physician, and therefore he very seldom asks it. The Prussian Penal Code, in unison with all the other German Penal Codes, provides that in cases of defensive illegal actions committed in " surprise, fear, or terror," all responsibility shall be excluded (*vide* the statutory regulations, p. 277, Vol. IV.), and places the "irritation of anger" by which the culprit has been "hurried into" (of itself a most significant expression) the commission of any offensive action on a level with " other" extenuating circumstances, that is, it practically provides that a diminished responsibility shall be held to exist in the case of actions committed in the heat of anger. In particular, as experience teaches, such a state of mind as excludes responsibility may be instantly produced by any sudden and unexpected attempt upon one's dearest earthly possessions, life, honour, or property, with which the whole soul is bound up. The individual is placed in a state of " surprise, fear, or terror," he becomes distracted. The harmony of his mental powers is disturbed and dissolved, he has not a word to say to the slanderer of his honour, he stares apathetically into the flames which have suddenly and unexpectedly swallowed up all his possessions, or, in this discordant condition, in which he has lost all knowledge of the results of his actions, he replies with a lethal weapon to the sudden attack upon his life or honour, and acts with the utmost rashness, as he probably would not otherwise have been capable of doing. Precisely in a penal respect we must not omit the consideration of a cause which is, under certain circumstances, so peculiarly fitted to produce "surprise, fear, or terror," and consequently distraction, in the production of which the helpless condition of the body must also be regarded as co-operative. I refer to *the act of parturition* in unmarried, lonely, and destitute women (*vide* pp. 186, 190, Vol. IV.). The state of distraction has a close psychological connexion with that of dreaming, therefore the statutory regulations of all times can only be regarded as justified. It is different with regard to passions so dangerous in their effects as anger and revenge. How powerful their influence is sufficiently exhibited in their purely corporeal effects. The pulse is quickened, the face and eyes are reddened, the temperature of the body rises, and the secretions and excretions are rendered more active. That a state of such great excitement may produce a restraining action upon the " free exercise of the will" (Penal Code) just as well as the perfectly allied condition of intoxication, is just as certainly to be supposed, *à priori*, as it has been actually proved by experience to occur, and the old proverb *ira furor brevis*, as well as the expression "*Zorntrunkenheit*," *anger-intoxication*, are more than mere similes. If it be thought right to adopt Plattner's term for this condition, and call it *excandescentia furibunda*, to this there can be no objection, so long as no attempt is made to construct from it a peculiar species of mental disturbance, which it by no means is, and against which precisely the same objections may be made as have already been made in regard to the similar scientific blunders, *amentia occulta, mania sine delirio*, and *mania transitoria* (*vide* pp. 194, 199, 204, Vol. IV.). In any given case, in which it may be doubted whether the accused was actually so much excited by anger as to be so "hurried into" the commission of the deed as that the free action of the will was excluded, or the reverse, the diagnosis will have to be made in accordance with the general diagnostic rules (§ 61, &c.), which are applicable to these as well as to every other case of disputed mental condition. In these cases, as well as in those of intoxication, further corroborative evidence which may enable us to guard against the mere lying pretence of a blind fury, may be gained, if, on examining the individual, we ascertain the existence of circumstances whose concomitant influence must of necessity have greatly increased the emotional excitement, such as a peculiar irritability generally, a so-called choleric temperament, bodily disease, particularly epilepsy, functional disturbances of the abdominal organs, &c., or, as is so frequently the case, a contemporaneous state of intoxication.

§ 87. ILLUSTRATIVE CASES.

CASE CXCIV.—INSANITY OF ANGER.

On the afternoon of the 29th of April, the writer B., came home drunk, as his wife deposed, and commenced to quarrel with his children. He behaved himself in the court, according to the statement of the witness R., "like one deranged," and, holding his child, one year and a half old, in his arms, he struck like a madman at the surrounding workmen. R. sought to quiet him, whereupon he held out his hand to him, called him his friend, and asked him to accompany him into his dwelling. When they got in, B. flung his child into the bed which stood from three to six paces off, and with the words, "What do you want in my house?" struck at R. with an instrument which

he had previously concealed in his sleeve ; R. parried this, and summoned two soldiers. These followed B., apparently quite quiet, to his
house-door, then he caused them to return into his dwelling with him,
when he wrested the musket from one of them, stood upon his guard,
and bit one of the bystanders, till at length he was bound and
carried off. R regards him upon this occasion as maniacal rather than
drunk. His landlord also states that, from similar previous occurrences, he was firmly convinced that B. laboured under " temporary
insanity," and according to the testimony of another witness was at
the time " in such a state of rage that his behaviour was like that of
a madman." His mental condition, of course, had to be inquired
into. I found a man, aged thirty-nine, of a short, powerful build,
slightly icteric complexion, otherwise in good bodily health. He
confessed that he was of an unusually violent temperament, and that
he could not endure contradiction nor an attack of any kind, because
he was thereby at once irritated to an extreme degree. He confessed
that upon such occasions he had often abused his wife, though not
with deeds, and that he had repeatedly broken the household goods
and furniture, and this his wife confirmed. A previous occurrence
was important. While taking a walk outside of the town he was
separated from his wife by a threatening storm. From a mis-
understanding she went home, whilst he thought that she was
to wait for him at the gate. After seeking up and down for a long
time and waiting in vain, during which he became always the more
excited, he also finally went home, when he found his wife had
arrived before him. On this occasion, he was so transported with
wrath, that he was no longer master of his violent actions. He
wished to unclothe himself, and he did so in a most unpurpose-like
manner. He wished to ease himself, and could not place himself
upon the night stool, so that he left his ordure lying in the room, &c.
He confessed that similar occurrences had often happened to him.
When quiet and unirritated, that is, in his ordinary condition, B.
was rational, behaved himself becomingly and properly, carried on
his business, maintained his family, and betrayed not the slightest
trace of mental disturbance. Consequently it must be assumed
that this was one of those rare cases in which an excited mental
emotion, such as chagrin, anger, &c., has risen for the moment to
actual mania, especially considering that in this case there was the
concurrence, if not of senseless intoxication, at least of tipsiness, and
after quoting the reasoning given in the text, the question put before

me in regard to the irresponsible condition of B. at the time of the commission of the deed was answered affirmatively.

CASE CXCV.—HOMICIDE DURING THE AFTER EFFECTS OF A SEVERE FIT OF INTOXICATION.—DIMINISHED RESPONSIBILITY.

A most instructive case, from the manifold concurring circumstances. It occurred sixteen years ago, under the former Penal Code, which, like our present Civil Code (*vide* p. 93, Vol. IV.), laid down the precise degrees of responsibility, a matter of which the present Penal Code takes no cognizance; but I shall return to this subject presently. The toolsmith, Loch, never previously punished, had on the 5th of August fatally injured Mrs. Bugges, his neighbour, by blows on the head with a smith's hammer. There had been a quarrel and altercation between them on the evening before the deed was committed, and from vexation at the low abuse which had been heaped upon him, he had gone late in the evening and drank several groschen's (one groschen = 1¼d.) worth of brandy, so that he came home so drunk that his wife had to take off his clothes and put him to bed. "Next morning when he rose," his wife states, " he was, as usual, after having been drunk at night, quite confused and stupid, so that he was by no means himself." A quarrel again fell out between him and the woman Bugge, who was speedily joined by her husband. Loch ran from the entrance of the house across the court into his cellar-dwelling and bolted the door, the man Bugge seized a besom shank, and struck with it against this door, while the woman Bugge continued to pour forth her abuse, so that, as widow G. has deposed, "the man must have had no gall, if he had not been excited by it and got into a rage." Loch finally opened his door, and now the man Bugge struck him in the face with the besom shank, so that it broke, demanding how he could so illtreat his wife? Loch replied, "How long will you speak so cunningly? now I will strike you all dead." Upon which he fetched a smith's hammer out of his cellar, and let drive with it at the man Bugge, but struck him but slightly. Mrs. Bugge hurried to the assistance of her husband, upon which L. turned upon her, and saying in a fury, "are you, cursed madwoman, there too?" he struck her with the hammer on the back part of her head; after this he went back to his cellar, and commenced to work at his forge, whistling at the same time. Mrs. Loch states in regard to his then state of mind, that "he was quite

out of his mind and senses, he spoke no word, and drank no coffee."
To the question of the police officer, who arrested him shortly afterwards, what made him act so? he replied, "Oh God, what does not
one do from thoughtlessness!" he was then, however, " very cool,
quiet, and not in the least excited." In regard to his general character,
every one of his acquaintances spoke well of him; and in regard to this
it is worthy of remark, that a document with many signatures from
his fellow-workmen was handed in, dated the 30th of November, in
which the leniency of the Judge was bespoken for him, and he was
described as an "honourable, upright, extremely good-tempered,
friendly, and sociable man." All the witnesses, however, agreed that
he was very fond of brandy, and that though when sober he was orderly, quiet, and industrious, when drunk he was always choleric and
quarrelsome, that then "he did not know himself," and that when he
had been drunk " for several days he could do nothing, and was quite
beside himself." The statements of the accused in regard to his mental
condition at the time of the commission of the deed completely agreed
with this evidence. " I know not what I have done," he said at his
precognition, " even if I were to be instantly hanged on the gallows
for it. I was so excited and maddened by the language of the woman
Bugge, that I do not know whether I struck her or not. At the instant the deed was committed all the colours were glancing before
my eyes, and my old complaint fell upon me :" in answer to a query
respecting this, he further stated, " Oh God! I will say nothing
about it, I am often seized with violent attacks of heat and congestion of blood towards the forehead." He further said, " I do not
know whether I had a hammer in my hand or not; if I had as much
sense as to know that I had a hammer in my hand, then I would
have had enough of sense to leave it lying." Finally, he declared
with tears that he bitterly repented of the deed, but that he had no
recollection of the circumstance attending it. — I stated, in my
report, " Loch has made precisely the same statements to me in my
private conversations with him. The accused is a very tall and somewhat slender man, thirty-nine years of age, but older in appearance,
very pale, with red and inflamed-looking eyes—probably caused by his
trade and by brandy drinking, which also explains a slight tremulous-
ness of the hands—and a mild and good-natured expression. His demeanour is quiet and composed, with a certain amount of seriousness;
his speech is slow, distinct, and gentle; his great disposition to
weep displays no small degree of irritability in his nervous system.

Even when brought upon the subject of the deceased woman
Bugge and her family, he never displays the slightest trace of violence, of irascibility, or of hostile feeling, but states with his
wonted calmness how wickedly that family have behaved towards
him. As to his bodily health, L. complains of a constant painful
feeling of distension in the superior part of the abdomen, which also
feels somewhat hard, of a great tendency to constipation, and of frequent headaches, which at times deprive him of his senses, or make
him confused. He has had these attacks tolerably often in the prison.
' Don't think, however, that I am therefore deranged,' he said to me
quite voluntarily and extremely characteristically, ' I have the full use
of my reason.' As if quite by the way—and his whole demeanour and
mode of expressing himself had the appearance of complete absence
of design—he stated to me, that the moon had a remarkable influence
upon him, and that in moonlight nights he was always restless and
sleepless. He also maintained to me, that his crime seemed to him
' like a dream,' and that if he had the use of his reason at the time
he would certainly not have done it.

"Authors worthy of all respect have assumed the existence of a
peculiar transitory condition, produced by the excitement of violent
and angry emotions by which the passions become so unbridled and
blind, that it is impossible for the excited party to preserve in his
momentary action the laws of morality. It would be an easy matter,
supported by these authorities, to refer the crime of L. to this so-
called *excandescentia furibunda,* and thus to declare him irresponsible.
But the task of a psychological physician is a loftier one than merely
to catalogue an individual mentally and morally under one of the
headings laid down by science; it has to enquire how in each individual case the crime came to arise in the mind of the perpetrator,
and whether the cause and its result bear a general, necessary, and
normal relation to one another, or not? If Loch have committed
this homicide in an ordinary fit of rage, such as he was so subject
to, or from a vindictive feeling against the woman Bugge, then he has
no more right to be declared to have been at the time the deed was
committed incapable of recognising the connexion between his action
and its statutory results, or irresponsible, than any other person who
has committed an offence in the heat of passion. But in this case, circumstances have been at work which seem to compel us to assume
that this has been a case of something more than a mere ordinary fit of
anger. A man of such an irritable nervous system as Loch is, indeed, of

such a morbid irritability that, according to his own statement, which there is no reason to doubt, he has for years laboured under a slight degree of that form of nervous disease which is termed somnambulism, a man who suffers from periodic headaches of such violence that they make him senseless and confused; a man, finally, whose fundamental character is predisposed to violent paroxysms of passion, to fits of anger, such a man gives himself up to drunkenness, an influence which more than any other enfeebles, irritates, and destroys the nervous system. The extent to which this destructive influence has been already exerted in Loch, is not only exhibited by the tremulous state of his hands, not only by his complaints of abdominal disease, which bear the stamp of truth, but also and quite peculiarly, by the unanimous statement of all the witnesses as to the condition in which he was after having been drunk, which he so often was. I do not now refer to that irascibility which is common to thousands when deeply intoxicated, but specially to that condition in which the accused was sworn by the witnesses to be for several days after a fit of drunkenness, in which he was incapable of work, and behaved himself in a most extraordinary manner. It is proved, that on the 4th of August, late on the evening previous to the commission of the deed, and, as is well worthy of observation, while already excited by anger and quarrelling, he had drunk 'several groschen's' worth of brandy, and that he, very naturally, had got so drunk that his wife had to take off his clothes, and put him to bed. Further, his statement, confirmed by his wife, that he spent that night, in which according to the calendar the moon was in her first quarter, sleeplessly, is quite credible; and it is equally credible that when he rose next morning he trembled over his whole body, and that his wife did not find him 'quite steady;' him, in whom the effects of intoxication lasted often for several days. In this extraordinary, morbidly irritable condition the quarrel with the Bugge family again commenced, and speedily resulted in such vile abuse, that the 'man could have had no gall' who could have peaceably submitted to it; this abuse does not cease when Loch sought to escape from it into his dwelling, it becomes now irritating and ends in actual assault; the man Bugge breaks a besom shank over his head, and Loch, now quite overcome with passion, and without having any definite object in view— for he struck first at the man and then at the woman Bugge—knocks the latter on the head with a hammer. I do not require to prove that this deed, done under circumstances such as I have just described, was not committed in a state of full and unlimited responsibility, because in regard to it there is a concurrence of so many circumstances, each one of which is enough to limit, or occasionally to destroy, the responsibility of a man. But I can just as little convince myself that Loch committed this deed in a state of absolute irresponsibility. Because, for example, he knew that he saw the woman Bugge, 'that cursed madwoman,' before him, he knew that he had a so-called lethal weapon in his hand, indeed he declared heedlessly his intention ' to kill them all with it,' and not one of the witnesses has been able to state that he was at that time in a perfectly senseless condition. Indubitably he had lost command of himself, he was beside himself, but he had not lost his senses; he knew to select an appropriate means, when he fetched the hammer to defend himself from the assaults and abuse of the Bugges, and thus, in conclusion and with respect to the facts just recited, I can only give it as my opinion, *that Loch committed the homicide while in a state of only diminished responsibility."* — This opinion was accepted, and a lenient sentence was passed upon him. The question is, however, forced upon our attention, how would a precisely similar case have to be treated in the present state of the Penal Code? If the Judge should now, as then, place the "Responsibility," of the accused in question, I should not hesitate for the reasons already stated (in § 60), to deliver precisely the same opinion (Diminished responsibility). Should, however, the query be put as to the "Insanity or Idiocy" of the culprit in accordance with the words of the Penal Code, then both of these diseases would have to be declared not to exist, the reasons for which are contained in my report; but, at the conclusion, it would have to be pointed out, that the influences proved to exist, especially when taken together, excluded the idea of unrestricted and perfectly free action at the time of the commission of the deed. Unquestionably the Judge on his part would then make proper application of the clause of the statute in regard to "Extenuating circumstances," with which the medico-legal report has nothing further to do.

CASE CXCVI.—ALLEGED STATE OF IRASCIBILITY.

This was the case of the perpetrator of the attempted robbery and murder of the young woman suckling her child, the description of whose wounds is related in Case CXXXVI. (p. 56, Vol. IV.), a case which is certainly a most remarkable psychological one. I have already stated that the woman was sitting quietly in her room, when the accused entered and requested her to show him the old metal, which he alleged he had bought from her husband. She asked him to wait, declined to accede to his repeated wish that she should go into the front room, and he sat down quietly beside her. Shortly afterwards he requested a knife to cut his cigar, and with it he suddenly attacked her, threw her on the floor and cut and stabbed her till she pretended to be dead, when he hurried into the front room as if to rob it. He fled, when the wounded woman got up, but was held fast and arrested. At the trial it was mentioned that six years previously he had committed a precisely similar attack upon a watchmaker in K., apparently without any motive. At that time he had stated as an excuse, that seven years previously a tile had fallen upon his head, and that since then he had frequent attacks of absence of mind, and that in particular he was quite deranged when excited by spirituous liquors or other wine. Shortly before committing the assault upon the young woman he made indecent proposals to her which were rejected, moreover he was tipsy, and both together had " put him in a fury." I give the most important points of my report. " Müller is thirty-four years of age, powerful, and in good bodily health. His look is lively, almost fiery, his complexion always high. In regard to his mental state, the following witnesses for the defence have deposed. F. that the accused ' was always somewhat short in his temper and very irritable,' that, however, he (the witness) could not remember any instance of his having been so excited as to be out of his senses. The police sergeant B. has deposed precisely the same. On the other hand, the witness S. testified that he knew Müller to be ' an extremely hot tempered man, who got quite beside himself for the merest trifle, and sometimes acted quite irrationally.' The witness B. also testified to his ' unrestrained irascibility.' Mrs. L. said he was 'quarrelsome, irascible, and vindictive.' To myself also the accused has described himself as hot-tempered and easily irritated, particularly when tipsy or otherwise excited. In regard to the accident from the tile, it is very remarkable that Müller could give me no particular information about it, not even as to the time of its occurrence, which he sometimes said was seven, at others ' more ' than seven, and again, ' about' seven years ago, whilst the documentary evidence states, that in the year 1848 he had spoken of the accident having just previously occurred, so that it must have happened more than nine years ago. Others of his former acquaintances have no knowledge of this accident. On a careful examination of his head, which is covered but thinly and sparingly with hair, I could not discover the slightest trace of any such blow from a heavy stone upon it, in particular there was no ' depression' as the accused asserted, nor was there any scar. All these circumstances render this allegation in the highest degree suspicious. Granted, however, that the accident did occur as the accused states; granted that the fall of a tile upon the head may exercise a disturbing influence on the mental powers of the party injured, though happily in most cases this is not actually observed; yet, as in the case of all similar alleged injuries, we have first of all to decide in each individual case the preliminary question, whether the existence of any mental disturbance, previous to or at the time of the commission of the crime, has been proved? And when this is answered affirmatively, we may then fall back upon the alleged cause in explanation of its origin. There is, however, not the slightest proof of the existence in Müller of any mental derangement capable of limiting or destroying the freedom of the will in regard to action. There is nothing in his whole habitus, nor in his speech or statements, which is at all remarkable, or calls to mind in the least the behaviour of a lunatic. I may remark, that of course I had no opportunity in the prison of observing Müller when artificially excited by drink or otherwise. From his whole behaviour immediately before and at the time, as well as after the commission of the deed, I must, however, entirely doubt that any such morbid excitement, produced by the causes alleged by him, had hurried him into the assault upon Mrs. N. It is something more than merely remarkable that he alleges two causes, intoxication and lust. 'I was senseless from drink,' he declared at the examination on the 8th of December. But Mrs. N. has stated that he was 'quite sober,' and it is not to be supposed that this young woman with a child at her breast, would have sat so long alone in her house with a perfectly unknown and powerful man 'senseless from drink,' would have asked him to wait, &c. The excuse with which he introduced himself; the wish expressed to get into the front room ; the request for a knife to cut his cigar, &c.; all are opposed to the idea of his being ' senseless from drink.' It is also just as little credible that libidinous desire had rendered him furious, for, besides that he himself only brought forward this idea at a late period, and still speaks hesitatingly and uncertainly in regard to it, Mrs. N. wholly denies that he ever made any indecent advances to her, which she repelled. Though I do not mean to deny that Müller's deed is, psychologically, both peculiar and remarkable, yet I am far removed from attributing it for that reason to some morbid impulse. It is unquestionable that, according to the statement of the witnesses cited, the accused is a man of an extremely irritable disposition; a man whom every opposition, every contradiction was capable of exciting to wrath and fury. Such an opposition he found at this time to his desire to be taken into the front room, where, we must assume with the indictment, he intended to commit a robbery, and, in a character such as his, it is far from improbable, that he then set about gaining his end by committing a deed of violence. How little he was 'senseless from drink,' or excited by lust, he proved immediately after the commission of the deed, by listening with the utmost calmness to the apparently dead woman, in order to convince himself that he had no further opposition to fear from her. But also his subsequent behaviour after the deed proved distinctly how well he knew what he had done. He was, as the witnesses examined have proved, that same evening and during the following days ' uneasy, restless, anxious, absent, thoughtless, sought to conceal his countenance,' and alleged that a wound of his hand received during the scuffle, was the result of a fall upon broken glass. All these facts agree in proving that there was no morbid emotional excitement at the time the deed was committed, and I must rather, for the reasons just given, declare that it is my opinion, that *Müller was a responsible agent when he committed the deed, and that he is so still.*" In the jury court there was brought forward as a proof of Müller's unbounded irascibility, that some years ago he wished to train a dog, and when he did not find it docile enough, he dashed it so violently against the wall as to kill it on the spot. One of the jury (a pensioned staff-officer) raised the question, how a man of such a character could have been promoted to be a non-commissioned officer, and requested to hear the certificates of the military authorities. These were read in court, and were found to be *favourable*. I proceeded to state that these certificates afforded a most important support to my opinion, because they proved that M. *could restrain* his irascibility, when he *must* do so, as was the case in military matters. He was sentenced to fifteen years' penal servitude.

§ 88. INSANITY.—CONTINUATION.—THE SO-CALLED MORBID PRO-
PENSITIES.

The *natural impulses*, the impulse for self-preservation, those of
hunger and thirst, and the sexual impulse, may be the occasion of
immoral and criminal actions, and may consequently come in question in regard to the mental condition of the culprit, and his responsibility. It is only in the rarest cases, however, that the assistance of the
physician is required, since the statutes, in their regulations respecting
" self defence," " passions and emotions," and " extenuating circumstances," everywhere present the Judge with a suitable basis whereon to
frame his opinion. Where the instinct of self-preservation has sought
to rescue the existence threatened at a forbidden price, all that the
Judge has to determine is the fact of its being a case of "self-defence."
When the long unappeased desire for nourishment, which is only a
modal development of the instinct of self-preservation, has hurried
its victims into the most horrible and detestable deeds, as has been
the case in those who have been shipwrecked and driven up and down
upon the ocean, or have been buried alive, or in any other way shut
up together and apart from the world, then we may leave it to the
Judge to determine whether he will regard the case as one of self-
defence, or extenuating circumstances, &c. ; it is certain, that in these
very rare cases he will very properly consider himself authorized to
decide the case alone. Often enough on the other hand, there occur
in practice cases of trifling crimes and offences, such as thefts of food,
alleged to be caused by hunger. But in these also the Judge finds,
in the circumstances of the individual case, *criteria* sufficient to
enable him to decide without any medico-legal assistance as to
the punishability (responsibility) of the culprit, and I myself have
only been called upon in a few such cases, where, from apparent imbecility, a more accurate psychological estimation of the accused
seemed to be required, the cases after that presenting nothing peculiar
as to their decision. It must appear more than doubtful whether,
as is so often asserted, the *sexual impulse* in healthy men may, when
long unsatisfied, exert a power compelling the commission of irresponsible penal actions. In men this cannot be supposed to be the
case, because nature, by the psychological involuntary seminal evacu-
ations, has set bounds to the increase of the sexual excitement, to
which may be added the indisputable fact, common to both sexes,

that by long continence the sexual desire is not generally increased,
but rather diminished and suppressed. For these reasons, therefore,
. very properly and naturally, in all the numerous and continually recurring cases of accusations against men for rape and other sexual
crimes which have come before me, it has never once occurred that even
an attempt has been made by the advocate for the defence to assert
the constraining influence of the sexual impulse of the accused as a
mitigating circumstance, or, moreover, that the case has been decided
in this sense.   In healthy women we see often enough in all classes,
from the noble and famous historical dames down to the lowest
grades, the most immoral debauchery from libidinous desires;
but none, except, perhaps, those who belong to that school which
refers everything to nerves and blood, &c., would recognize in' this
a blind, transporting power. It is different in regard to the diseases
of the male and female organs of generation (*priapismus*, *satyriasis*,
nymphomania), which may produce insanity of an erotic character,
but on the one hand this is so well-marked that it cannot be mistaken, while on the other it cannot be recognized as a peculiar
species of general insanity, and it does not require the laying down
of any peculiar rules as to responsibility.

It has not, however, been considered sufficient to uphold views
diametrically opposed to these in regard to the natural instincts, and
to attribute to them the possibility of producing such an overpower-
ing influence as shall exclude responsibility in healthy men, but also,
and that alas! by the overwhelming majority of medical authors, the
unscientific and untenable doctrine of *morbid impulses*, and their constraining power in the causation of crime has been sought to be
introduced into forensic medicine and practice for the benefit and
advantage of frail mortality,—a theory which ought to be extir-
pated root and branch. There is no chapter of the science which
displays more of the ontological tendency I have so often had
occasion to oppose, none which exhibits more of the desire to genera-
lize and to force the clear expressions of mental life into a nosological
scheme, in none is the influence of incomplete and defective observation of individual cases, the want of thorough criticism, a more
disgraceful abuse of the word " experience," or, finally, the effect of
ignorance of forensic practice and the criminal world in theoretic
teachers more clearly evident or dangerous than in this. Inventions
like those of a *mania sine delirio*, *amentia occulta*, &c., are not to be
compared in their danger with that which would place the meanest

u 2

crime under the shelter of some morbid propensity in the accused, who was thus as it were predestinated to be a thief, a fire-raiser, a ravisher, or a murderer! For we have received into forensic medical literature, as "confirmed by experience" (!), a propensity for theft, a propensity for fire-raising, a propensity for sexual crimes (*Aidoiomania*, Marc, &c., *vide* Case CLXXX. already related), a propensity for murder; and with this tendency and absence of criticism, which I have so often opposed, there is no reason why we should not also have a propensity for cheating, from which might be deduced the morbid constraint to swindle, forge, and embezzle. Meanwhile, I, along with others, my honourable predecessors, few alas! in number, must lift my voice against this mode of treating forensic psychology, against this tendency to turn the Penal Code into a mere catalogue of morbid propensities. And in the first place, I beg to refer to the reasons bearing upon this subject, which have been already related in considering the question of mysterious voices said to have impelled the culprit to the commission of the crime, as well as in examining the expressions *causa facinoris* and monomania (§ 61, p. 103, § 63, p. 111, § 79, p. 234, Vol. IV.).

§ 89. CONTINUATION.—PROPENSITY FOR THEFT.—KLEPTOMANIA.

Mathey defines his Klopemania (subsequently termed Kleptomania) as a "propensity for theft without necessity, without being driven to it by the urgent wants of poverty."[*] Not only the criminal jurist or experienced forensic physician, but every subordinate officer of police, every shopkeeper in a large town, would laugh to hear this so oft-copied definition of the "propensity for theft," this complete misapprehension of the term *causa facinoris*, as if it was only " *le besoin pressant de la misère*," that was the cause of theft! That this is *not* the case in five-sixths of our common thieves, that the swell mob of every European capital immediately squander what they steal, are facts too well known to require proof. From the records of the Berlin police courts, and from my own experience, I have come to know more than *one lady of elevated position* who has stolen silk stuffs, &c., out of the shops, not from " *misère*," but from a love of dress, which could not be satisfied in an honest way. For the same reason a poor, but not needy woman, stole firewood placed in her charge. A woman of property living in a good position, repeatedly

---

[*] Nouvelles recherches sur les maladies de l'esprit. Paris, 1816, pp. 134, 146.

stole, till she was apprehended, roasts from a butcher's shop where she went to buy, not indeed from "*misère*," but because her means were not sufficient to supply as many roasts as she wished to consume (Case CC.). A young and highly-educated foreigner, many years ago, had been received into the highest circles in Berlin, and had been much talked of. At length I also made her acquaintance in a ——— criminal prison. She had committed a great robbery by breaking open the secretaire of a·female friend (*vide* also Cases CCII. to CCIV.). The inclination which collectors of objects of art or nature, curiosities, &c., have to gratify their love of collecting by illegally appropriating any remarkable objects which cannot be purchased, and which they see in any museum or collection which they visit, is everywhere well known. After so extensive an experience of this character, Marc's* advice, that "the first circumstance to be considered is the social position of the accused, and the value of the object stolen in proportion to his means," seems perfectly worthless. Will the other criteria laid down by Marc, the great advocate for "Kleptomania," stand the test any better? "The chief proof of this monomania," he says, "is the voluntary confession of the thief, and especially the restoration of the object stolen, or, at least, the speedy reimbursement of any loss caused to another." Thus the voluntary confession of an evil-doer is made the proof of the existence of mental derangement, of a morbid propensity?! Was the confession then always so voluntary, and particularly, was there not an intimate connection between the arrest of the kleptomanic thief and his "restoration of the object stolen, or the speedy reimbursement of its loss," which would be explained much more naturally and consistently with daily experience by the desire to avoid exposure, disgrace, and punishment, than by assuming the existence of a morbid propensity. "To this we must add," says Marc, "the trifling value of the stolen object, which is either thrown away, or presented to some other person." The "presenting to some other person" of the article stolen would be very properly reckoned by any simple and unlearned Judge as a perfectly valid *causa facinoris* of any such common theft. "The trifling value of the object stolen, which is thrown away," seems a more inexplicable criterion. But I must in the first place point out that neither Marc nor any one else, among all the many cases given, has quoted one in which the object was little valued or thrown away;† this must then be presumed to be the case

* Die Geisteskrankheiten. Berlin, 1844, S. 176.
† My own case, CXCVII., which will be given immediately, and in which

in the following instance, detailed by Marc in the following words:—
" 114 Observation.—It is well known that Victor Amadeus, King of
Sardinia, constantly stole articles of trifling value."

If this be an "observation," and if psychological theories are to
be based upon such observations—only compare this with the anecdotes in regard to kleptomania related by medical authors—then the
critic has easy work.   He has only to reproduce them, in order to
display their nakedness.

It is, however, quoted in favour of the supposition of a morbid
propensity for theft, that lunatics steal while insane and confined in
an asylum.   This fact, which is consistent with experience, must be
granted, without thereby affording the slightest aid to the hypothesis
of a kleptomania.   The pleasure experienced in possession, the inclination, indeed, the ardent desire to increase his possessions is
congenital and deeply rooted in man, as the conduct of a child at
once shows.   I need not detail here how this desire in moral individuals becomes a spur to productive activity, proving also nourishment for the passions of avarice and covetousness, and in the
immoral the cause of theft, robbery, and murder for the sake of
robbery.   It is certain that this desire is born with every one of us,
but reason and morality can and must regulate and keep in check
this, as well as all our other desires.   It is otherwise with lunatics, in
whom both of these checks are loosened and relaxed.   When in them,
therefore, this desire, as it were instinctively, asserts itself, this arises
through precisely the same mental process as in the case of the
assertion in lunatics of any of the other congenital propensities, which
are no longer regulated, such as sensuality.   Theft by lunatics is
consequently a *symptom* of their mental derangement, but is not the
mental disease itself.   Those always very rare cases already alluded
to, in which men have purloined, apparently without any object, and
have returned the objects stolen as alleged, sometimes even with
additional compensation, require a different psychological explanation. If we consider, quite apart from the motives already alluded to,
quite apart, moreover, from mere jokes, to which I do not now refer;
if we consider, I say, how much skill, cunning, craftiness, and courage
is required for this secret stealing, how requisite it is to watch for
and seize the proper instant, and then skilfully and rapidly to carry

the objects stolen were certainly thrown away, must not be adduced in opposition to my statement, because the subject of it was actually insane, and
laboured under hallucinations, &c.

out the plan conceived; how much pleasure is attached to every successful attempt under these conditions; how much gratification is
thereby afforded to personal vanity, we can then freely acknowledge that some may be, indeed are, found, who enjoy a certain kind
of pleasure in thus hunting after the property of others, quite
apart from its value. I have called it hunting, because, when the
keen sportsman risks his neck in a steeplechase, or in chasing a fox
or a hare, or for a similar purpose stands watching for hours in a
wood in wind and snow; when the angler stands half the day in
the water, as so many are fond of doing in England, and, that not
only without weariness, but watching his hook with the keenest excitement; we have in these examples of precisely a similar psychological process. The fact that the love of secret stealing, as here
defined, is of rarer occurrence than the love of hunting and fishing,
does not invalidate the explanation. Moreover, I have been for long
convinced, that even among those whose trade is theft, this peculiar
psychological process asserts its right besides other and more criminal motives. Otherwise it would be difficult to explain the fact, of
daily occurrence, that we see thieves just discharged from a long and
severe imprisonment, take at once to the picklock and the crowbar,
though conscious that a fresh discovery would only be followed by a
double punishment. I scarcely require to add one single word to
say, that the pleasure just referred to as being found in exercising
one's skill in stealing, may, like every other mere desire, be perfectly
easily kept in check, and wholly suppressed by the moral sense, and
that it has not the very least in common with a so-called morbid
propensity for theft, which irresistibly and instinctively hurries one on
to steal, and for that reason excludes responsibility,—a supposition in
support of which there are neither well-observed facts, nor tenable psychological theories; " a supposition," which a criminal-psychologist *
admirably says, " would lead the criminal law, which has so much
to do with thieves, into a pretty mess, if these individuals, together
with their organ of theft, were not to be hanged at once without
more ado immediately after their first theft !"

* Von Weber, Handbuch der psychologischen Anthropologie. Tübingen
1829, S. 345.

§ 90. ILLUSTRATIVE CASES.

CASE CXCVII.—DISPUTED MENTAL CONDITION OF A THIEF.

The merchant's clerk W., aged thirty, was accused by his master, and pleaded guilty to having stolen at various times sundry articles of ivory, &c., and his conduct at the precognition occasioned an inquiry being made into his mental condition. He was in excellent bodily health, but his appearance and demeanour struck one at the first glance as remarkable, somewhat conceited and like a dancing master, and his manner and speech was correspondingly affected. He mouths all his words and grimaces continually, using *g* for *y* with other affectations, and smiling constantly without any occasion. Even when brought upon the subject of his accusation and imprisonment, he smiles and says that is "only a trifle and of no consequence." If this may be looked upon as only a peculiar manner, yet much more important facts came to be considered. It was proved that repeatedly in broad daylight he had thrown the stolen articles, long before any suspicion was attached to him, into the bushes in the Thiergarten, and when questioned about this, he expressed himself quite irrationally, as "I, to be sure, why not?" &c. Still more. He declared that he had a certain Bertha H. for his sweetheart, and that he was to be married to her, and it was ascertained that this Bertha was a purely imaginary person, a phantasm. The erotic excitement thus indicated was more indubitably displayed when he was committed to prison, when he at once seized a fellow-prisoner by the genitals. Finally, he asserted that he was followed everywhere and watched by "an unknown gentleman," and his whole expression in making this as well as all his other statements completely excluded the idea of any simulation. From these facts and statements W. had to be declared to be insane in the medical sense of the word. I may point out that this was another of those by no means rare cases (§ 73, p. 192, Vol. IV.) in which the insanity assumed a two-fold character; here we had a combination of erotic mania with the "mania of persecution."

CASE CXCVIII.—A SIMILAR CASE.

Information was lodged that the merchant's apprentice N., aged twenty, had repeatedly stolen small sums out of his master's till, that at length when seized in the act and dismissed, he had from dread of punishment fled from his parents' house and attempted to drown himself. It was very well known to his master "that N. was of weak mind and incapable of acting with judgment," and he had only taken him into his business on trial. The family medical attendant also certified that he had long known N. to be of "weak mind," and that he could not depend upon his ability maturely to consider the results of his actions; the criminal commissary B. also "found him to be of so childish and simple a character, that he could not avoid doubting his responsibility."—"In fact," I stated, "a very short observation of this subject is all that is requisite to see that this generally expressed opinion is correct, since N. at the first glance is observed to be an imbecile. He labours under a so-called converging squint, which, of itself, is deserving of attention, and carries his head resting on his chest. His answers are childish and silly, and in regard to the offence of which he is accused they are perfectly incomprehensible, because he constantly cries like a child. His petty thefts have been committed for no evident selfish end, as he himself confesses, since there was no want of anything in his father's house, and he himself was most lovingly cared for because of his helplessness, which was very early apparent and rendered him unfit for any business. More important, however, than this, and extremely remarkable, was his conduct at his attempted suicide, when in broad daylight in the Thiergarten, when from its publicity he could not but expect his design to be frustrated, he, quite suddenly, stripped himself to his shirt before plunging into the water. He has not a word of explanation to give in regard to this extraordinary behaviour. It is not wonderful that such a man should commit petty thefts, since experience teaches that those decidedly deranged in their mind frequently yield to the inclination to appropriate secretly the property of others."—(Here follows a statement similar to that already given in the text.)—"No importance can, however, be placed upon the fact, proved by his dread of punishment, that N. knew that he had done a punishable act, because psychological experience also teaches us that mental derangement by no means excludes an obscure idea of the difference between good and evil, without there being any decided power of voluntary action, or any capacity for distinctly perceiving the results of their actions to their full extent." Accordingly I declared N. to be "incapable of considering the results of his actions," that is, an idiot in the statutory sense of the word (Gen. Common Law, § 28, I., 1. Penal Code, § 40).

CASE CXCIX.—A SIMILAR CASE.

This was another merchant's apprentice who was placed at the bar because of an attempted robbery of his master. I was summoned to attend the trial, without having ever seen the man or looked into the documentary evidence, in order that I might give my opinion as to the disputed mental condition of the accused, in accordance with the facts brought out during the trial. According to my usual custom, I declined this, and requested leave to make a preliminary medical examination of the accused, as well as a sight of the documentary evidence and of certain writings in his hand, of which legal possession had been taken, for my information. All this was granted, and I thus obtained the unexpected information contained in the following report :—"The accused, Falk, aged nineteen, confesses that on the fourteenth of July, while at work in the ware-room of his master, he threw a packet of silk goods under the clothes press; this was observed by his master from an adjoining apartment, and he at once taxed his apprentice with the attempted theft and gave him in charge. At the trial, however, the accused pleaded in bar of judgment that he is of feeble and absent mind, and that at the time of the commission of the deed he was not himself. His appearance, demeanour, and statements made at his last trial in regard to a collection of poems intended for the press, excited doubts in the minds of the subscriber in regard to the mental condition of the young man, and these doubts have since been fully confirmed. In the first place, in regard to the writings, which were seized at my request, these were, for the most part, only humorous poems, collected and transcribed from the works of various poets, which the accused intended to publish under the extraordinary name of 'Victoria:' yet, amongst them there are pieces—written of course without any intention or idea that they should be subsequently submitted to an official scrutiny—which supply the most convincing proof of an anormal mind. I refer to, e.g., a poem full of the vilest obscenities, which cannot be quoted (Falk says he wrote this as a joke), in a letter in which he signs himself ' Auerhahn,' and which is also addressed to a ' beloved Auerhahn ; ' a letter written in Hebrew, and specimens of printing types, apparently in the Hebrew character ; the accused (an evangelical Christian) stated to me that he had learned to write Hebrew for his own instruction, ' it was so easy to learn,' &c. ; and also I may add, a letter to myself, dated the eleventh of this month, regarding the abstraction of his papers, in which he calls the official concerned in it a 'footpad,' 'who had snapped up everything in his room,' and he begs me to send them all back again by 'this way.' This letter to me was signed 'Louis Refalk,' a signature which he had also adopted in several of his former writings. On my representing this to him he explained, laughingly, that his Jewish friends were always in the habit of calling him Redolph (instead of Rudolph), and therefore he sometimes signed himself Refalk ! The accused is small and feeble. His head is always bent forwards and his demeanour is quiet, though he frequently smiles without any apparent occasion. He complains of a tendency of blood to the head, which is confirmed by his unusually hard and full pulse, as well as by his unusually high complexion and the remarkable redness of his forehead. Otherwise his bodily health is good. The fact that he lives with his parents in comfortable circumstances, and has no motive for such a (paltry) theft, seems to confirm his statement to me that he did not know what he could have been thinking of at the moment when he flung aside the packet. His behaviour in regard to the commission of the deed is, however, more important, for he carried this out in a most unquestionably injudicious manner, namely, in broad day-light, and while his principal was in the next room, and might have been expected to observe him, while it is evident that a man of clear judgment in his position, if he intended to commit a theft, might have carried it out otherwise in the easiest possible manner, and so as not to be instantly discovered. The accused explains his hazy mental condition, which is well known to him, by the compulsion which was put upon him to make him learn a trade, while he had an irresistible impulse for a theatrical career. At present he is decided to devote himself to the stage. I found him busy studying the character of 'Mortimer,' and he says he will make his first appearance in a small theatre of this city. Even this apparently unimportant trait is significant for the psychological unravelling of Falk. It must, at least, be regarded as proof of no small amount of over-estimation and self-esteem, when a small, insignificant looking, feeble man in sober earnest believes himself qualified for the representation of a character such as that of 'Mortimer,' and this all the more that he must know that one of Schiller's great tragedies neither could nor would be likely to be represented in some paltry booth or suburban theatre, but only in

some one of the larger theatres, which indubitably would not admit
any such 'actor of heroes.' All this is, however, not understood by
the accused, and he avoids all rational remonstrance on the subject
with a smile or an absurd exclamation. According to what has just
been stated, the accused exhibits himself as a man who lives in a
world of fancy surrounded by hazy ideas, without moral firmness,
and with, what is not uncommon at his age, the notion that he
possesses a poetic and dramatic talent, which has already led him to
play absurd pranks. That this excited fancy has already carried
him beyond the limits of mental soundness is evident from the facts
already detailed, his Hebrew letter, his signing his name as 'Auerhahn' or 'Refalk,' &c., which are the deeds of men living in
delusions. From all that I have just stated I am convinced that the
accused is affected with a delusory mental disturbance (§ 40,
Penal Code), and that he was also in such a condition at the time of
the commission of the deed." He was acquitted.

CASE CC.—THEFT COMMITTED IN A STATE OF ALLEGED IMBECILITY.

This was one of those cases alluded to at p. 292, Vol. IV., in
which the theft was not committed from "*le besoin pressant de la
misère*." During the spring and summer of 18— the flesher R. had
observed that pieces of meat were frequently stolen from his shop.
There was then a cessation for two months, after which thefts repeatedly recurred. It was subsequently ascertained that the person
accused, M., a woman of property, had been absent from Berlin
during these two months. Suspicions connecting this woman with
these thefts began to arise, especially after a piece of veal was stolen
from the shop on (Wednesday) the first of April, when the shopkeeper knew that M. had just returned from a visit to a watering-
place. The Sunday following R. arrested the accused. She had
come into the shop and was bargaining for a piece of meat, when R.
remarked that a piece of a pig's flank, which he had intentionally
left lying on a block, was missing, whilst M., who had on a shawl
which hung over her right arm, was going towards the door as if
leaving the shop. R. went up to her and accused her, in plain
terms, of theft, at the same time throwing back the shawl and
taking from her the missing piece of meat, which M. was carrying
wedged in between her right arm and her body. The accused
declared that she only wished to see the meat by better light at the

door, and then to buy it, and exhibited a thaler (three shillings) with which she wished to pay it. The eyewitness K. disposed of this allegation by stating that the occurrence happened in the morning during clear weather, and that the shop was quite light, so that it was not requisite for any one to go to the door to examine anything. To repel the evidence sworn to by the witnesses, the woman M. at her precognition denied all intention of committing theft, and asserted that she laboured under occasional fits of "momentary loss of consciousness," in consequence of a previous fit of apoplexy, a kind of "partial imbecility," so that she readily forgot even things that had just happened. "At my examination she, with rare consistence, maintained this assertion. It is, indeed, quite impossible to get anything out of M. but that she knows nothing, not a thing, her head is so weak, that she could have no motive for theft, since from her host and relation X. ' she got as much to eat as she wished,' &c. She scarcely remembered going into the shop on Sunday, and the stolen meat she says she did not even see, &c. The whole conduct of M. is not that of a weak-minded person, still less that of one become imbecile from a previous attack of apoplexy, which the accused says she suffered from years ago. Besides that there is not a trace of apoplectic paralysis observable about her, she is extremely lively, which patients of that character are not accustomed to be. Her alleged weakness of memory only affects her recollection of the theft of which she is accused. Thus, she knows accurately the names and addresses of three medical men who have had her under treatment; she knows her exact relationship to the family of X.; she knows that she left her basket standing upon the shop-table, and this fact she adduces in support of her statement that she had no intention of going off with the meat; in one word, like all similar subjects, she has an accurate memory for exculpatory facts, and an alleged reverse for everything inculpatory, a state of mind not usual with those actually imbecile. To this I may add, that the crime of which she is accused is not deficient in an appropriate *causa facinoris*, inasmuch as to an immoral individual it may seem desirable to have roasts upon the table, even by forbidden means, oftener than his means afford, as is often enough proved by many more striking examples than the present : I may also add— without considering the former thefts from R.'s shop, which are not proved to have been committed by the accused—that not one other fact, except the crime of which she is accused, is to be found in the documentary evidence which could prove the occurrence of absence of mind or momentary loss of consciousness in the woman M.; and this is extremely remarkable, seeing it is alleged that this condition has lasted for many years : and I may also add, that even this deed was undertaken and carried out with an amount of cunning and judiciously contrived endeavour to avoid discovery, which clearly pointed to a consciousness of the illegal nature of the action, so that I have not been able to avoid the conviction which I now express, that the woman M. was neither insane nor imbecile (§ 40, Penal Code) when she committed the crime of which she is accused, but a perfectly responsible agent." At the time of the trial, while seated at the bar, M. behaved in the coarsest and most suspicious manner, for she firmly maintained that because of her great mental infirmity she knew nothing. She did not even know her Christian name, nor her age, and every question of the Judge was answered with "I don't know." For formal reasons the trial had to be postponed, and only after several months M. again appeared at the bar. Her behaviour was precisely the same. There was not the slightest reason for swerving from my former opinion, which was only confirmed by her present conduct, and the woman M., whose well-to-do condition was only now accurately ascertained, was condemned as a common thief.

CASE CCI.—THEFT IN A STATE OF ALLEGED MENTAL DISEASE.

This case is analogous to the one just related, and concerned a woman, aged fifty, the wife of a subordinate official. While buying bread in a baker's shop she stole pastry goods to the value of three silver groschen (3½d.), carrying these off secretly, after glancing sidewards at the shop-woman to see if she were observed. When arrested, she at first attempted to flee, and then gave up the goods, begging that she might not be further annoyed, that she would willingly pay, &c. A few days subsequently she called in a medical man, to whom she stated (what was also proved) that five years ago she had a severe fall upon her head, and further declared that since then she had laboured under a weakness of memory, and that at times she did not know what she did. At the time of the trial, as well as in the course of my previous examinations, no symptom was discoverable from which any mental infirmity could be deduced. Nevertheless, she declared in court that she had not the slightest recollection of any of the circumstances connected with the crime of which she was accused; but (as usual in all such cases) she could remember very well everything which could exculpate her. The medical man referred to considered her justified in assuming that a "compression of the brain," which still existed, had resulted from the accident alluded to, and that this had produced in her a state of "imbecility, mixed with insanity." Upon me devolved the easy task of confuting this opinion, as well as the supposed existence of "cerebral compression" for years in a perfectly healthy woman, and of proving the perfect responsibility of the accused. The result was that the woman was condemned to eight days' imprisonment with costs,—Both of the two cases last related were, as already remarked, cases in which "*le besoin pressant de la misère*" (p. 292, Vol IV.) was not the occasion of the theft, and where "the worth of the articles stolen in relation to the means of the accused" was certainly absurdly trifling. Both of these cases ought, therefore, according to the Marc-Mathey theory, to be reckoned as examples of "Kleptomania!"

CASES CCII. AND CCIII.—THEFTS COMMITTED BY MEN OF EDUCATION.

CCII.—In the spring of 18— the student v. Z. saw upon the street a man clad in an overcoat, which he had lost in a coffee-house one evening in February, and it was ascertained that Edward, a student of medicine, had carried off the coat upon that evening and had sold it to the clothes-dealer Isaac, with whom he had often similar "dealings." The accused alleged in exculpation that he was senseless from drink at the time he committed the deed, and, further, that he had sold the coat for four thalers (twelve shillings) because he could not remember the place whence he had removed it. The student from whom the coat had been stolen stated, that when Edward saw that he was discovered, he came to him, and urgently begged him to leave the matter alone. The testimony of other witnesses was much more favourable for the accused. His brother, a medical man, declared that Edward "had for a long time laboured under absence of mind and mental disturbance." Another physician, who had served with Edward in the army, testified as to his "eccentric" actions; for instance, that he had once gone out in uniform with a civilian's hat on. The surgeon of the company, R., confirmed this, adding that Edward was "very often lost in thought, indeed, almost quite absent in mind:" thus, once when asked by his superior as to the condition of a patient, he recited, quite seriously, to him a passage from Schiller. (It was not stated whether this was *àpropos* or not.) This evidence as to the conduct of the accused during his period of military service was, however, completely opposed to the very favourable character of his certificate of discharge from the superior military medical board, which is, nevertheless, very well informed as to the conduct and actions of its subordinates, and which could not have remained in ignorance of continued and remarkable conduct indicative of mental derangement. R., just referred to, moreover, subsequently deposed that he had gone with Edward to the coffee-house in question on the evening alluded to, but that he heard him talk such "absurd nonsense" that he left him. It seems very remarkable, I said in my subsequent report, that Edward did not speak to his friend in the course of the next few days as to what had happened at the coffee-house, the strange coat, &c., but had at once proceeded to sell the coat without further enquiry. It was, however, further ascertained that the parents of the accused were in a good position and let him want for nothing. A barmaid stated that his "apparently senseless tricks" often caused much laughter; that he had once come into the shop with one boot on and one slipper, that he frequently hid things, &c. I myself found him to be a well-formed, powerful young man, twenty-six years of age, who exhibited not a trace of any mental disturbance, but behaved himself like a man of perfectly sound mind. To me also he alleged that he had got drunk in the coffee-house, but he could not explain the contradiction pointed out to him, that his statement that next morning, and subsequently, he could not remember into what house he had gone in the evening referred to, when sober, did not tally with experience in regard to drunkenness. From the report given in I only quote the following concluding passages :—"The testimony of several of the witnesses seems to prove that E. was absent in mind and fond of being peculiar. Such a mental condition is of no rare occurrence; but there is not the slightest reason to deduce from it any want of mental freedom; because the consciousness of good and evil is far too deeply rooted for that, and is thoroughly recognised in many low degrees of mental derangement, which by no means exist here. The behaviour of the accused after the commission of the deed proves also that the accused had not lost this consciousness. The extreme degree of drunkenness alleged·by him to have existed on the evening in question would, most probably, have caused him to be much remarked by the guests in the coffee-house; nevertheless, the student robbed declares that he never saw him. I have already stated my opinion as to the alleged forgetting of the locality after the drunken fit was over. Further, there is no sign of mental disturbance and thoughtless action in the accused, after some days, taking the coat to a clothes-dealer and getting hard cash for it, nor in his going, when discovered, to the man he robbed and begging him to let the matter drop, asserting to the court the untruth that he had replaced the coat. The apparent absence of any motive for the deed is the only point capable of raising any doubt in the matter, and it seems to be quite permissible, in estimating psychologically the culprit and the deed, to devote a few concluding paragraphs to the elucidation of this point. It has been certified that the parents of E. are well to do, and suffer him to want for nothing, and yet all that he got for the paletot was only four thalers (twelve shillings). But the trifling character of the motive can never be accepted as a proof of the irresponsibility of the culprit, as I do not need detail here more at large. Moreover, I may point out in this case, that of a young man living in a large town, far from his friends, where pecuniary matters are in question; the amount is a very relative idea; and in regard to this we must at least not leave altogether out of view that statement of the clothes-dealer, that E. often had 'dealings with him.' As I was asked as to the 'responsibility' of the accused at the time of the commission of the deed, in accordance with what I have just stated I assumed it to have then existed."

CCIII.—Such a decided opinion could not be given in the following not quite usual case. "On the afternoon of the 4th of November, K., a candidate for the ministry, was in his land-lady's room along with her daughter, aged ten years. He sent her out to get blotting-paper, and he was then left alone with the son of a neighbour on the same flat, aged eight years, whom the girl had left in the apartment, to take care of it during her absence. On the girl's return, K. met her on the stair, took the blotting-paper from her, and left the house and Berlin, after having, as alleged, under the eye of the boy and in the absence of the girl, taken ten thalers (thirty shillings) from an unlocked chest-of-drawers, and a silver watch from off the wall; wherefore proceedings were taken against him for theft. Meanwhile, however, the advocate employed by the family for the defence, on the 11th of this month sought to show that at the time of the commission of the deed K. was 'wholly irresponsible.' The causes of his mental disturbance, which had lasted for a long time, arose partly from religious doubts, partly from the unfortunate result of an examination, and partly from his impecuniosity in contrast with his boundless vanity. The advocate for the defence further declared that K. on his arrival in Hamburgh, whither he had gone to enter upon a situation, had taken up his abode in one of the first hotels, and had lived in it in as foolishly extravagant a manner as if he had inexhaustible resources at his command. On the other hand, he was so bold as to return hither, going from house to house buying old clothes when he got money from his relations, and again hawking these old clothes and cigars through the country, &c. The advocate for the defence also lodged two medical certificates, which in the main certified what he himself had stated, and declared K. to be 'perfectly irresponsible.' I have repeatedly examined him, and have not been able to convince myself of this. The circumstance that K. is now in the lazareth of the town prison, labouring under the windpox, made this examination of his mental condition appear quite unintentional. K. does not impress one with the idea of being so insane as his advocate states. He seems old for his alleged age of twenty-nine years, but his look is clear, his features bear the impress of whim and frivolity, and his appearance is unpleasant, because he is ugly, with a strongly marked Jewish countenance and black, bristly hair; but, as I have already said, a first glance at him by no means gives an expert the impression of one mentally deranged. And such a condition is just as little to be suspected from his speech and behaviour. The plucking at his hands, mentioned by his advocate, I have never once observed. At each of my visits, K. raised himself courteously in bed and sat there quiet and composed during our conversation. I spoke to him intentionally about the crime for which he was arrested; but also to me he denied it decisively, and declared he knew nothing about it. His statements in regard to it were quite distinct and connected, such as could only be expected from one perfectly innocent or from a consistently lying thief. He said that the evidence of two children was worth little; he had left Berlin to get his legitimation papers, which were lying in the neighbourhood of the city. He knew nothing at all about either the watch or the money, just as he had deposed at his first judicial precognition. Of his residence in Hamburgh he said, laughingly, that he had certainly gone into a hotel, because he could get no other lodging, but that he had never dined there, and his entire bill amounted only to fifty marks (£3 15s.) for coffee, &c., so that it was non-sense to talk of his foolish extravagance. He obstinately denied the alleged hawking of clothes and cigars, and added that he might, perhaps, have sold one of his own old coats; but there was nothing remarkable in that. He acknowledged the unfortunate result of his examination, and he took this occasion to explain the number of scientific prizes which he had taken, unmistakably exhibiting in this a certain amount of self-esteem. This, however, did not exceed in amount that usually to be found in vain men, and was not in the least obnoxious to me either upon this or any other occasion, and if I had no knowledge of the *anteacta*, it would never have raised the slightest doubt in my mind as to the mental sanity of K. Neverthe-less, we have still the perfectly opposite statements of the advocate for the defence, as well as the two medical certificates, and I would not be justified in accepting K.'s statements as the more truthful or the reverse, and this all the less that it is consistent with experience that there are certain mental conditions which exactly occupy the boundary line between health and disease, or which may even have overstepped it, and in which even the most experienced psycholo-gist is unable, from the want of a long and intimate acquaintance with the respective individuals, to see clearly the state of their mind, because they still have sufficient power of will to command the delusions which enchain them. Such a condition is specially apt to be found in men whose morality is lowered by drunkenness and dissolute living of every kind, and in whom want, passion, defective character, love of idleness, and the tendency to vice and crime gradually disturb the normal mental life, and who, finally, though often only after the lapse of years, fall into actual insanity. There exist certain data in this case which justify the assumption that K. may belong to this class. I regret, however, that at present (at the commencement of the preliminary investigation), and before being more clearly informed from the statements of witnesses, &c., as to the former life of K., I cannot conscientiously give any more certain or decided an opinion in so important a matter. Nevertheless, from the result of my own examination I am convinced, and, with due regard to what I have just stated, I am able to give it as my *provi-*

sional opinion that K. is *not a perfectly irresponsible agent."* The case was publicly tried during my absence from Berlin, and I have not been able to learn the result.

CASE CCIV.—THEFT ALLEGED TO BE PRODUCED BY THE LONGING
OF A PREGNANT WOMAN.

This was a case of extreme interest, in which the longing of the pregnant woman was determined and unquestionable, and all that had to be medico-legally decided was whether the repeated thefts perpetrated by the culprit, a lady of a certain rank, were to be ascribed to this cause. Frau von X. had in January, 18—, ordered from a goldsmith, a present for her husband, and being left alone for one instant in the shop, she had employed it in bending over the glass cases containing articles of ornament, and these she "meddled with." While doing this she was surprised by the assistant; she became very pale, asked for a glass of water, from which, however, she scarcely drank any, and went away hastily. At that time she was in the fifth month of her first pregnancy, which had set in immediately after her marriage. The shopkeeper immediately missed several articles of ornament from the case referred to, among others a seal and a kind of medallion. In the beginning of May, four weeks previous to her confinement, Frau von X. went to a second goldsmith, and selected earrings at the price of three thalers (nine shillings), and offered in payment old ornaments, in particular fragments of the medallion just referred to. The request that she should select some more goods for them, she parried by stating that at present " she required nothing," she nevertheless took some silver teaspoons and ten thalers (thirty shillings) in hard cash. About the same time she appeared in a third goldsmith's, asked for a silver teaspoon, and offered in payment the lower part of the signet referred to. Meanwhile, the favoured goldsmiths had begun to suspect the Frau von X., and the second one who was robbed went to her under some pretence to recognize her. Scarcely had he returned, when the woman, now close to her confinement, came to him, and entreated him " with the appearance of the utmost anxiety to tell her, by all that was holy, why he had come to her?" On the twenty-ninth of May she was confined, and on the thirtieth she received a summons to appear before the criminal court, which of course astonished her husband extremely. " As if awakening from a dream," he stated before the court, "she made the following confession, that during her pregnancy she had been seized with a hitherto unknown and unconquerable desire for everything bright, especially for bright articles of gold and silver, and that her greatest wish was to possess them." Thus she had in a state of "complete insanity" taken these things out of the shops. At another time she had declared for certain that she had no recollection of these occurrences; and again, that she had gone out with the intention of returning the articles, but had become convinced by the way that they were her own property properly acquired. She expressed herself in the same contradictory manner at her own precognitions. The documentary evidence contained the following remarks as to her character :—This woman, aged twenty-two years, was of good family. She had early exhibited "a ridiculous amount of vanity and love of dress," but this her husband would not allow, and he stated that she was always "thoughtful, quiet, thoroughly honest, and religious." It was, however, indisputable, and proved by the unanimous testimony of a great number of witnesses, from the domestics of her own house up to her noble relations, that very soon after the commencement of her first pregnancy, she had exhibited a most remarkable alteration in her mind. She was absent, forgetful, and exhibited a most remarkable love for bright, glittering articles, which she gratified in the most extraordinary manner. Thus she was, in spite of all objections, continually cleaning the household utensils of brass; she played with new bright thalers, &c., and her husband deposed, that she had often complained to him that frequently when visiting acquaintances who had bright silver or other articles, she was seized "with such an inclination and desire to take them forcibly, that he must not take her there any more, for she was afraid of herself." Numerous facts were deposed to in regard to this forgetfulness and desire for bright articles; of these, I only quote that she had taken from a relative staying in her house a knife with a mother-of-pearl handle, and when in company had carried off before the eyes of the players the glittering whist markers, and this had not been looked upon as a joke, but rather as a sign that "her head was quite turned" at present. Her medical attendant, in the course of the trial, declared her to be irresponsible, while my very experienced predecessor in office did not even assume her responsibility to be diminished. In this contradictory state of matters a superarbitrium was required from the college. This we drew up and it was accepted and presented. This extremely detailed and minute report was based upon the doctrines already evolved in the text, particularly in relation to the fixed idea, the longings of pregnancy, and monomania, and therefore it need not be reproduced here. It was assumed as indubitable, that the accused was affected with the described longing, also, that her mental faculties were otherwise obscured by her pregnant condition, that she had been careless in her household matters, forgetful, and absent. It is perfectly consistent with experience, we said, that she should be conscious generally of this longing, as appears from the deposition of her husband, that she requested him to take her no more to visit friends who had glittering silver. "Every one acts precisely so who is partially affected by some longing, or fixed idea, from which he cannot escape, but which he still masters with his reason, because he recognizes it." It is however, very remarkable, that when she was aware of this longing she did not, as in regard to visiting her friends, avoid and shun all magazines full of "glittering articles," but rather visited them without any urgent occasion, or any decided wish to purchase, which, moreover, in her position, she might readily have managed by means of a messenger; indeed, even when far advanced in pregnancy she went long distances through the city to visit various shops, whose dangerous contents could not have been unknown to her. Her suspicious behaviour to the shopkeepers whom she robbed was now considered, the important circumstance that once instead of taking-really "glittering articles," she preferred taking hard cash, observing, that she "required nothing at present," the fact that she had kept her thefts a profound secret from her husband; the very important circumstance was brought forward that she herself had confessed that she had once gone out to return the articles which she, nevertheless, omitted to do; further attention was directed to the fact, not coming under the head of her longing, that each time she broke and rendered unrecognizable the ornamental articles formerly stolen, also that she had gone each time to a different goldsmith. Her numerous lies and contradictions at her precognitions were placed together, and from all these details the conclusion was drawn, that the alleged longing caused by the pregnant condition of Frau von X. was not irresistible, that it had not morbidly impelled her to commit her threefold theft, but these presented all the characteristics of responsible actions, and that the accused must be regarded as a responsible agent.

The result was the punishment of the lady, her divorce, and—— —

after a lapse of years, and at a time when she was not pregnant, a new theft from a shop of a piece of black silk !

§ 91. CONTINUATION.—THE MORBID PROPENSITY TO FIRE-RAISING.
—PYROMANIA.

We need not dally long with this obsolete question, now that Flemming, Meyre, Brefeld and Richter have proved by psychological reasons, and I myself also with the same, but also with facts from criminal statistics, that there has seldom been any doctrine in psychology which has been taken less from nature, or from life, or which has been more purely evolved at the desk from superficially examined and irrelevant facts, and which has thus become a tradition, than the notorious doctrine of the morbid propensity to fire-raising.* Who can reckon the cases in which the application of the penal law in one of the vilest of crimes, because most secret, most difficult to investigate, and most dangerous to the public, has been misdirected by the physicians from the assumed existence of a morbid propensity in the criminals, which impelled them to fire-raising as the starved are impelled to eat ! The actual facts of the case are essentially limited to these, that, first, as the advocates for the existence of a pyromania assert, the cases in which young individuals of both sexes, particularly females, have committed acts of fire-raising, are of " very frequent " occurrence, which must strike one as remarkable ; second, that no visible reason for this crime can be scarcely ever ascertained, but rather that the young culprits themselves have not been able to allege any, or, at the most, have declared that they felt as if they must do it, or, as if an internal voice drove them to the deed. Then around these " facts " the garment of theory was hung. These young criminals were about the age of puberty, and at this epoch of life "a state of venosity prevails, and the eye instinctively seeks for light and flame to oxidate the hypercarbonated blood," &c. ! It did not, however, seem anything remarkable to extend the epoch of puberty from the eighth to the twenty-second year, in order to exhibit the frequent occurrence of the probative facts. Nevertheless, at least, the frequency of these cases is a fact, and Friedrich, about eighteen years ago, collected sixty-nine of such " observations " out of periodicals, &c., most of

* " On the Hobgoblin called the Morbid Impulse to Fireraising," in Casper's Denkwürdigkeiten zur medic. Statistik und Staatsarzneikunde Berlin, 1846, S. 257-392.

them to be sure only consisting in three words: the eight, ten, or seventeen years' old N. N. has set fire to so and so.* I do not intend to deny this, but it would be better in the first place to examine this constantly referred-to frequency of this crime, statistically and in comparison with the relative frequency of the other offences of juvenile criminals. In regard to this I have—*loc. cit.*—shown by means of irrefutable official figures taken from the tables of the Prussian criminal statistics for twelve years, that in *a hundred thousand* young lads and girls, one—*one fire-raiser* I say, but *thirty-nine* thieves and receivers of stolen goods were brought up for examination! This of itself at once disposes of the "morbid propensity to fire-raising at the age of puberty," for the crime occurs, as we see, during this epoch not "very frequently," but *very rarely*, whilst the "hunger for light," strangely enough, impels young lads and girls very frequently to—theft! It seems not to have appeared remarkable that this "new disease," as A. Meckel calls it, should only be found in Germany and in no other country (because foreign physicians have given such cases a more correct psychological explanation), indeed, that almost all these pyromaniacs have lived in the country and not in the towns, seems also not to have been thought surprising, although the influence of the epoch of puberty is not confined to German maidens nor only to peasant girls! And, finally, in regard to the repeated statements as to the alleged want of any motive for a responsible crime, which "*per viam exclusionis*" has led to the assumption of a morbid impulse, in this we have another instance of that frequent and dangerous error so often found in forensic psychology, and to which I have already directed attention (§ 61, p. 103, Vol. IV). When the young boys or girls knew not what to say for having committed an act of fire-raising, or when they have declared (as they have certainly very often been cross-questioned into doing, since "pyromania" has forced its way into science), that they merely did it to please themselves, indeed, when they have stated, as in many other cases, that an abusive word from their master or mistress, a supper withheld as a punishment, or the refusal of permission to revisit their parental home, &c., has driven them to incendiarism, the disproportion between the cause and the effect was found to be so inexplicable as to

* Henke, Abhandlungen u. s. w. III. 2 Aufl. S. 226. Friedrich, Syst. d. ger. Psych. 2 Aufl. 1842. S. 272. *Vide* also H. E. Richter's instructive Table of Juvenile Fireraisers, 1844, in which attention is already directed to the abusive extension of the period of puberty.

force one to take refuge in the hypothesis of a morbid propensity. I need not, however, repeat what I have already said at p. 104, Vol. IV., as to the necessity of placing one's self in the same position as the culprit in considering the *causa facinoris ;* in this case, without exception, these are partly truly silly, partly lazy, or petulant, ill-natured subjects, certainly often only half children, who, when they are impelled by their psychological tendencies to any malicious act, of course are led to attempt such as requires neither bodily nor mental skill or exertion, but only to be one instant unwatched and a match or burning coal, which are everywhere to be had. In regard, however, to what we frequently hear as to the influence of internal voices in the examination of such subjects, I may refer to what has been already said on this subject (§ 63, p. 111, Vol. IV.).

In the "Denkwürdigkeiten" I have related thirteen cases of so-called pyromania observed by myself, and in my "Vierteljarschrift" (Bd. III. p. 34, &c.) I have detailed a fourteenth very remarkable case, which was most seductive for the hypothesis of a pyromania ; I now add a few of the most remarkable allied cases which have come before me since then.

§ 92. ILLUSTRATIVE CASES.

CASE CCV.—A FIFTEEN-YEARS'-OLD INCENDIARY.

Augusta S., a servant in the country, aged fifteen, had confessedly set fire to the dwelling house and hostelry of her master on the eleventh and thirteenth of November. At first she stated that she felt an internal and irresistible impulse to the deed, subsequently she alleged as her motive, that she hoped thus to escape from her service. Having been requested by the district court concerned to examine her mental condition, I would not have quoted the case here, if I had not had an opportunity of ascertaining most indubitably that the above mentioned motive (so usual in such cases) was distinctly supported by the other evidence. But there existed also another unusual circumstance. Having been formerly open and courteous, she had become since her entry upon this her first service reserved and lazy. Her genitals and breasts were of only half the usual virgin development, she had menstruated three times previous to the deed. But I found the girl S. already deflowered, and she confessed that she had a sweetheart at home with whom she had repeated sexual connexion! It was therefore very easy to apply the above doctrines to this individual case, in which another physician had already assumed the existence of a pyromania. Setting aside all and every idea of pyromania, I declared for the above-mentioned reasons before the jury court, that I regarded her as a responsible agent, coupling this statement with the addition, which often forces itself on our attention in such cases, that I meant the responsibility of a stupid girl of deficient intellect, who was still half a child. The jury accordingly gave a verdict of guilty, but recommended the girl to the royal clemency.

CASE CCVI.—A SIMILAR CASE.

Caroline St., aged fifteen years and six months, was accused by her master of having set fire to his mill. She had confessed the deed both to the gensdarmes who arrested her, and also to myself, in precisely similar words; but, on the other hand, at her first precognition after her arrest she denied the deed, and ascribed the fire to accident. Fourteen days subsequently, however, at her second precognition, she made a full confession, as follows:—"I was busy in the kitchen in the evening. It then occurred to me that I might set fire to the pine trees in my master's garden, and without thinking that greater damage might ensue from the fire, and without being rightly conscious what I was about, I went with a couple of matches, which lay open in the kitchen, into the coach-house, in which one of the panes of the window was broken. The pine twigs, with their leaves, reached close to this window. I struck a match, and stuck it among the dry pine leaves, which at once took fire. After this was done, I returned to the front of the house, and placed myself in the front door." I found the girl St. large and powerful for her years, and so far advanced in sexual development, that her virgin genitals were covered with hair and her breasts were already somewhat developed, whilst, according to her statement, not even the *molimina* of menstruation had appeared. She carried her head somewhat depressed, looking to the ground or sideways, seldom looking the questioner in the face, so that at the first glance the accused appeared somewhat shy, silly and stupid in her behaviour —her speech corresponded. "To indifferent questions she gives her answers pretty quickly and fluently, but at once becomes shy and reserved whenever the conversation is turned towards the crime of which she is accused. In particular, I have repeatedly failed to get her to confess the cause of her incendiarism, and she persistently refused to confess what she had previously stated to the police officials, that she had done it to revenge herself of her master. This is all the more remarkable that she makes no secret of the fact that she had, from her point of view, sufficient reason for revenging herself. She asserts naïvely that she was very ill-treated in her service, both in respect to her food and the amount of work expected from her; indeed that she had to submit to actual blows. For these reasons she had given warning that she would leave her place; but this warning was not accepted. On repeating her warning about three weeks before the deed, not only was it not accepted, but blows were also given her. This confession, and the known fact that she earnestly desired to return to the village in which she was born, gave foundation to the suspicion that her fire-raising had been carried out with the most carefully considered intention, and that vengeance was its psychological basis. The accused, notwithstanding repeated inquiries as to this point, maintained a persistent silence, without, however, presenting any appearance of obstinate refusal, and our conversation was thus each time forced to be broken off. To this must be added, that the mistress of the accused deposed that the latter had 'never been orderly and diligent,' that her master had never, any more than myself, ' observed any traces of melancholy or mental disease' in her; and there would then have been no reason for doubting her responsibility as to the crime committed, had not medico-legal science during the past ten years permitted the assumption of a peculiar form of monomania, called pyromania, under the influence of which precisely such subjects as the girl St. are supposed to be impelled to incendiarism. Such a peculiar species of mental disease has, however, no existence," &c. "Even the criteria by which the advocates for this hypothesis have vindicated its existence are absent in the accused. She states that she has never had any peculiar fondness for fire and flame; her development is not anormally advanced; she has never had any of the so-called *molimina menstrualia*, palpitation of the heart, giddiness, heavy dreams, congestion, &c.; and the data described do not in the least give to her deed the character of a blind and instinctive impulse, nor is there any want of an appropriate motive as a basis." It was easy to show that the desire of revenge existed in this case. But attention was also directed to the entire conduct of the girl St. and her half-childish behaviour, and it was then assumed that she was responsible for her crime, but that the deed was to be regarded as the result of a half-childish desire of revenge.

CASE CCVII.—ANOTHER INSTANCE OF THE " INTERNAL VOICE" IN
A JUVENILE INCENDIARY.

I shall give the full details of this case, because of its peculiarity;
my certainly severe, but not unreasonable, opinion in regard to it was
not accepted by the jury. "The journeyman joiner, Voigt, now in
his nineteenth year, is accused, and has confessed to having set fire,
on the morning of the 11th of Nevember, to his mother's clothes-
press, in which his own and her clothes were hanging, by which the
clothes were all burnt. In the first place, and particularly when
lodging information with the police, which he did himself, and at
once after committing the deed, he alleged his motive to be the
desire of obtaining vengeance on his mother, for her having given
him a good scolding some days before for his laziness and repeated
desertion from work. This statement, however, he subsequently
denied at his judicial precognition, when he, as also to myself, made
the following statement in regard to the deed :—' The thought came
into my head that morning that I must set fire to my mother's clothes-
press. It was, as it were, an internal voice calling to me that I must
set fire to the press. At first I resisted the impulse, afterwards,
however, I went into the kitchen and split some very resinous firewood, then I lighted a match, kindled the firewood with it, and put
it into the open clothes-press. It now occurred to me that I was
doing wrong, and I took the burning firewood out of the press, and
blew the fire out. I then sat down on a stool in the room. In about
five minutes the thought came again into my head, I felt the thought
buzzing through my head, that I must burn the press and the clothes.
I therefore again set fire to the same piece of firewood, and again
placed it, when alight, in the clothes-press; but again my heart mis-
gave me, and I took the firewood out before anything was burned,
and put it out. I then walked several times up and down the room,
seeking to suppress my inclination to fire-raising, but in this I was
not successful, for the internal voice kept continually calling to me,
you must do it, you must do it; you must put the burning firewood
into the press. Therefore for the third time I put the burning firewood into the press, and shut it,' &c. Upon this he left the house;
but he had scarcely reached the street when his conscience smote
him, and he went to the police to lodge information against himself.
He declared then that he regretted what he had done. The accused

is a man of powerful build, and sound bodily health. His pale complexion is to be ascribed to his imprisonment, which has already
lasted for several months. He complains of an occasional headache,
but this purely subjective statement cannot of course be subjected to
any proof, it is, however, of no consequence, since an 'occasional'
headache can neither explain a crime, nor deprive any one of his
voluntary freedom in action. His look is dull and inexpressive, and
certainly seems to indicate no special development of his intellect.
His skull is perfectly normal in form. He complains of weakness of
memory, but his statements do not reveal it. On the contrary, his
answers exhibit a correct memory, are rapid, accurate, distinct, and
perfectly coherent. There is no trace of any peculiar gestures, faces,
distortions of the features, or anything else remarkable to be observed
about him, except this, that, though in his nineteenth year, and of
full sexual development, he has as yet not a trace of a beard. In
regard to his disposition, the deposition of his former master is very
significant, as he gives the most unfavourable testimony regarding him.
He says he is lazy—which the accused has proved by repeatedly
running away from his work under different masters, having twice
wandered as far as Brandenburg and Friesack—obdurate, cunning,
and vengeful, so that his (the master's) wife was afraid of him, and
he was glad to be quit of Voigt. On the other hand, not one of the
witnesses examined ever observed a trace of mental disease about
him. And I also have not been able to discover a single trace of the
existence of any mental alienation about him. Nevertheless the circumstances attendant on his peculiar crime have raised doubts as to
the integrity of his mental powers; and it will have to be shown
whether these doubts are justified by general psychological experiences.
I have already stated that his crime did not want a true *causa facinoris*, by which I mean an impulse to the illegal gratification of a
selfish desire, and this desire was that of obtaining vengeance on his
mother, who had scolded him well on the day before the commission
of the crime—indeed also very shortly before it—and had threatened
him with the workhouse. A malicious deed might very well be expected of one of his disposition as described, indeed—and this is of
importance in deciding as to its nature—such a deed is by no means
inconsistent with his character. On the other hand, his assertion
that an 'internal voice' drove him to the deed, always calling to
him 'you must do it,' is extremely deceptive, as it leads to the conclusion that he was impelled thereto by the blind, irresistible impulse

of a morbid mind. Nevertheless, this ' internal voice ' is a phenomenon very well known to every expert as of extremely frequent occurrence in similar cases. The thought of incendiarism gets into the head of the accused. I have already related how that has happened. He proceeds to carry it out, since he is alone and unobserved, and the deed is one easily executed, with little expenditure of either mental or material energy. His conscience smites him, and he strives to undo his work. *Consequently he knows* that he has done what is wrong, and ought to be repented of, and he thus proves that at this moment he possessed *the power of perfect discrimination.* Now comes the time in which, as so often happens with criminals before the completion of the deed, he wrestles with himself, and wrestle he does on the internal question, ' shall I, or shall I not ? ' When at length the evil principle gains the victory, and *his own internal voice, with a* ' *do it !* ' *hurries him on to the completion of the crime ;* and Voigt is therefore perfectly justified, and makes no evasion when he states that he felt as if a voice were calling to him, ' You must do it.' Though I do not require to enter into further detail to prove that this so-called ' internal voice' is not to be regarded in his case as a delusion of the senses, or as a symptom of fully developed insanity, since he is not insane at all—another circumstance that might excite doubts is, that the accused exposed his own clothes to be burned. A similar thing has been observed in many similar cases. Voigt, the lazy, beardless lad, who has hitherto been supported by his mother, as yet lays no such value on his own property as is done by a thoughtful, orderly man, who has laboriously acquired his goods. His explanation to myself in regard to this is extremely characteristic. ' I had,' he said, ' a good coat, trousers, and vest upon me, and those in the press were too tight to be put on over them.' In the face of such a declaration the doubts referred to cannot be looked upon as of any value, and this circumstance only further confirms how little confused or unconscious the accused was at the time when he committed the deed, and shows how considerate and deliberate he was, consequently that he was neither ' completely deprived of the use of his reason,' nor ' incapable of considering the results of his actions,' that is in the statutory sense neither a lunatic nor an idiot (Gen. Com. Law, §§ 27, 28, Tit. 1, Part I. Penal Code, § 40). In regard to the latter statement, and in answer to my question, whether he did not know that he exposed himself to a severe punishment by executing such a deed, the accused replied that at the moment he did not think of it. To deduce from this, that he could not have foreseen the results of his deed, would be to deny this deliberate crime to be such, for every criminal (responsible evil-doer) knows the 'results of his actions;' at the instant of committing the crime, however, he does not think of them, because he is impelled to the deed by a more powerful though momentary excitement. Even the immediate self-denunciation, which seems to be somewhat remarkable, is explained by Voigt in the simplest manner. He says, and rightly, 'as how that' he must have been recognized as the perpetrator of the deed, because he was quite alone in the house, and fire cannot get into a clothes-press in broad daylight unless put there intentionally. Only the reproaches of his conscience, and perhaps the thought that, by a voluntary confession, he might mitigate his punishment, impelled him —still with a perfectly rational consideration of the circumstances— to denounce himself. I have not considered it superfluous to ask the accused if he ever had any predilection for playing with fire, if he frequently dreamed of fire, or if he had been guilty of any sexual excesses? Though a decided opponent of the notorious hypothesis of a so-called morbid impulse to fire-raising, I dared not evade in this particular case any possible objections based upon this doctrine, · now happily set aside, which might subsequently be brought to bear upon it. Voigt answered all these questions in the negative, and his behaviour showed that he regarded them as perfectly irrelevant, and that there is nothing he has had so little feeling of as an inexplicable, morbid pleasure or fancy for fire and flame. It is not the part of the forensic physician to point out the distinction between a villainous crime and a mere boyish prank; and the drawing of this distinction must be left with a different judge. In regard to my own duty, however, I think I have shown sufficient reason for stating it in conclusion to be my opinion, that Voigt neither is at present, nor was at the time of the commission of the deed, insane or imbecile, or deprived of the free use of his will by any other internal cause." (§ 40, Penal Code). The jury found him guilty of fire-raising, but denied his responsibility.

CASE CCVIII.—RESPONSIBILITY OF A FEEBLE-MINDED YOUNG
INCENDIARY.

" On the sixteenth of June, 1846 " (previous to the introduction of the new Penal Code), " a fire broke out in the hay-loft of a stable

belonging to a plasterer called Appel, by which the greater part of the
roof was destroyed. Suspicion was directed to the youngest son of
the proprietor as the cause of the fire, and two days subsequently he
repeated before the commissary of police the confession he had
previously made to his father, that he had raised the fire because he
was kept too hard at work by his father and his elder brother, and in
particular had too far to walk. Therefore, he had long thought of
getting work in the neighbourhood, and it seemed to him most convenient to burn down his father's stable, which must be rebuilt. For
this purpose he bought some matches, took them into the hayloft on
the evening of the sixteenth and set fire to the hay with them.
When it began to burn he rapidly descended the ladder and ran off.
The police official did not hesitate after this short communication to
remark, in his statement, that it appeared from the declaration of the
accused that he was an idiot in the statutory sense of the word (!).
He made, however, subsequently quite other and different depositions. According to that made at his first judicial precognition on
the twenty-third of June, he stated that on his return home from the
puppet-show he found the stable on fire, and he appears to think that
the fire arose from the carelessness of the coachman, asserting that
the confession made to the police was only extorted by the dread of
being punished if he denied the deed. Also after this precognition
it is registered, ' that A. exhibits evident traces of mental weakness.' At the precognition held in my presence upon the eleventh
of this month, as well as in my subsequent conversation with him, he
could not tell the year in which he was born; he only knows that he
is twenty-one years of age and that the thirteenth of February is his
birthday. In all conversations he repeats the questions put to him
first before answering them; for instance, What is your name?
' What is my name,' &c., a very characteristic, and, according to
experience, a most usual procedure in all weak-minded people, who
seek, as it were, to make themselves fully comprehend the question
by a previous repetition before answering it. In regard to the cause
of the deed, the accused gave various answers both at the precognition referred to, and also to myself subsequently ; they substantially,
however, amounted to this, that he had done it that he might for
once give them at home a right good vexation, for they had so often
vexed him. And he makes this statement with an obvious feeling
of mischievous pleasure, and with a stupid, roguish smile. In the
precognition referred to, he made the astounding statement that he

had 'a bride' (who has never been discovered by the police, either at
the address given by him or elsewhere), but that he had never had connexion either with her or any other woman; on the other hand, he made
the unsolicited confession that he practised onanism 'every evening' in
the prison, and I thus took occasion subsequently to return to these
points.   He confesses that he also practised onanism daily at home,
and on my asking him if stains did not thus arise, and if the attention of his mother was not thus attracted, he stated, laughingly, that
he had a little dog at home, and that he had made his mother believe
that it had made the stains in his bed.   A. knows the Ten Commandments; he also knows that though fire-raising is not expressly
mentioned in them, yet that it is a wicked action.   To my question,
whether he did not see this clearly before committing the deed, and
that he would be punished for it, also that in destroying his father's
property he destroyed his own with it, he returned a negative answer,
and stated that he had not thought of that, that he only wished 'to give
them a thorough good fright.'   A few days subsequent to the commission of the deed, the father of the accused applied for a civil-law
declaration of idiocy against his son, and this application he supported by the certificates of two of his son's teachers, and by a number
of facts out of the previous life of the accused.   The father ascribes,
what he states to be the evident weak-mindedness of his son, to a faint-
ing fit which happened in his fifth year, and describes many traits
which he thinks ought to prove his assertion, such as, for instance,
that his son does not know the different kinds of coin, that up to
this day he has not advanced in the art of masonry beyond being a
hodman; that he, an adult, often plays for half-a-day at a time with
quite little children, that he has spent upon himself the money en-
trusted to him for the purpose of procuring victuals, and stated in
excuse that it was all the same who spent the money; that once in
the house of an acquaintance he took off all his clothes and went to
bed in broad daylight, &c.   After examining A. personally, I can
readily believe him guilty of all these stupidities.   He is a man of a
flabby habit of body, obviously enfeebled by onanism, with a stupid,
inexpressive look, never directed towards his questioner, but always
sideways, and he is fond of interspersing with a stupid smile his
answers, which, as I have already said, are expressed slowly and after
repeating the question.   It is impossible to carry on any continuous
conversation with him.   Childish indifference, childish tendencies,
and a childish absence of shame, are the most prominent features in

his character. 'His ten years' (penal servitude) he says, are certain, but he says this with an indifference not observed in the most hardened criminal; in ten years, he says quite quietly, he shall be thirty-one years old, but still young enough to commence something. Childish tendencies, I say, for his father states that 'he borrows money everywhere, to take him to the puppet-show.' That he is devoid of all sense of shame, like a child, and not like a young man of twenty-one, is evident from the manner in which he has repeatedly expressed himself to the Judge and to myself in regard to his daily sexual excitements, of which he speaks with an indifference for which there is no comparison. As a child, however, knows that it has done wrong when it has broken something, &c., so also A. feels that he has done wrong, but he also feels it only thus. He, whom no one would call a hard-hearted miscreant—for of this there is not the slightest proof—neither expresses nor exhibits the slightest trace of grief for having caused his parents so much sorrow, although from his speech it would appear that at least he loves his mother, if not his father. It is also remarkable, and it remains to be considered, that he has sought to evade the punishment due to his crime by denying having done it, and that there is no want of motive for his deed. But his lies were not persisted in, and the first confident remonstrance brought him to a confession, and thus, as already shown, he certainly feels that he has done something wrong, and only will not allow it, just as children deny in like cases. Further, while it is confessed that a desire of vengeance against his father and brother, consequently an easily recognizable and most usual *causa facinoris*, led him to commit the deed; yet, we must not overlook the facts that in this case it was only a roguish trick that was intended, as he himself says, he wished to give his father a 'fright,' and that the cause and effect were so disproportionate, as to stamp him either as a monstrous criminal, or as a silly, childish being. From the documentary evidence it appears as clearly that A. is not the former, as I think I have shown in the details just given, from which I indeed deduce, and hereby give it as my opinion, that A. is of very defective mental development, and, therefore, only a very much diminished amount of responsibility must be accorded to him." At the public oral trial, Appel was acquitted. The public prosecutor, however, appealed, and in the court of appeal I had to maintain the opinion I had given, since no new facts had been brought to light. The most unusual question was now put to me, whether I considered any appropriate punishment applicable, and

what I would prefer? Upon this I declared that corporal chastisement was the most fitting punishment. Accordingly the court
condemned him to receive "twenty blows at twice."

CASE CCIX.—ATTEMPT AT POISONING AND FIRE-RAISING BY A
YOUNG APPRENTICE.

This case was peculiarly remarkable for the extraordinary malicious
pranks of the culprit, as well as for the complication of crimes and
offences of which the "impulse to fire-raising" was by no means the
solitary one. I first became acquainted with this subject at the jury
trial. He was an apprentice named Möller, aged fifteen, who had been
for five months learning the trade of a grocer and had committed the
following deeds. 1. On the twentieth of June he had poured sulphuric acid into a pot in which his master's coffee was being kept warm;
by chance, no harm arose from this. 2. In the same month he one
evening threw a piece of burning tinder, which he had lighted at the
lamp, under the stair of the butter cellar, where he then happened to
be on business; no fire resulted. 3. When he rose at five o'clock
on the morning of the fifth of July, to go to his business, he threw a
burning sulphur match upon a cane chair upon which a dressing-gown
of his master lay. The coat was burned. 4. About the middle of
the month, he poured sulphuric acid into a cask in the cellar containing cherry brandy for sale, and he afterwards sold some of it to a
woman. At first he denied the fire-raising and confessed the attempt
at poisoning, but subsequently he only persisted in asserting that he
intended nothing malicious to the person or property of his master,
and the latter declared that there was always a perfectly good understanding between them both, so that he could not account for the
actions of the accused. The accused himself said that he "must have
had an irresistible impulse to do something wicked;" it came out, however, that the police official who first precognosced him, had supplied
him with this motive (!). His last as well as his former master, and
the shop assistants, had nothing to say prejudicial to his character.
His father stated that he was "thoroughly good tempered, but
childish, so, that, for example, he liked to play with little children,
surpassing in this his own little sister aged five." I found him small
and beardless, his genitals were, however, developed and covered
with hair; he denies having ever practised onanism. His head was
bent forwards, his look dull, without life, and gave a decided impres-

Y 2

sion of stupidity. I detailed before the jury the views I have already repeatedly stated, and assumed a "diminished responsibility." The jury accepted my views and declared "that M. is to be regarded as guilty of having committed the various attempts at fire-raising, the attempt at bodily injury, and the damage to the property of others from wickedness" (? not so, rather from mere wantonness), and condemned him to be deprived of the national cockade, to three years' penal servitude, and to pay the damages. The culprit was satisfied with this sentence, but his father appealed ; I do not, however, know the result of this appeal.

APPENDIX.—Closely allied with the so-called morbid impulse to fire-raising in young people, who so often commit criminal actions only from a half-childish wantonness, are similar boyish pranks, which also appear somewhat mysteriously enigmatical, when attention is only directed to the deed, and especially when the motive is erroneously apprehended, and which have been frequently thus explained. I have already, in my explanation of the expression *causa facinoris* (§ 61, p. 104, Vol. IV.), directed attention to such "inexplicable" crimes, which are often kept up by imitation, and for whole months keep a whole people in terror. Such were the *Piqueurs* in Paris during the twenties, and their imitators, the "girl abusers" of Augsburg, in the years from 1819 to 1832, who waylaid young girls to wound them with stilettoes ; such were the miscreants who poured sulphuric acid on the clothes of women wholly unknown to them, whom they met in theatres, &c.; such were the South German "queue-cutters" of the year 1858, who, in the dark, cut off the hanging plaited tresses of the women (it was not proved whether this was done from thievish intentions or not), &c. All these contemptible offences have one and the same psychological source, are based upon natural mental phenomena, and by no means require the supposition of some obscure demoniac impulse, and an irresponsible condition of the agent. I shall now relate two of these wonderful cases belonging to this category, in relation to the first of which I gave an accurate account of the evolutions of such phenomena, which was accepted by the court, and made use of as the basis of their sentence. It may, perhaps, be found useful in connexion with the occurrence of similar cases.

CASE CCX.—A JUVENILE DESECRATOR OF GRAVES.

" According to the report of the police official Q., on the forenoon of
Sunday, the 30th of April, the flower-dressed mounds of five graves
in two churchyards were destroyed and trampled smooth. On the
10th of May, a fast-day, in another churchyard two family-burial
places, surrounded with a high railing, were also climbed into and
desecrated; the mounds, ornamented with flowers, were destroyed,
and the flowers and flower-pots trampled to pieces. Again, on the
forenoon of Sunday, the 14th of May, in another churchyard, the
mounds of four children's graves were destroyed. The perpetrator
of these repeated misdemeanours has been ascertained to be the
journeyman weaver, Carl Müller, aged twenty-six, a native of Berlin,
who, when first accused of it by the police, confessed to the intentional
desecration of one of the family burying-places; but when further
asked regarding the further desecration of other graves, he rejoined
interrogatively, what graves? and after long hesitation he remarked,
that he could not exactly remember just now. Upon this he was
arrested, and at the first precognition he made the following
confession:—' In the course of the spring I have been in the habit
of frequenting the churchyards outside of the Halle gate. Why I
did so, I can give no proper reason. It was on Sundays, when I
was not working, that I went thither. In these churchyards I have
torn down the flowers and other ornaments from several graves,
trampled and destroyed them. I have never stolen anything from
the graves. I do not know what induced me to commit such excesses. I cannot account for it myself. I do not know the families
to whom the graves, which I destroyed, belonged. I have not acted
therefore from any malicious motive. I was not drunk, nor out of
my senses, but perfectly rational. Nevertheless, I cannot account
nor give any reason for my acting thus. I was not affected by any
religious excitement; and if you ask me ever so often the reason of
my acting thus, I must still continually repeat that I cannot tell.
I understand that my deeds were unlawful and punishable. The
damage has been done by me, and I must now smart for it. I stand
this day, for the first time, before a court. I have always supported myself decently and honestly, and earned as much as I require.' These latter statements are not contradicted by the documentary evidence. The witnesses examined, in particular the grave-

diggers, have given no material evidence in regard to the present object. Only the glazier M. deposes what may be here quoted, that in regard to the misdemeanour committed on the 10th of May, he saw the accused climb over the churchyard railings, and after looking carefully all round him, get into the family burial-place. There he trampled the graves with his feet, and then ran off from the witness, who was pursuing him; but he was speedily overtaken by the grave-diggers. Müller, upon this, was condemned to six months' penal servitude, and to pay the damages. The accused's advocate, however, in his defence, threw doubts upon his mental condition, and the undersigned was requested to examine into this.

" Nevertheless, little of any consequence has been ascertained. Müller is a feeble, middle-sized young man, with an extremely pale complexion, remarkable for a dull, inexpressive look, and with a vacant, stupid physiognomy. He denies having ever practised any abuse in *venere*. He declares himself to be perfectly sound in body, and this is also confirmed on examination. In regard to the outrage committed by him, he gave me repeatedly the same answers as he had done before the court, and declared that he could not in the least say how it happened that he had desecrated the graves. Any statements in regard to this he made monosyllabically, and with a certain amount of embarrassment; while, in regard to other matters, his trade, his manner of life, &c., he was open and communicative, and expressed himself as always perfectly consistently, clearly and rationally, so that I have not been able to discover the slightest deviation from the normal mental state.

" However remarkable the present case may seem at the first glance, it may yet be comprehended under general psychological laws. The accused has declared that neither covetousness, hatred to the dead, nor religious enthusiasm, has induced him to commit the outrages described, and there exists no reason for doubting the truth of this statement; since the grave-diggers have not discovered any theft from the graves, which he denied, and hatred of the many dead whose resting-places he wantonly outraged is just as little to be thought of, as that a man, so silly, and with so little mind, could be the subject of an exaltation of any kind, such as religious enthusiasm. By this apparent absence of all *causa facinoris*, we are certainly at first disposed to imagine that some blind impulse, the excitement of some mental disturbance, had driven him to these deeds, for it is true, though it has been disputed, that where no actual *causa facinoris*

can be recognized, no crime has been committed, for a man, so long
as he has the free use of his mental powers, is only induced to action
by motives in accordance with the general laws of human thought
and feeling. Nevertheless, for the discovery of the *causa facinoris*,
the first requisite is that we place ourselves in the same position as
the culprit, and then we will always find whenever the moral and
penal laws have been broken by a responsible agent, that a motive
does exist, rooted in the mental and moral nature of the culprit, and
which has impelled him to commit the deed, the punishable character
of which was not unknown to him, although the same motive would
have proved insufficient to induce thousands of other men to act
similarly. It will not be difficult by applying this dogma to the
accused to explain psychologically and unconstrainedly his apparently so extraordinary crime, without falling into the very frequent
error of deducing an irresponsible state of mind from the mere remarkable and unusual character of a deed, and the absence of any
obvious cause for it.

"The desire of exercising and asserting his energy is deeply rooted
in man. The child is actuated by this impulse when he destroys his
toys after they have lost the charm of novelty. The more this
impulse is kept in check by reason and morality the more it becomes
partly ennobled in character and partly suppressed. In the educated
and rational man it becomes a spur and goad to his endeavours to
distinguish himself among his peers; but even he disdains not
when walking in his idle hours to strike with his stick among the
weeds, &c. But he does not sing or shout aloud on the street, nor
does he, if otherwise in a good frame, break lamps like a wanton
youth or silly street boy. The less a man is ennobled by the culture of
his reason and moral feelings, and the less expenditure of bodily or
moral energy is required, so much the more readily is this impulse
exerted in low, vulgar, and detestable actions; and in this alone is to
be found the key to many apparently inexplicable crimes and offences:
I may refer to the so-called *piqueurs*, to the cases in which sulphuric
acid has been thrown upon persons wholly unknown to the culprits
and unconnected with them, and, finally, to a long list of acts of
incendiarism by juvenile culprits belonging to this category, and to
nothing else, one of whom (in my own official experience) once
upon a time actually declared—after an ordinary *causa facinoris* had
been sought for in vain—that he had raised this fire 'because while
lying idle in the sheep's stable, the thought came into his head, *that*

he should go and do something!' This desire to exercise one's mettle, this mettlesomeness ought to be and may be curbed by reason and morality, and is therefore very properly punished by the laws of morality, when, as unbridled wantonness, it reveals itself in illegal actions.

"Every congenital impulse is capable of being repressed, though only temporarily in those low in a mental and moral position, by employment, because the mind is diverted by the work, and the proverb that 'idleness is the root of all evil' is as generally true as it is particularly applicable in the present case. The accused, a young man from the lower classes, and employed in the most mechanical work of weaving, and whose physiognomy, as already remarked, betrays his deficient intellect, confesses that he frequently on Sundays 'when he was not working' went to the churchyards alone, and it is proved by the documentary evidence that his mischief was only perpetrated on Sundays and feast days. While here alone and with both mind and body wholly unemployed, idle, it might very readily occur to him that he might vindicate his personality most remarkably in the most simple manner, by a trifling exertion of his hands and feet, and so procure the great satisfaction of destroying—probably saying to himself all the time, I alone have done this—what it had cost others an expenditure of time, trouble, and money, to create. That he is no longer conscious of such a train of thought is no proof, even if we do not suppose him lying, of the erroneousness of this deduction, since even the recognition of such a motive presupposes an, acuteness of intellect which is not to be expected of Müller, or any similar individuals, who generally, in similar cases, express themselves in precisely the same manner in regard to the cause of the deed. He knew very well, however, according to his own confession and according to his conduct as proved by the evidence, and he knows very well yet, that his deed was a punishable one. According to the evidence of the eye-witness R. 'he looked carefully round about him' before he clambered into the K. family burial place, doubtless he did the same at other times, though he did not happen to be seen, and he ran off when he found himself pursued; these circumstances prove that he was well aware of the punishable nature of his deed, a fact easily reconcilable with the statement made by him, and which may be regarded as true, that hitherto he had conducted himself well, maintained himself honestly, and had never been punished, nor even suspected of crime. Finally, there is no reason to suppose

that M. was prevented by any temporary or persistent mental disturbance from pretermitting the execution of this deed, which he knew
to be punishable, since neither the documentary evidence nor my
own examination have revealed the slightest trace of any such
derangement, and a state of irresponsibility can never be presupposed. Accordingly, I declared it to be my opinion that the
journeyman weaver Carl Müller was a responsible agent when he
committed the crime of which he is accused, and that he is still to
be regarded as such." At this second trial he was condemned to six
months' imprisonment for "damage done to the property of others
through wantonness."

CASE CCXI.—A YOUNG SWINDLER WITHOUT ANY APPARENT
MOTIVE FOR THE DEED.

This case was indeed peculiar, and the Judge thought it requisite
to have his mental condition and responsibility determined. H. was
a Jew, aged eighteen, a commercial traveller, at present in Berlin on
a business journey; quite recently he had gone into several shops
one after the other, and in them all he had given himself out to be a
certain Count Bernitzki, had spoken broken German, and had
ordered goods to be sent to him at a hotel where he was not
residing, and where no one knew anything about him; and while
ordering these goods he had taken in one shop a cigar, and in
another a few bonbons. He never for one instant denied having
done this, either at his precognition or to myself. But he expressed
himself as completely in the dark as to any motive for the deed, for
which he knew no other reason than merely "to make fools of the
people." He stated that he was intimately acquainted with a Count
Bernitzki, and it occurred to him, for the reason just mentioned, to
play his part for a few moments, without having the slightest
intention of swindling the people, which was also, he said, proved
by the fact that he neither did nor indeed could take possession of
any of the goods ordered. His appearance, his demeanour, his conversation, and manner of speech presented nothing remarkable, and
in the course of repeated conversations with him I could not discover
any trace of delusion, or anything anomalous in his mind. In fact,
the motive alleged by him, the mere wanton desire to banter people,
and thus to gratify his own vanity, must be regarded as true, and also
one psychologically perfectly sufficient and admissible; for the same

reasons, therefore, which have been already detailed in the foregoing case, he was declared to have been a responsible agent at the time of the commission of the deed. He escaped with a slight punishment.

§ 93. CONTINUATION.—HOMICIDAL MONOMANIA.

The observation that men occasionally commit murder without any of the usual motives, indeed under the most remarkable circumstances, and apparently on the most sudden impulse, and that frequently on persons most dearly loved by them, is not of recent date. Even Felix Plater quotes the case of a mother who had a desire to murder her own beloved child, and in the works upon demoniac possession, and the like, are included similar cases of the olden time. But to gather these facts into a scientific category, and to clothe them with the mantle of theory in order to construct from them a peculiar species of mental derangement has been the work of recent French psychonosologists, particularly of Esquirol, who was speedily followed by Marc and others, till at last this new homicidal insanity, *"monomanie homicide,"* was adopted. If the mere number of bare facts be sufficient to shut the mouth of criticism, then the existence of this morbid impulse is indisputable. There exists in special works and medical periodicals a vast number of cases of men who have executed the most bloody deeds in apparently the most inexplicable state of mind. Mothers, for instance, have been seized with the most irresistible impulse to kill their children whom they dearly loved, or they have actually killed them in the most horrible manner. But is the supposition of an instinctive impulse to murder a sufficient psychological explanation or interpretation of such cases, or is it not rather *explicare obscurum per idem obscurum?* If we analyze this accumulation of cases we distinguish quite distinctly these different categories which have no connexion with each other, and the combination of which into one species has produced a confusion, which is distinctly evident in the writings of the original authors themselves. For example—1. Under the head of homicidal monomania there have been reckoned cases evidently of the most barefaced and ordinary criminality; thus that brutal eight-year-old girl of Esquirol's,* who, with frantic hatred, persecuted her stepmother, whom she heard continually abused by her grand-parents, and of whom the stepmother says "no day passes in

* Die Geisteskrankheiten, aus d. Französ. Berlin, 1838, II. S. 62.

which she does not strike me; when I stop before the fireplace, she strikes me on the back to knock me into the fire; she gives me blows with her fists, and seizes scissors, knives, and other utensils, saying, I would like to kill you, I wish you were dead," &c. She had the same hatred for her little brother, while she declared at her precognition that she did not entertain the same wish towards her father or her grandmother. A veritable instance of the old proverb, "wickedness makes up for age!" But also a case which, like all similar ones, has nothing in common with one of monomania.

2. But, by far the larger portion of those cases cited as proofs of this monomania have been of a very different nature. The "impulse" displayed by such men to kill (themselves or others) unquestionably existed, and the most horrible deeds have been frequently executed from the point of view of this impulse. But these men were mentally diseased, afflicted with melancholia. Long before the discovery of "homicidal insanity" it was known that there was such a thing as a *raptus melancholicus* (Metzger*), a "furious melancholia" (Chiarugi†). I have already detailed several cases (CLXXIII. to CLXXVI.) of this character, and could easily relate many more. The labourer, whose case is related at p. 215, Vol. IV., who loved his children passionately, cut all their throats one morning, without any one having ever suspected him to be capable of doing such a horrible deed. But the investigation proved that he had lapsed into melancholy. A few days before the commission of the deed, after he had come to the conclusion to kill himself, he wrote out a perfectly insane testament, from which I only quote that he had named the president of the ministry, who was wholly unknown to him, as his executor, and had instructed him to take care that the diseased foot of his youngest child was bathed every week with chamomile tea, &c. With these words: "I have killed my child and cut my own throat," a young woman came before the magistrates and exhibited the body of her child, aged one year and a half, which she carried in her apron, and whose skull she had smashed with an axe; to do which deed "the thought suddenly came into her head" as she sat quietly by her child's cradle. The inquiry revealed that she had become melancholic, because her seducer had long deferred his marriage with her, and had become possessed with the thought, that "it would be better both for the child and herself to be out of the

* System der gerichtl. Arzneiwiss, § 427.
† Ueber den Wahnsinn, § 422.

road. In prison she subsequently became perfectly insane. Another mother, a day-labourer's wife, who was sitting by the cradle of her youngest child, was, as she alleged, "suddenly seized with such an aversion to the child." She first put on all her clothes, then fetched her husband's razor from the fireplace, took the child into her lap and cut its throat. She then went to her sister's, where her elder child was dry-nursing, in order to fetch it, and, as she subsequently confessed, to kill it also. But this poor woman also, formerly a most worthy mother and a happy wife, had fallen into melancholy after her last confinement, which continued to develop itself more and more fully : seven months previous to the commission of the deed she had for fourteen days, alternative with maniacal outbreaks, attacks in which she wept, wrung her hands, declared she was forsaken of God, feared she would be sent to hell, &c. During repeated similar attacks she declared to several witnesses that she would murder her children, "the devil was already within her, and she was in hell even now," &c. A whole collection of similar cases observed by others are contained in the various annual series of Henke's Zeitschrift and of the Annales d'Hygiène, to which I need not at present do more than refer. Even the notorious and much-quoted case of Henriette Cornier, who suddenly cut the head off the child of an acquaintance,* belongs to the category of melancholia, as well as the most recent cases related by Ideler † and Maschka.‡ The former was that of a woman, who, in consequence of a severe bodily disease (chronic inflammatory enlargement of the uterus with pelvic abscess, opened by puncture), for some time previous to her death "fell into a state of great mental disturbance, accompanied by the idea that she would injure herself and others. This restlessness increased with the lapse of time, produced sleeplessness, manifested itself in weeping, wringing of the hands, roving about, and by the dread expressed by the patient that she would murder herself and others. One morning she awoke with the idea that she had murdered her mother," &c. It is scarcely possible in this case to avoid recognizing the portrait of melancholia. In Maschka's case, the moral and God-fearing Anna P., aged thirty-eight, killed the eighteen-months' old child of her brother by cutting its throat, in order that she herself might get out of the world ; and the medical faculty of Prague assumed with unquestionable justice, in

* *Vide* the details of the case, in Marc, *op. cit.* II. S. 48.
† Lehrbuch, &c., S. 307.
‡ Sammlung gerichtsärztlicher Gutachten, &c. II. Prag. 1858, S. 260.

accordance with the circumstances, that she had done the deed
" during and in consequence of mental derangement." That, however, in certain forms of insanity, particularly in mania, when the
passions rage tumultuously, and the patient is hurried into a great
variety of violent actions, not only towards men but also towards
things, to murder, inflict injuries, and smash everything at random,
&c.; and also in melancholia, in which, from the deepest mental depression, man is disgusted with life and all its pleasures, and regards
death as a much-longed for deliverance from all grief, both for himself and all those whom he loves as his own self; that from and in
these mental derangements the most horrible bloody deeds have been
done, has been experienced and known for at least so long as these
forms of mental diseases have themselves been known. In these
cases, therefore, also, we have nothing specific, nothing to justify
the assumption of an "inexplicable impulse" existing isolated,
like a dark stain in the pure and healthy mind. The "homicidal
insanity" is in such patients only a form assumed by the disease in
many of them, only a symptom of that general mental derangement
which will not be difficult to ascertain in each individual case of the
kind, if we only examine it carefully, and on every side (§§ 61, 62),
and do not permit ourselves to be blinded by what is remarkable in
the deed. All these cases here referred to, which comprise, I may
repeat, the larger number of those cases included under the head of
" homicidal monomania," must therefore *be struck out of this
category.*

* Marc (*op. cit.* II. S. 158) has collected (exclusive of the blackguard
child already mentioned, and two anecdotes related in a couple of lines!)
eight cases of so-called homicidal mania. There is, however, not one among
them in which general mental disease did not indubitably exist. Cazauvielh (*Annales d'Hygiène publ.* T. XVI. p. 121) has collected as many as
four-and-twenty French cases, among which there are several cases of newly-
delivered women who felt an impulse to murder their children. This, of
course, was no permanent monomania, but rapidly passed off, and only one
of these, about to be related (§ 94), falls to be considered here as coming
under this category. All the others, without exception, refer to lunatics.
Of these, I will only quote the following case, to show the careless manner
in which facts have been gathered into the category of an isolated "instinctive impulse," which have nothing in common with it:—" Jeanne
Desroches took a knife and went to her sister, where she found two little
children and an old woman. She killed her two-years' old niece, by
stabbing her with the knife. She then went into her mother's house, bade
her good day, threw her down, gave her a few stabs with the knife, and

§ 94. CONTINUATION.

3. There are still other cases, whose actual existence I am all the less inclined to deny, as I myself have had occasion to make similar observations. These pure cases, that is, those in which, without the individual having laboured under any form whatever of insanity, or having been from any bodily cause suddenly and transitorily affected by mental disturbance, those cases, therefore, in which there coexisted with otherwise mental integrity an "inexplicable something," an "instinctive desire" to kill (Esquirol, Marc, Georget, &c.), are extremely rare, or rather, there are extremely few of these cases published; for I am convinced that such pure cases actually occur far more frequently than their literary history would seem to show. The following are a few examples. Esquirol (*op. cit.*, p. 357) quotes from Gall the case of a mother who, particularly during menstruation, laboured under an indescribable anxiety, and was tempted to murder herself, her husband, and her children, whom she dearly loved. She had not courage to bathe her youngest child, for an "internal voice" was constantly saying to her, "let him drown!" Often she had scarcely time to throw from her a knife, with which she was tempted to murder herself and her children. If she went into the apartment where her family slept and found them asleep, she shut the door hastily behind her, and flung the key far from her, that she might not be led into temptation. There is nothing said as to the existence of any mental derangement in this woman, or in the subject of the following case, and we have no right to presuppose anything of the kind. Frau H. (Cazauvielh, *op. cit.*) had at times (*par instans*) thoughts which incited her to kill her four children. She dreaded

smashed her head with a hatchet. She then went upstairs into a room, where she smashed everything she could lay her hands on" (*sic!*), "went thence into a neighbour's, whom she stabbed several times with the same knife, so that she died in three days. Jeanne then betook herself to another woman, whom she called into the street, slipping meanwhile into the house and murdering her seven-years' old child. The mother, who then hurried up, she stabbed several times with the knife, and finally ran off to her mother, and concealed herself in the cellar. At the precognition this woman confessed all the particulars of her murderous deeds. Her answers, however, most unquestionably proved her mental derangement" (which no one can possibly doubt who reads even these fragmentary details of this case, which is only an ordinary example of furious mania!).

lest she should do some fearful deed, she wept, she despaired, she desired to throw herself out of the window. She fled from her family, was intentionally much from home, and hid all knives and scissors. In the confessions of a so-called hypochondriac* we find the following: "By chance I once had a sharp knife in my hand as I was busy about one of my children whom I dearly loved. Suddenly the thought came into my head how miserable I would be, if I should plunge this deadly instrument into the child's breast. This idea constantly returned in the same form. Diversion, active exercise, &c., were not spared, but all in vain; nothing could free me from my fixed idea." Finally, from my own experience, I can relate the following. A young lady of position, aged twenty years, living at the country seat of her mother, a widow and most estimable woman, very irritable and easily excited, but perfectly healthy both in body and mind, had for long before I was consulted on the matter, allowed the idea gradually to take root that she should murder her former governess who lived in the house, and was on the best of terms with herself as well as the rest of the family. This impulse continually recurred with ever-increasing force, and her struggles against it became always more severe. Her letters breathed the utmost despair as to her misfortune. She herself finally advised all knives, scissors, &c., to be hid from her, which was done ; she could not even trust herself with knitting needles. She begged, what was also conceded, that she should no longer share by night the apartment of her governess, as she had done from her childhood, &c. I advised a long journey through France and Italy without the governess, and this was carried out with the most happy results. A man nearly connected with me, in his sixties, has for at least twenty years had the extraordinary notion, which continually seizes him when he uses a razor, that he should cut out both his eyes. It has never happened that he has turned this to earnest, but continually upon the occasions referred to the question emerges within him, "If you were now to walk into your family with blind and bleeding eyes—what a misfortune it would be!" It cannot, therefore, be doubted that the most unnatural thoughts as to the commission of violent actions, in particular as to the murder of persons dearly loved, may arise and take root in the mind. This process, however, presents, apart from the wonder of a mental life in itself, by no means anything so "inexplicable," as is asserted, as to require a peculiar disease

* Reil's u. Hoffbauer's Beiträge, &c. I. S. 588.

to be made of it. It is the product of the excited fancy alone, and is perfectly analogous to the mental processes already referred to in considering the other "morbid impulses" (§§ 88 to 91). The imagination of the horrible and the dreadful has an acknowledged attraction for the fancy. Criminal *causes célèbres* are frequented with avidity both by the educated and uneducated, and tales of highwaymen, melodramas full of horror, &c., are always and everywhere sure of meeting with ample attention. But the imagination itself has its own independent fancies, and is fond of entertaining grand phantasms to break, as it were, the common-place monotony of life. When standing on the top of a lofty mountain, at the brink of a precipice, upon a tower, &c.—a voice whispers, " If you were now to cast yourself down, what would people say ? " In crossing a bridge in which there is nothing capable of exciting a fear of its fall, we hear again, "what if the bridge were to fall, and men and bridge were now to be hurled to a common destruction?" A solemn, religious service has collected a vast multitude, and the thought arises, " What if you were suddenly to fire a blank cartridge from a pistol into the air above your head, what an uproar there would be, what consternation, what a running and crowding ! " The children are so good, so dear, "what or how would it be if you were to murder them ?" Thus these thoughts arise, often suddenly, as pure freaks of the imagination, and they have their own peculiar attraction, especially for men of general excitability, for those afflicted with bodily disease, hypochondriacs, hysterical persons, and, in regard to sex, particularly for women. Once arisen, however, these thoughts continually reproduce themselves, according to the laws of the association of ideas. That mother, who, when bathing her child, has once been seized with the idea, what if you were to drown it ?—very naturally recurs to the same idea during subsequent baths; each time of setting the razor recalls, in a perfectly natural and psychological manner, the cutting out of the eyes, &c. Thus the idea gradually becomes rooted in the mind, and may under certain circumstances finally attain the force of a fixed idea, and exert a disturbing influence upon the vital relations ; indeed, it may produce all the more unhappiness and despair, the more horrible the original idea is, such, for instance, as the murder of beloved children, and the more the person is otherwise in perfect moral sanity, and able to command his fixed idea, inasmuch as he recognizes it as such (§ 79, p. 234, Vol. IV.). And how he finds in his fundamental morality, that aid which makes him victorious in this certainly often severe struggle, we learn from experience, and is exhibited in these pure cases just detailed, in which none of the fancy-born misfortunes actually happened, no horrible deed was actually executed.

I think I have given a psychologically true explanation of those cases in which the impulse to hurt or murder has developed itself into a fixed idea, and which belong purely and exclusively to that category, and each individual case of which is to be decided according to the rules applicable to it (p. 234, Vol. IV., &c.). To construct from these cases a peculiar species of insanity, is just as unscientific as it would be to take a hundred other similar fixed ideas, of which, probably, far more cases could be collected than of that under consideration, and construct from them a hundred similar " morbid impulses " and monomaniæ; the attempts to do this being already only too numerous to the great injury of the administration of the criminal law. I may repeat that both of the other two categories of cases reckoned as examples of homicidal insanity do not belong to this head at all, and I have already given the reasons for this opinion; thus we arrive at the dogma that *there is no such peculiar species of insanity as is termed homicidal insanity, homicidal monomania,* and that forensic medicine neither can nor ought to recognise any such. In a purely practical respect an opposite opinion is also perfectly superfluous, since, irrespective of that, each individual case must be unfolded before the Judge in all its diagnostic relations.

SECOND SECTION.

IDIOCY.

§ 95. General.

The Prussian Common Law defines Idiocy as "the inability to consider the consequences of an act." It has been often said, and it is daily felt by Prussian physicians, that this definition, even if we interpret it as referring only to the "statutory" results of the act, is extremely defective, since our Common Law definition of insanity (and mania)—a "total deprivation of reason"—is far more applicable to the state of idiocy. The various scientific appellations which have been given to this condition, *amentia, fatuitas, imbecillitas, idiotismus,* exhibit the various modifications and gradations in which it occurs in nature, from mere defective intelligence, stupidity, up to complete negation of every mental faculty, actual idiocy; I do not say up to cretinism, for this condition, that of the cretin, who is only a a human caricature (Heinroth's "man-brute"), is no longer an object for forensic psychology. But these gradations, which certainly do occur individually in nature, can never be separated by any determined-limits from one another, and all attempts systematically to separate from one another, defective intellect, stupidity, silliness, imbecility, idiocy, have been thwarted by those innumerable transitional forms by which nature so often derides attempts at scientific systematizing. The more easy, however, it is in general to diagnose this condition, which is readily done not only from the mere deportment of the patient, but at once from his answer to the simplest question; for instance, from the impossibility of solving the simplest question in arithmetic, &c., and the less recognition of these various modifications is made by the statutes, so much the less practical interest is there in any systematic gradation. The ease with which the diagnosis is made in general, contrasted with the various forms and individual cases of insanity, is also increased by the circumstance that, as experience teaches us, mere simulations of idiocy are extremely rare, and when they do occur are usually so coarse and so mingled with every possible symptom of insanity, that the discovery of the deceit is generally speedily made. On the other hand, unfounded accusations of defective intelligence, imbecility or idiocy, are often made, which do not always arise from the malevolence of interested parties, but are often based upon deception, since certain bodily conditions, as St. Vitus's dance, deafmuteness, even a high degree of stammering, may produce the appearance of seriously defective intelligence, cases which a slight experience in the matter enables us readily to diagnose. Moreover, the idiot—taking the word in its most comprehensive signification—is not able to conceal his mental condition from the observer, as is so often done by the lunatic, and thus the physician has in each individual case the advantage of having a pure and unadulterated object for his observation, inasmuch as the idiot, the mental child, steps before him and exhibits himself as he is.

§ 96. CONTINUATION.

In by far the larger proportion of cases it is the civil responsibility of idiots that has to be judicially considered, and only rarely does their criminal responsibility come in question. In regard to civil responsibility its amount is usually readily ascertained. A man who knows the year in which he was born and the current year, but who cannot tell his own age (a very common example!), is not, of course, in a position to look after his property himself, or to undertake any office or employment, &c., and how much less would this be the case if he were mentally a cipher in even a higher degree. But criminal responsibility is not to be so unconditionally and absolutely excluded in all degrees of idiocy, for the "power of discrimination" (Penal Code), the consciousness of the difference between good and evil is inherent in the mind, and not in the intelligence, as has been already fully shown (§ 63, p. 111, Vol. IV.) and need not be repeated here. Mrs. D. was so imbecile that she was fit for nothing but sweeping out the room, and had to be combed and washed by her friends. She gave a negative answer to my question, whether she loved her nine-years' old son? But to the question, whether she would then strike him dead? she replied, with a stupid smile, "oh no, why not!" Cases certainly do occur,

though rarely, in medico-legal practice, in which the criminal responsibility of imbeciles and idiots is concerned (Cases CCVIII., CCXIII.,
and CCXIV.), for observation teaches us that even such individuals
are capable of emotion, and that the original human passions, particularly anger, revenge, and childish petulance, are not completely
extinguished in them, and may impel them to the most violent
actions. Thus, incendiarism, murder, and dangerous violence, have
been perpetrated by idiots, who, therefore, are not quite so harmless
as they are usually reputed. The less, therefore, the limits of the
various gradations between the merely weak-minded and true idiots
can be generally defined, so much the more is it necessary in every
individual case of disputed criminal responsibility in an idiot, who
has perpetrated an illegal action, to fall back upon the circumstances
attending this individual case, and their interpretation according to the
general diagnostic rules (§§ 61 to 63), and to decide the case accordingly. The external phenomena of idiocy as exhibited in one
affected by it, must, as well as the physiognomy of insanity, be
presupposed by forensic medicine, to be well-known by the forensic
physician. The physiognomy of idiocy is masterly delineated true
to nature in a few lines by Esquirol, *op. cit.* (Die Geisteskrankheiten, &c.), § 165.

§ 97. ILLUSTRATIVE CASES.

CASE CCXII.—CREDIBILITY OF THE EVIDENCE OF AN IMBECILE.

Sophia, aged eighteen, was said to have been assaulted and
violently ravished; at her precognition she seemed so stupid, that it
was considered necessary to examine into her mental condition, and
this was expressly required for the purpose of ascertaining the amount
of credibility due to her evidence. "I found her to be a decided
imbecile. She was of short stature, apparently but little developed
corporeally as yet, the hinder part of her head was characteristically
flat, her forehead was low, and her look stupid and lifeless. She never
looks at any one speaking to her, but always sideways away from him.
She smiles without any occasion. Her deportment is negligent and
slovenly. She can neither read nor write, and, according to the
statement of her mother, she is even unfit for the simplest household
work. It is impossible to carry on any connected conversation with
her, apparently because she does not understand the questions.
When brought upon the subject of the present enquiry, at first, she

always merely said, 'he sits—will be punished,' and when asked why he would be punished, she merely said, 'has played with pussy.' It appeared also as if a certain amount of natural shame prevented her from confessing the alleged rape. But when she perceived from my questions that I knew everything from the documentary evidence, she related that B. had thrown her upon the bed, and after her fashion, she expressed by hints more than words the union of the genital organs. It was impossible to get anything more accurate out of her, because then she constantly repeated 'has played with pussy—will now be punished.' In regard to the question as to her credibility, her reply to my enquiry as to the nature of the foreign body hinted at, was very characteristic, 'that it was as long as my stick!' At the same time, in all her statements was discoverable that absence of mind always to be found in imbeciles and weak-minded persons, because they are not able to fix their attention upon any particular subject for any length of time. Accordingly, I do not hesitate to declare that Sophia *is an imbecile.* From this alone it follows that *full credibility cannot be accorded to her statements;* because, like all her other mental faculties, her memory is also necessarily feeble, and therefore she is not in a position accurately to remember and describe the circumstances accompanying the alleged assault; but at every attempt to ascertain these she continually recurred to the expression already recorded, 'has played with pussy,' &c. Therefore, in conclusion, I declare it to be my opinion, that Sophia is affected with (congenital) imbecility, and that the full amount of credibility which would be given to the statements of any other mentally sound and moderately gifted maiden of eighteen, cannot be accorded to her statements."

CASE CCXIII.—CRIMINAL RESPONSIBILITY OF AN EPILEPTIC IMBECILE FOR RESISTANCE TO THE LAW.

During the night between the 6th and the 7th December, 1841 (consequently under the former Penal Code), the tailor Marsch had violently resisted and injured several watchmen. The medical evidence in regard to him rendered a medico-legal examination necessary, and this I carried out so as to supply an answer to the question, "whether, and in what degree is M. to be considered a criminally responsible agent?" "Marsch," I stated, "presents at the first glance an extremely remarkable appearance. His forehead is lofty, but flat, his look is lifeless, his gait is irresolute, his demeanour vacillating. On the right parietal bone he has an old cicatrix, one inch in length, and above it a fresh scab, alleged to have arisen, with great probability, from a recent fall. A constant silly smile during his examination, and still more the manner in which Marsch expressed himself, in combination with the known origin of his ailment, left no doubt as to the state of his intelligence, which will be presently detailed. Marsch declares that he has from childhood suffered from epileptic attacks, and that he is dull of hearing. The latter is confirmed by our conversation with him, and in particular he is quite deaf of the right ear, and very dull of hearing of the left. The medical certificates leave no reason for doubting the fact of the long duration of his epilepsy and of the severity of his disease. In particular, the resident physician of the workhouse certifies 'that Marsch has daily attacks of epilepsy, especially towards night.' It is known how readily epilepsy of many years' standing, particularly when severe, as a case must be in which the fits occur 'daily and especially towards night,' induces mental weakness or even actual idiocy in those affected by it, and the individual before us affords proof of the truth of this experience. Marsch was not able to tell me his age accurately, he reckoned himself only forty-two, although he stated that he was born in 1798, and he knows that the current year is 1842. To my representation that his calculation was not correct, he replied with his usual smile, constantly repeating the question put to him (as is customary with such people), 'that is not his fault, he is unable to calculate it.' It is, moreover, difficult to make the simplest question comprehensible to M., and it is not easy to understand his obscure and stuttering answers. It could not be ascertained how old his two dead children had been; and when I asked him as to the length of time he had been married, and the years in which the children had been born, he involved himself in the most extraordinary contradictions, at last always, once more in a very characteristic and usual manner, assuring me that in regard to these matters his wife would be able to give me every information. When asked as to the sex of his children, he at first replied, 'they were only quite little children,' and only when I asked more distinctly whether they were boys or girls did he answer the question after some consideration. Also, he could not give the maiden name of his wife; but stated that if she were present I would soon learn it. When asked how it happened that he had resisted the watch, he declared that 'when any one attacked him, he always resisted:' in regard to the more immediate incidents of the deed of which he was accused, he could give no information, nor could he recognise in it anything illegal. In accordance with all these details, I had no doubt that Marsch was a perfectly helpless, very weak-minded, and imbecile man, who by a longer continuance of his epilepsy would, most probably, in the course of years lapse into the lowest degree of mental weakness, true idiocy, and who even now, and extremely probably for a long time past, has been incapable of considering the results of his actions (Gen. Com. Law), and I, consequently, answered the question put before me as follows :—That Marsch is now (three weeks after the deed) to be regarded as an irresponsible agent, and that extremely probably he was also irresponsible even at the time when the deed was committed."

CASE CCXIV.—THEFT COMMITTED BY AN IMBECILE.

This was no common case, and no one would certainly, à priori, have supposed that its subject, lame in body, and degraded in mind, could have performed such a deed, and in such a manner as was, nevertheless, done by him. The labourer Hoffmann had, on the 5th of February, offered to pledge an alleged life assurance policy with the tradesman R., and took the occasion to steal out of the coat of the latter, and behind his back, a pocket-case containing cigars and valuable papers. R. did not hear any irrational talk from Hoffmann at this time, but his "staggering gait" struck him. After his suspicions had been directed to the accused, he sought him out, and charged him with the crime, which, however, H. denied. His wife begged him "on her knees" to tell the truth; the accused, however, folded his hands and replied, "dear little mother, if I had them, I would give them back to Mr. R." At the same time, he asserted that the cigars found by him had been previously in his possession, and that he had bought the ham, also found by him in a neighbouring shop, which he pointed out; this, however, was not confirmed. After some urgent exhortation, however, he confessed his crime to the party robbed, and declared that he had hid the paper-money in a place which he pointed out, but had thrown the bills, &c., into an apartment, which he also pointed out. Both of these statements were found to be true. " At his precognition, after his arrest, H. declared that he did not know the names of his father and mother, nor the names of his deceased or present wife, nor the number and age of his step-children; he confessed the theft of the cigars, 'because he wished to smoke a cigar,' and attempted to explain what he was further accused of as an accident. His behaviour during the precognition, however, was of such a character that it was recorded 'that he leaves the impression of a wholly irresponsible agent.' As often as it was mentioned that he had done wrong, he burst into tears, and called out that he was quite losing his reason: he interrupted the proceedings with questions which proved that his mind was occupied with other matters : he answered the question as to the name of his wife very angrily in these words, 'mother is her name,' &c. H. has repeatedly behaved in a precisely similar manner before me. He is thirty-eight years of age, very tall, has a high complexion, bristly hair covering the forehead, and a staring, decidedly stupid look. He speaks with a half stammering tongue, and cannot take two steps without staggering. As this evidently not simulated walk and speech pointed to some previous cerebral lesion and consequent semi-paralysis, and as H. declared that he had been treated in the Charité Hospital, I requested a sight of the sick reports. From these it appeared that H., about Christmas, 1855, had lost his property, and fallen into a state of 'melancholia,' in which he also had visions of devils, mistook known persons, 'and could no longer form any opinion in regard to the affairs of life.' At his admission into the Charité Hospital on the 26th of January of last year, the paralysis of the extremities was already very distinct, the patient was very irrational, lay constantly and slept much, gave stupid, unconnected answers; he improved, however, under treatment which, according to his statement to me, consisted chiefly of douche baths, the remembrance of which he preserves. In April he was transferred to another station, and in June he was dismissed. Though, according to this, it is indubitable that H. has suffered from a cerebral affection, which has injuriously affected both body and mind, and his whole appearance made it not doubtful to me even before I obtained a knowledge of the anteacta, and that no simulation is to be supposed in this case, yet a deed such as the one described, committed by a man who is half paralyzed and wholly imbecile, must seem remarkable. Still on further considering it psychologically, this case loses its unprecedented character. It is a well known fact, often exemplified in lunatic asylums, that lunatics steal, and do so with a certain amount of cunning. It is also well known that lunatics, though they no longer possess any actual 'power of discrimination,' yet have a certain instinctive foreboding of what is allowed, and what is forbidden. From this never distinct knowledge, such lunatics are observed to conceal the stolen goods, deny the deed, and exculpate themselves till they are palpably convicted. When H. states, that he observed the cigar-case in the, according to the statement of R., 'front pocket hanging outwards' of the coat, and that he took a fancy to smoke a cigar, such an extremely simple mental process is even by him very possible. Even the concealment of the paper money, and the throwing away of the bills, subsequently discovered in the pocket-case, is no more than what a child would have done under similar circumstances. Since now the examining Judge very properly declared his doubts as to the mental condition of H., since the remark of the physician in the Catholic Hospital, in which H. was treated for some bodily ailment, 'that his behaviour there in the period from the 20th to the 28th of January of this year' (*that is but a few days before the theft*), 'was of such a character, that he must in the most decided manner deny his responsibility,' is of much value, since the report of the condition of the accused while in the Charité Hospital is of great importance, so also my own examination of H. leaves no doubt upon my mind that H. is no feigner, but is an individual seriously diseased, as the result of a depressing mental affection and a bodily disease of the brain, and that now, as well as before, and at the time of the commission of the crime he was not in a position to 'consider the results of his actions.' Since the legal terminology calls this condition 'Idiocy,' I therfore declare that H. has perpetrated the deed of which he is accused when in a state of Idiocy (§ 40, Penal Code)." The case was therefore let drop.

§ 98. DEAFMUTENESS.

STATUTORY REGULATIONS.

GENERAL COMMON LAW. Part I., Tit. 9, § 340. *Lunatics and idiots, as well as deafmutes, enjoy similar prescriptive rights* (*vide ibid.* § 595, and Part II., Tit. 18, § 346, where minors, lunatics, or idiots and deafmutes are placed upon the same level in regard to the appointment of trustees).

IBIDEM. Part II., Tit. 18, § 15. *Congenital deafmutes, as well as those who have become so before attaining their fifteenth year, must be taken under the guardianship of the state, whenever they cease to be under paternal control.*

IBIDEM, §16. *Those who have become deafmutes at a later period of life, must be placed under guardianship only when they cannot express themselves by signs generally understood, and are therefore incapable of looking after their affairs.*

IBIDEM, § 818. *The guardianship of deafmutes ceases when upon examination it is found that they have attained the power of taking charge of their own affairs.*

IBIDEM, § 819. *Therefore, even if the defect in the hearing and speech should be removed, an examination must yet be made, lest imbecility or weakness of mind should make the continuance of the guardianship necessary.*

IBIDEM, § 820. *Both of these examinations referred to must be undertaken with the assistance of the persons named in* § 817 (namely the guardian, an expert named by the court, the relations, &c.).

(In regard to the testamentary capacity of deafmutes, *vide* Part I., Tit. 12, §§ 26 and 123. In regard to their power to receive presents and legacies according to the Rhenish-French Law, *vide* the Civil Code III., I. Art. 936.)

The Penal Code knows no deafmutes.

Not only the Prussian, but also every other German statute book places the deafmutes in identically the same position as to legal rights with (minors) lunatics or idiots, particularly the latter, and with perfect justice. For these unfortunates, whether the deafmuteness has been congenital, or acquired during the early years of childhood by the accidental loss of hearing, after which, the scarcely learned speech is speedily forgotten, are quite peculiarly idiots in the verbal sense of the word (ἴδιος, *solitarius, privatus*). They stand alone in the world, seeing that two of the most important means of communication with them are closed, and they crawl through life in a state of social starvation, as it were, in the train of their fellow-men. In the larger proportion of cases they are, indeed, originally furnished with every mental capacity, and can, therefore, not only do good work in all simple mechanical occupations, maintain themselves very well, and become useful members of society; but some of them even have talent and become artists, to say nothing of those certainly very rare cases in which they possess actually much higher endowments. But the mental powers are not developed, and remain in the lowest stage, because the enlivening mental intercourse with their fellow-men, enjoyed by the meanest peasant lad, is cut off from the deafmute, or reduced to its lowest ebb. All statutes, all authors, therefore, set a value upon the instruction which the deafmutes may have enjoyed, and of course I do not mean to deny, that special instruction may and does act beneficially, even if it only provides the deafmute with a certain amount of acquaintance with the elementary sciences, and some degree of understanding in religion and morality. How much or how little, however, even the best educational institutions for deafmutes, or even the most renowned teachers can effect in the education of these unfortunates, what invincible obstacles have been opposed to their endeavours by the natural helpless condition of the deafmutes, I have, alas! during my continually recurring investigation of the mental condition of deafmutes had only too many occasions of discovering, and of this I shall presently give a few examples.

In almost all cases, these investigations are connected with an enquiry into the civil responsibility of deafmutes, in accordance with directions given in the statutory regulations already quoted, and refer specially to the removal of the curatorship prescribed by statute; this removal is very frequently petitioned for by a deafmute who has long been of full age or by his curator, and can never be legally agreed to without hearing the opinion of the expert. Deafmutes are men, and are not devoid of the original passions and emotions of men, particularly anger, hatred, and revenge. They have, therefore, been even the objects of criminal accusations, and even murders committed by deafmutes have been reported by Alberti, Hoffbauer, Itard, Marc, Jendritza, &c. My own experience, however, which includes very many cases in which the civil responsibility of deafmutes was in question, comprises but a very few in which their criminal responsibility has had to be enquired into after the commission of some illegal act. I cannot regard this relative proportion as purely accidental; taken in connexion with the fact, that only very few cases are known where crimes have been committed by deafmutes, and with the favourable testimony as to the disposition which I have so often received in the course of my examinations from the relatives and acquaintances of deafmutes, we are rather driven to the conclusion, that the nondevelopment of the mental powers by the internal isolation of the deafmute is accompanied *pari passu* by the nondevelopment of his passions.

§ 99. CONTINUATION.

In regard to the manner of investigating such cases, in the first place, every attempt at communication with, or questioning the party to be examined by means of the ordinary form of speech, is wholly insufficient. I have frequently convinced myself, even in the case of deafmutes who had enjoyed many years' instruction in the admirable Royal Institution for the Deaf and Dumb of this city (Berlin), that questions uttered with the most careful and distinct slowness, and with the clearest definition of each syllable, &c., may indeed be understood, but, even after a long and troublesome conversation, our object is never thus attained. Still less can our end be gained by means of pantomime. It is often extraordinary to observe how much skill in this mode of communication is attained by those who constantly associate with deafmutes, as members of the same family, trade masters, &c. But, besides that the same practice and skill is not possessed by others as the physician, or the Judge, this mode of communication is limited solely to simple, ordinary household inter-course, and the assistance given me in this respect by the members of the same family was, therefore, almost always very far from being suffi-cient for my object. The only sufficient mode of examination, and one which is, therefore, always recommended by all experts, is by means of written communications, presupposing, of course, that the party to be examined can write and read writing. This has fortu-nately been, without one exception, the case with all the deafmutes of this city (Berlin) which have come before me. When the reverse is the case, and even in the case of deafmutes who can read and write, when they are accused of serious crimes, and it is needful to obtain as much insight into their mental state as possible, the physician must decline undertaking the investigation by himself, and the necessity of obtaining the assistance of a teacher of deafmutes must be pointed out to the Judge; this has been of the greatest assistance to myself in cases of this character. For even conversation by means of writing can, as I can testify, only be carried on within very narrow limits. It is self-evident that in these communications we must commence with the very simplest question; questions as to his name, age, family re-lations, &c., then short arithmetical tasks may be given, such as would be given to a child ; questions as to matters universally known, the King's name for instance. But the utter helplessness of the deafmute, their actual mental weakness is very speedily shown. It is affecting to see how, when they have carefully studied a written question, with that vivacity which is peculiar to most deafmutes, and often with joyful excitement at having at length comprehended it, they quickly seize the slate pencil and write down their answer. How erroneous this, however, often is, and how the deafmute deceives himself as to his understanding of the question, I shall presently exhibit by giving a few examples. The further, however, we proceed with our queries, so much the more does the enfeebled mind of the deafmute give way, partly from the unusual exertion, and partly because it is no longer able to follow and comprehend the sense of the questions, and a stop must be put to the examination, because to urge it any further would only be a useless annoyance of the unfortunate individual. Fortunately such a necessarily superficial examination is usually practically sufficient. For as the teacher, when a boy cannot tell him correctly the genitive of *pater*, justly considers that he has made a sufficient examination to enable him to give evidence as to his scholar's knowledge of Latin, so also the physician will consider himself entitled to give a conscientious opinion as to the civil responsibility of a deafmute, and his capacity for taking charge of his own affairs, &c., when he finds that he is not capable of giving correct answers to trifling questions in arithmetic, which a boy after two years' instruction could readily solve. And this is the case, I may repeat, in, alas! the larger proportion of such cases, and I can only remember one case (CCXVII.) in which I was able to give an opinion favourable to the removal of the guardianship which was applied for. From the mental organization of the deafmutes, however, we must in every case exercise the very greatest caution in declaring them civilly responsible, because to be under trust is the greatest good fortune for them ; is the necessary supplement of their existence, without which they would fall a prey to the first swindler who came across them. In regard to the disputed criminal reponsibility of deafmutes, in any case which may occur, the materials collected during this examination with the assistance of an "expert" teacher of the deaf and dumb, must be duly considered, in accordance with the general diagnostic rules (§§ 61 to 63) already laid down, which are generally applicable to deafmutes as well as to other men, and in doing this the consideration of the circumstances of each individual case must, even in regard to deafmutes, constitute the most important requisite for the formation of an opinion in regard to it. I have already (p. 90, Vol. IV.) referred to the simulation of deafmutism.

§ 100. Illustrative Cases.

Case CCXV.—Attempt at Rape and Homicide by a Deafmute.

Shoemaker Nitsch was the accused deafmute examined with the
assistance of R., a teacher of deafmutes. At the first glance his very
defective intelligence was betrayed by his flat forehead, and his perfectly stupid and inexpressive look. And also, by the remarkable
circumstance, that N. had none of the great vivacity and excitability
of gesture and pantomimic speech which is usually so characteristic
of deafmutes. Whether and in how far *excess in venere*, to which
N., according to his own statement—and this is interesting as regards
the present inquiry—was much addicted, had occasioned this general
mental and bodily flabbiness, must remain undetermined. After a
few introductory questions, N. was brought upon the subject of the
deed of which he was accused, and it was repeated to him that he
had forced his way into the house of a man unknown to him, the
sexton, Sch., and demanded that he should give him a girl (the sexton's
daughter) to go to bed with, and that on a second attempt, after he
had been threatened with being kicked out, he had drawn a knife,
and with it attacked the sexton. With the ingenuousness of a little
child, he smilingly confessed all these deeds. On the possible
results of his act for Sch., as well as for himself, being represented
to him, he stated, as he had previously done when precognosced, that
he certainly might have killed the sexton, and that this would have
cost him his head. After long consideration, he also remembered the
Ten Commandments. But all these statements, in which, as already
remarked, the relative passiveness of his gestures was remarkable,
and which were taken *solely* from the interpretation of the teacher R.,
who himself seemed to have some trouble to make him understand,
and get anything out of him, all these statements were made after a
fashion which left no doubt that in regard to all these matters, and
to the difference between good and evil, Nitsch had only an obscure
idea, but no distinct understanding. In accordance with what I had
seen, and the results of the examination, I could not do otherwise
than state, "that the deafmute Nitsch was, from great feebleness of
mind, unable to consider the results of his actions, and that, therefore, he was to be considered an idiot in the Common Law sense
of the word." (§ 21. i. 1).

CASE CCXVI.—CIVIL RESPONSIBILITY OF A DEAFMUTE.

She was thirty-two years of age, blooming and healthy, with a lively look, and, according to the statement of her mother, had lost her hearing when eight months old from the effects of a draught of air, and that she could still hear with her right ear but very feebly. Her civil responsibility was in question. She had a hundred thalers (£15) lodged with a woman Lehmann, who paid interest for it; some of the questions following have reference to this, and the answers, which lie before me in her own handwriting, I have here copied literally. I may remark, that this woman had been instructed in a deaf and dumb institution for *seven years*. We shall see how much of this instruction remained.

Have you a father still? "Yes; he is dead."
Have you a guardian? "The proprietor G."
What is a guardian? No answer.
How many Commandments are there? No answer.
Have you never heard of the Ten Commandments? "I know no commandments."
In what year were you born? "1809." (Correct.)
When will you be forty?
She misunderstood this question, thought I considered her to be forty years old, and wrote down rapidly "not yet thirty-two," giving me to understand by the most lively gestures that she felt herself aggrieved by my thinking her so old! ("Vanity, thy name is woman!")
And what year will it be eight years after this? "1850." (Correct.) Upon this she wrote, "What is your name?" I wrote down my name, and asked, what is a doctor? "The disease."
What is the name of our King? "William Frederick the Fifth, King of Prussia, to-day is his birthday." (The latter statement was correct.)
Do you get interest from the woman Lehmann? "Every three months, one thaler and twenty-five silver groschen" (five shillings and sixpence).
How much does that make in a year? No answer, and signs that she did not understand the question.
How many times three months are there in a year? The same.
How many months are in a year? "January, April, July,

October." (The quarters when the interest was paid were thus impressed upon her !)

Are there no more months in a year? "Before the year 1838." Her attention was again directed to the question, and after some consideration she wrote, "a year has twelve months."

Therefore, when you get every three months one thaler and twenty-five silver groschen (five shillings and sixpence), how much do you get in a whole year? "Five thalers" (fifteen shillings).

Are you certain that the woman Lehmann will give you back your money? "One hundred and seventy thalers, seventy thalers" (£25 10s—£10 10s.).

After a little explanation, she gave me to understand that she now understood the question correctly, and that she did not doubt the woman Lehmann.

Why do you think so? No answer.

If she does not pay you, what will you do? No answer.

When will you again receive one thaler and twenty-five silver groschen (five shillings and sixpence)? "In October."

What month are we now in? (It was precisely the 15th of October). After long consideration, "October."

Therefore, you have just lately received money? No answer, &c. It is evident that no civil responsibility could be accorded to such an individual.

CASE CCXVII. — CIVIL RESPONSIBILITY REACQUIRED BY A
DEAFMUTE.

An equally rare and pleasing case, in which a subsequent examination after the lapse of nine years gave quite a different result from the first, which I carried out in 1842, after the guardian had applied for the removal of the curatorship; "since his ward N., now of full age, had been nine years in a deaf and dumb institute, and could now make himself understood in writing by everyone." How far this understanding went is exhibited here in a few of his answers —literally transcribed—to my questions.

Where were you born? "I was in Berlin on the 4th of April, 1812." (Correct.)

Have you property, and how much? "Four hundred and forty-one thalers." (£66 3s).

Where have you lodged the money? " With the house-proprietor."

Would you give the money to any house-proprietor? No answer.

Do you require any security from the house-proprietor? " I require four thalers, fifteen silver groschen per cent." (thirteen shillings and sixpence.)

Must he give you any writing upon the matter? "Yes."

A bill for instance? "I can also write other things."

Do you require the mere promise of the man to pay you every year four thalers fifteen silver groschen (thirteen shillings and sixpence)? "Four thalers (twelve shillings) per cent."

If the house-proprietor will neither pay you the four hundred and forty-one thalers (£66 3s.) nor your four per cent., what will you do? " I can let it remain, and three per cent., nine thalers, fifteen silber groschen (twenty-eight shillings and sixpence)," &c., &c.

Accordingly, I could not recommend the removal of the curatorship. Nine years subsequently, in the summer of 1851, the application was renewed before the court of guardianship, and in support of it a number of certificates were handed in, some from the Directors of the Royal Institution for the Deaf and Dumb, and others from one of the first printing establishments in Berlin, in which N. had worked for a long time, from the proprietor, from the assistants, &c., all of which were remarkably favourable as to his powers. From the great interest of the case, I examined him repeatedly, because at the first I found a material and surprising advance beyond his former state. N. had so improved in the power of speaking, that he now spoke so that he could be understood tolerably well, and it was supposable that those who were in the habit of associating with him could understand him sufficiently well, and this was also confirmed to me. Thus one great medium of communication with the external world was opened up, and its result was visible. His look was now clearer and opener than formerly, and his eye lively. His punctuality and skill at his work were not only much praised by the experts, but also every question which I put to him in regard to his trade, the way to manage it, his savings, easy arithmetical questions, &c., he now answered in a manner which could only be termed pleasing. Accordingly, I could declare that N. was now capable of looking after himself and taking charge of his own affairs, and that he no longer required to be under curators. He has not since appeared before the court of guardianship, a proof that there has been no necessity for placing him again under curators.

CASE CCXVIII. — RESTRICTED CIVIL RESPONSIBILITY OF A
DEAFMUTE.

In regard to the deafmute journeyman book-binder, St., aged forty-
four, the following question, formulated according to the statutes,
was laid before me—"Does he possess, 1. the power of expressing himself rationally, and 2. the capacity for looking after his own affairs?"
He had been formerly for ten years a pupil of the Royal Institution
for the Deaf and Dumb, and had, as I stated in my report, "in it at
least learned to write currently and almost quite orthographically; of
this I satisfied myself. Questions put to him regarding his profession,
his gains from it, and the amount required for his maintenance, he
answers in writing with ease, showing thereby that he is not unacquainted with the elements of arithmetic. It is also to be assumed,
as his sister states, that he transacts his business orderly and effici-
ently, especially as his profession is a still and quiet one, and does
not concern those senses of which St. is devoid. It is different, however, in regard to the question, whether he is capable of looking
after his own affairs in the full statutory sense? I must answer this
question negatively in the interest of the party himself. My examination has shown that he has no idea in regard to any of the only
moderately complicated affairs of civil life, involved in the management of property, such, for instance, as lending money upon security
of mortgage, &c. Nothing, therefore, would be easier, on the part of
a swindler, than to cheat St. out of all his property, as, according to the
evidence of his brother-in-law, on the 12th of this month, he has
already lent money inconsiderately. Experience has, moreover, made
me sufficiently aware, that only a small proportion of all deafmutes
attain a higher development of their mental faculties. For all these
reasons I must declare, that St. certainly possesses the power of
making himself understood (in writing), but that he is not capable
of looking after his affairs himself."

CASE CCXIX. — ABSENCE OF CIVIL RESPONSIBILITY IN A
DEAFMUTE.

This was a similar case; it concerned Caroline R., of full age, who
had been born deaf and dumb, and, like the man in the former case,
had been instructed for ten years in the Deaf and Dumb Institution

of this city (Berlin). The judicial question put to me will appear from its literal answer, which will be presently given. She was brought to me by N., a teacher of deafmutes, who was proposed as her curator, and who once more rendered material service in the examinations. Caroline R. was a person of small but compact growth, of healthy complexion, normal formation of skull, and good bodily health. Her eyes were clear and lively, and betrayed no evidence of imbecility. She spoke well, or rather she forcibly uttered hard, articulate sounds; without the assistance of the teacher, I would not have understood them, and long practice was certainly needful to enable one to understand this so-called speech of R.'s. I was just as little able to understand her language of signs, which was the usual pantomimic language of deafmutes. I therefore carried on my conversation with her in writing, and as the basis of my decision I may employ a few samples of this.

Question. After she had stated that she possessed three hundred thalers (£45)—What would you do with the money if it were handed over to you ?

Answer. " I would save the money " (tolerably legibly written).

Question. Would you not buy anything with it?

Answer. " No."

Question. If you put your thalers in a box, you have then nothing for them ?

Answer. " No."

Question. Do you sometimes visit your acquaintances, what do you do then ?

Answer. " A person of property."

Many other questions in regard to other subjects resulted in similar unsatisfactory answers.—I then detailed that mental powers such as hers, " were insufficient for her proper guidance in any of the more complicated relations of life, such as those concerned in making a contract, a will, &c.," and in conclusion, I answered the question put to me as follows :—" 1. That Caroline R. is certainly capable of making herself tolerably well understood in writing, but not by signs generally understood ; 2. that she is not capable of making full use of her mental powers, or of taking charge of her own affairs; and 3. that the appointment of curators is rendered necessary, not by the existence of idiocy, but by mental weakness."

CASES CCXX. and CCXXI.—TWO BROTHERS BORN DEAFMUTES.

Both of these had enjoyed, with interruptions, five years' instruction in the Institution for the Deaf and Dumb already referred to. In the course of a long and troublesome examination, it was not possible to get one single pertinent answer from either of them, except in regard to their age, which they correctly stated to be twenty-eight and twenty-four years. To every other question, even of the simplest character, as, When were they born? What was the current year? Had they any money? &c.—they wrote the most preposterous answers, and it was impossible to enlighten them as to their errors. They were otherwise, according to the statement of their elder brother (who could speak), perfectly useful as assistant-gardeners. It is evident, that civil responsibility had to be denied to both of them.

INDEX.

Dissection of the body; abdominal cavity, I. 210
Docimasia pulmonaria - III. 41
„ „ a. vaulting of the chest, III. 41
„ „ b. position of the diaphragm III. 48
„ „ c. the liver test - III. 49
„ „ d. volume of the lungs III. 50
„ „ e. colour of the lungs III. 51
„ „ f. consistence of the lungs III. 53
„ „ g. weight of the lungs III. 56
„ „ h. the floating of the lungs III. 62
„ „ post-respiratory sinking of the lungs - III. 76
„ „ incision into the pulmonary substance - III. 78
„ „ uric acid deposit in the ducts of Bellini, III. 80
„ „ the remains of the umbilical cord - III. 83
„ „ obliteration of the fœtal circulatory canals, III. 85
„ „ bladder and rectum test III. 86
„ „ ecchymosis - III. 87
„ „ probative value of, III. 89
„ „ when it is superfluous III. 90
„ „ cases of, III. 94, 99, 103, 107
Driving over, cases of death by, I. 250, 251, 252, 254
Drowning, death from, I. 219; II. 229
„ „ diagnosis of, II. 232, 237
„ „ cases of, II. 245, 248, 249, 250, 252, 253, 254, 255, 256, 257
„ has it been homicidal or suicidal II. 258
„ „ „ cases illustrative of this question, II. 261, 265, 266, 267, 269, 270, 271, 273, 274
Drunkenness, in a forensic point of view - - IV. 262
„ cases illustrative of - IV. 270
Ductus Botalli, in new-born children, III. 11, 85
Ductus venosus, in new-born children III. 11, 85
Ducts, fœtal circulatory - III. 11
Ecchymoses, capillary, on the lungs and heart of those suffocated II. 126
„ „ cases of, in children, II. 143, 144, 145, 154
„ „ „ adults, II. 187, 219
„ in new-born children, III. 87

Embryo, the development of, according to months - - III. 15
Emotions and passions - IV. 277
Emphysema pulmonum neonatorum, III. 68, 72
Epiglottis, upright position of, in those drowned - - - II. 237
Epilepsy, simulation of, - IV. 85
„ as a cause of mental disease IV. 188
Epispadia - - III. 247, 251
Ether, vide Chloroform.
Ether, chloric, vide Chloroform.
Ethyle, nitric oxide of, vide Chloroform.
Exhaustion, death from - II. 2
„ „ cases of, II. 9, 10, 11
Excandescentia furibunda, vide Insanity of Anger.
Exhumation of bodies - I. 66
„ „ cases of, I. 68, 69, 70, 71, 75, 76, 78, 80, 81
Fall of the child's head on the floor at birth - III. 129
„ „ cases of death from, III. 137, 138, 139, 140, 141, 142, 143, 144
Fellare - - - III. 337
Firearms - - I. 134
Fire-raising, morbid propensity to IV. 311
„ „ cases of - IV. 313
Fœtus, age of - - III. 5
„ definition of a - III. 5, 13
„ development of, vide Embryo.
„ signs of maturity of - III. 18
Food of Convicts III. 195, 197, notes.
Foramen ovale in new-born children III. 11, 85
Frogs, vomiting of - - IV. 83
Fungi, poisonous - - II. 61
Gas, carbonic acid, case of suffocation by - - - II. 154
„ carbonic oxide, state of the blood in those suffocated by it - - II. 131
„ „ cases of suffocation by II. 147, 148, 149, 150, 151, 152
„ carburetted hydrogen, vide illuminating gas.
„ sulphuretted hydrogen, case of suffocation by - II. 154
„ ordinary illuminating, case of suffocation by - II. 158
Gestation, protracted - III. 363
„ „ diagnosis of - III. 369
„ „ duration of - III. 369
Gonorrheal infection as a proof of rape III. 285, 299
„ „ cases of - III. 323, 325
B D 2